[A]grade

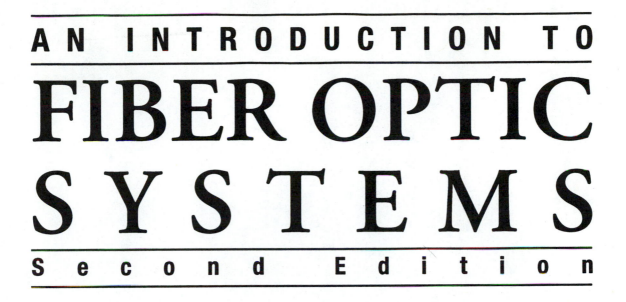

AN INTRODUCTION TO

FIBER OPTIC SYSTEMS

Second Edition

10/2007

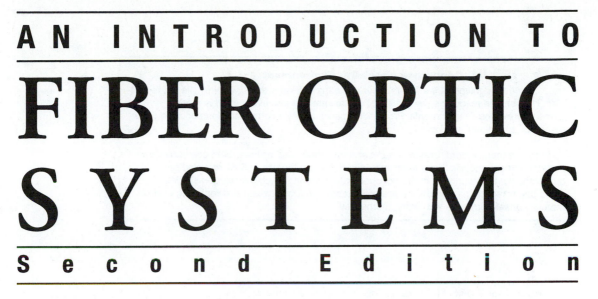

AN INTRODUCTION TO
FIBER OPTIC
SYSTEMS
Second Edition

John Powers

Professor of Electrical and Computer Engineering
Naval Postgraduate School
Monterey, California

IRWIN

Chicago • Bogotá • Boston • Buenos Aires • Caracas
London • Madrid • Mexico City • Sydney • Toronto

© Richard D. Irwin, a Times Mirror Higher Education Group, Inc company, 1993 and 1997

Publisher: *Tom Casson*
Senior sponsoring editor: *Scott Isenberg*
Editorial Assistant: *Anne O'Day*
Director of Marketing: *Kurt L. Strand*
Project supervisor: *Margaret Rathke*
Production supervisor: *Bob Lange*
Designer: *Matthew Slomka Baldwin*
Director, Prepress Purchasing: *Kimberly Meriwether David*
Typeface: *10/12*
Printer: *R. R. Donnelley & Sons Company*

Library of Congress Cataloging-In-Publication Data
Powers, John P.
   An introduction to fiber optic systems/John P. Powers.--2nd ed.
    p. cm.
   Includes index.
   ISBN 0-256-20414-4
   1. Optical communication systems. 2. Fiber optics. I. Title
TK5103.59.P68   1997
   621.382'75--dc20   96-33103

Printed in the United Sates of America
1 2 3 4 5 6 7 8 9 0 DO 3 2 1 0 9 8 7 6

This book is dedicated in memory of my parents and with thanks for the loving support of my wife, Anne, and sons, Matt and Dan.

On Bell's Photophone . . .

"The ordinary man . . . will find a little difficulty in comprehending how sunbeams are to be used. Does Prof. Bell intend to connect Boston and Cambridge . . . with a line of sunbeams hung on telegraph posts, and, if so, what diameter are the sunbeams to be . . . ? . . . will it be necessary to insulate them against the weather . . . ? . . . . . . until (the public) sees a man going through the streets with a coil of No. 12 sunbeams on his shoulder, and suspending them from pole to pole, there will be a general feeling that there is something about Prof. Bell's photophone which places a tremendous strain on human credulity."

New York Times Editorial, 30 August 1880
Source: International Fiber Optics & Communication, June, 1986, p. 29

Preface

This textbook is written for the beginning user of optical fibers in communications. My purposes are ...

- to introduce the terminology used in optical fibers,

- to describe the building blocks of an optical fiber system,

- to facilitate the initial first-order design of optical links, and

- to provide an entry to the research literature of optical fiber system components.

As a result, the book is more pragmatic than most texts on fiber optics. Few derivations of formulas are given; the formulas are introduced to support design applications and to aid the reader's understanding of the physical phenomena being described. Detailed discussion of advanced topics is left for more advanced texts and the research literature.

The assumed prerequisites are (1) an introductory knowledge of the electromagnetic theory of waveguides (especially the existence of modes in waveguides, mode cutoff, the concepts of phase and group velocity, as well as reflection and refraction at an interface), (2) an introduction to communications theory (especially amplitude modulation, pulse modulation, data rate, and bandwidth concepts), and (3) fundamentals of electronics (including amplifier frequency response, comparators, transistor circuits, and logic gates).

A major goal of the text is to bring the reader to the point where he or she can intelligently read and use the information presented on the data sheets to incorporate the information into the system analysis. (It is suggested that the instructor provide the students with an up-to-date collection of representative manufacturer data sheets of typical fibers, sources, receivers, connectors, couplers, and other appropriate components to supplement and illustrate the material presented in the text.) Another goal of this text is provide the reader with the capability to evaluate the potential of a device for use in the synthesis of an optical link.

The book is designed to progress from the description of the components in a fiber link (the fiber, the connections, the sources and receivers) to the interconnection of the components into a link or a network. Chapter 1 introduces the advantages and disadvantages of optical fibers and reviews their history; Chapter 2 presents an introduction to the structure of the optical fibers and their ability to guide light. Many of the terms used to describe the physical attributes of fibers are introduced in this chapter. (It should be noted that the treatment of the electromagnetic modes in a fiber is different in this text than most other texts. A detailed mathematical description of the modes is not included (there are many lucid treatments referenced). The results of these electromagnetic analyses are described only in terms of how they impact the fiber performance in communication systems.) Chapter 3 presents a description of the optical properties, including the important concepts of fiber loss and fiber dispersion. It also includes a description of optical-fiber cables. It also introduces the effects of nonlinearities in the fiber, a rapidly developing area of fiber optics since we can now generate high power densities and travel extremely long lengths in fiber links at high data rates. This combination allows nonlinearities to affect the fiber link performance. Chapter 4 applies the properties of the fiber to splices and connectors. The dependence of the joint loss on fiber properties and physical misalignment are described. The chapter includes a description of some of the passive fiber-optic components, such as couplers (used to split and combine optical signals) and wave-division multiplexers and demultiplexers (used to combine and split signals of differing wavelengths). It also includes a description of the recently introduced technology of fiber gratings, optical devices incorporated in the fiber that act as optical filters.

Optical sources (semiconductor lasers and light emitting diodes) are described in Chapter 5 with an emphasis on the devices' output characteristics and the factors that limit the performance as a communications device, such as modulation response times and reliability factors. Chapter 6 describes optical detectors and their performance limits and, then, goes on to a discussion of the performance of these detectors when combined with preamplifier and equalization amplifiers. Emphasis in this chapter is on the signal-to-noise performance of the receiver with the goal of estimating the power required at the detector to provide a required signal-to-noise ratio or to achieve a desired bit error rate.

The rest of the text brings the optical-component building blocks together in a complete optical links. Chapter 7 is the climactic chapter that integrates the material introduced in earlier chapters into a method for the design of an optical link. Both power budgets and timing budgets are introduced to show that links can be either attenuation-limited or dispersion-limited. Dynamic range considerations, which are of special importance for shorter links, are also introduced. One of the revolutionary concepts being introduced to fiber links is the optical amplifier (that eliminates the need for electronic repeaters). These amplifiers can be used to boost the signal power as it leaves the source, to amplify the signal before entering the detector, or as in-line amplifiers embedded periodically within the link. The optical fiber amplifiers are described in this chapter with analysis done on their role as in-line amplifiers, where the link distance can now be several thousand kilometers before the optical signal is converted into electronic form. Just as optical amplifiers allow the power budget to be extended to long distances, dispersion management (or dispersion compensation) can reduce the total link dispersion, while the local dispersion is relatively high. These techniques allow us to extend the limits of link distances due to dispersion. Another technique for overcoming dispersion is to allow the effects of the fiber nonlinearities to cancel the dispersion effects. As described in this chapter, this can be done if the optical pulse has the right temporal shape and amplitude. Optical amplifiers can be used to regenerate the amplitude of the signals. Such waves are called *solitons* and offer the promise of long-distance high-data-rate communications.

Chapter 8 describes the application of optical fibers to modern data and digital telecommunications networks, including star and linear topologies as well as introducing the proposed standards of the Fiber Distributed Data Interface (FDDI) for computer network applications and the Synchronous Optical NETwork (SONET) for telecommunications applications. The combination of asynchronous transfer mode (ATM) technology and fiber links using SONET is also covered in this chapter. Chapter 9 describes the use of multiple signals at differing wavelengths (wavelength-division multiplexing) to increase channel capacity and the technology required to implement this multiplexing. It also introduces some WDM network architectures to implement interconnections.

Most of Chapters 1 through 7, along with selected topics from the later chapters, can be covered in a one-quarter course on optical fiber communications, offered at the senior-year level.

The addition of other subjects from the later chapters or topics of the instructor's choice (e.g., the electromagnetic mode analysis) leads to a one-semester course.

In all cases, references to the literature have been included for a more detailed or alternative treatment. The literature cited is heavy on tutorial papers on the subjects and offers a gateway into the research literature of fiber optics for those who seek more detailed information. Inevitably, omissions will have occurred and I apologize to those whose work has not been included.

This second edition has added material on fiber nonlinearities, fiber gratings, dispersion compensation techniques, solitons, ATM network technology, and nonlinear effects on WDM systems. To make room for these topics, coverage of fiber strength issues, coherent detection, fiber sensors, fiber fabrication, and the measurement of fiber parameters (other than fiber attenuation) that appeared in the first edition has been dropped.

I want to thank my thesis students who have sparked my interests in fiber optics, my past research sponsors (ARPA, the Space and Naval Warfare Command, and the Naval Underwater Warfare Engineering Station) for providing the means to allow the interest to flourish, and to the students of my fiber optics course at the Naval Postgraduate School who have tested this manuscript, making wise and helpful comments. Their efforts are greatly appreciated.

I would also like to thank the many reviewers of the text who offered suggestions for improving the text. Their comments have definitely helped to improve the material.

John P. Powers
Pacific Grove, California
jpowers@nps.navy.mil

USEFUL CONSTANTS

Speed of light (vacuum) $c = 3.00 \times 10^8$ m/s
Planck's constant $h = 6.63 \times 10^{-34}$ joule·s
Boltzmann's constant $k = 1.38 \times 10^{-23}$ joules·K^{-1}
Electron charge $q = 1.60 \times 10^{-19}$ coulombs

Contents

USEFUL CONSTANTS

Speed of light (vacuum) $c = 3.00 \times 10^8$ m/s
Planck's constant $h = 6.63 \times 10^{-34}$ joule·s
Boltzmann's constant $k = 1.38 \times 10^{-23}$ joules·$(K)^{-1}$
Electron charge $q = 1.60 \times 10^{-19}$ coulombs

Chapter 1

Introduction

This text focuses on the application of optical fibers to the task of carrying information from point to point at high data rate-distance products and with high data integrity. It will omit imaging applications of fibers, even though they historically antedated the communications application.

Figure 1.1 on the following page shows a representative fiber link. In the top line of the figure, an optical source, such as a semiconductor laser or LED, is modulated by a signal. (The modulator can be either external to the source, as shown, or the source can be modulated directly.) The modulated output light is introduced into the fiber link through a set of connectors or through a permanent fiber splice. If the link is long, the light intensity diminishes because of attenuation in the fiber, and the optical signal may need to be regenerated by a repeater, as shown in the second line of the figure. This device consists of an optical detector that converts the light into a voltage, some electronic circuitry to detect the signal, and an optical source that regenerates the detected signal. Thus strengthened, the optical signal proceeds on its way through more fiber. An alternative to repeating the signal is to optically amplify the light, as shown in the third line of the figure. The "WDM" box represents an optical wavelength-division multiplexer which combines the signal beam and a pump laser beam (at two different wavelengths) in an optical-fiber amplifier. The fiber amplifier strengthens the signal beam using the power from the pump laser beam without requiring optical/electronic and electronic/optical conversions. Finally, the optical signal arrives at its destination (as in the bottom right of the figure) and the information is recovered and converted back into its original format.

The design of a fiber-optic link similar to the one shown in Fig. 1.1 is an interactive process. To illustrate some of the decisions that are required, we consider some of the choices that must be made. Selections in one area will impact the choices in the other area.

- **Signal to be transmitted:** Is the signal digital or analog? What is the signal's (analog) bandwidth or (digital) data rate? What is the dynamic range (i.e., what are the maximum and minimum signal strengths)? What is an acceptable (analog) signal-to-noise ratio at the output of the receiver? What is an acceptable (digital) bit error rate?

- **Source:** Should one choose an LED or laser source? At what wavelength? What modulation format should be used—AM or FM for analog signals; pulse amplitude modulation (PAM) or pulse position modulation (PPM), etc. for a digital waveform? What is the source cost, reliability, output power level? How stable is the source in the face of temperature changes? A more powerful source will allow longer distances—how far can the signal go in the user's application?

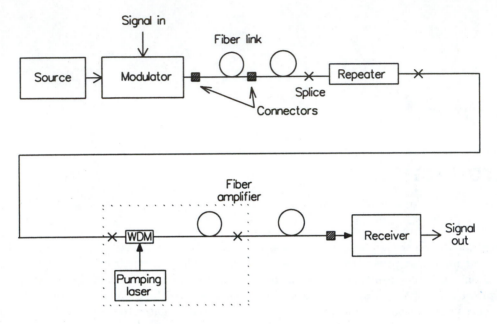

Figure 1.1 Typical fiber-optic communication system.

- **Detector:** Of what material should the detector be made to provide the maximum sensitivity at the wavelength of interest? At what cost? Does the user require enough sensitivity to justify the extra expense and complexity of an avalanche detector? How stable is the performance of the detector in the presence of changes of operating temperature?

- **Fiber:** What fiber attenuation is required to meet the system design objectives? What bandwidth-distance product? Is single-mode fiber required or will multimode fiber suffice? What cabling of the fiber is required—strength members, power conductor, size, weight?

- **Connectors and splices:** What splices and/or connectors will be required to meet the required attenuation limits? How easy is it to make a splice under operating conditions? Does the connector have to keep out water or gases?

- **Repeaters and amplifiers:** Should the link use electronic repeaters (where the signal is detected, cleaned up, and retransmitted) or optical amplifiers (where the optical signal is amplified)? Where should these devices be located to maximize the signal-to-noise ratio? What is best spacing?

These and other questions must be answered to accomplish a successful design that meets the link objectives [1]. To the beginning worker in fiber optics technology the choices can be overwhelming. It is the purpose of this text to allow the user of fiber optics technology to accomplish a first-order design of a link for an application, while appreciating the tradeoffs required to achieve the desired performance goal.

Name	Data rate	Number of voice channels
DS-0	64 kb/s	1
DS-1 (T1)	1.544 Mb/s	24
DS-2 (T2)	6.312 Mb/s	96
DS-3 (T3)	44.736 Mb/s	672
DS-4 (T4)	274.176 Mb/s	4,032

Table 1.1 Standard United States telecommunications data rates.

Japan	Europe
1.544 Mb/s	2.048 Mb/s
6.312 Mb/s	8.448 Mb/s
32.064 Mb/s	34.364 Mb/s
97.728 Mb/s	139.264 Mb/s
396.20 Mb/s	565.148 Mb/s

Table 1.2 Standard data rates in Japan and Europe.

1.1 Bandwidth and Data Rate

Historically, bandwidth requirements for communications have shown an increasing trend, roughly doubling every two years. The traditional way of meeting this requirement has been to increase the carrier frequency, as the information bandwidth is constrained to be, at most, equal to the carrier (for baseband transmission) or some fraction of the carrier. Hence, to meet the increased bandwidth demands, information carriers have transitioned from HF to VHF to UHF to microwaves to millimeter waves and, finally, to light waves.

The bandwidth of a copper-based medium is tied to the losses in that medium. In coaxial-line technology, the cable losses (in decibels per length) increase linearly with the carrier frequency. Reduction of the losses at any particular frequency can be achieved by increasing the diameter of the cable, but eventually the cable becomes too large and bulky. Fiber-optic cables do not exhibit this linear increase in loss with frequency, and the losses can be made quite small by the proper fiber construction and the proper choice of operating wavelength.

In considering data rates, some benchmark data rates might be useful. The telephone industry has established *standard data rates* for various applications. Table 1.1 lists these data rates for the North American system, and Table 1.2 lists them for the Japanese and European systems.

1.1.1 Data Rate and Bandwidth

Let us consider the relationship between the *analog bandwidth* BW of a signal and the *data rate* DR required to transmit a digitized version of that signal.

Suppose that the signal extends from 0 Hz to an upper frequency of BW Hz. (For example, an audio signal extends from dc to 20,000 Hz; a video signal in the United States extends from 0 to 5 MHz.) Such a signal, before it is modulated onto a carrier wave, is called a *baseband signal*.

- **Sampling:** The *Nyquist criterion* tells us that, to faithfully reconstruct an analog signal, we must sample the wave at a rate equal to or greater than twice the highest frequency (i.e., we must sample at a rate greater than or equal to $2 \times$ BW). Allowing for a sampling factor of S (where $S \geq 2$), the sampling frequency is given by $S \times$ BW. Reasonable values of S range from 6 to 10.

- **Digitization:** Once sampled, each sample of the waveform must be digitized. The number of bits N per sample depends on the accuracy required. Eight bits allows the data to be divided into 256 (2^8) *quantization levels*. More accuracy requires more bits. Twelve bits allows the sample to be represented as one of 4096 levels; sixteen bits would give 65,536 levels. Currently eight bits usually represents the low end of acceptable accuracy and sixteen bits represents the high end (except in cases calling for extreme accuracy).

So, we find that the data rate of the digitized signal (in units of bits per second) will be

$$DR = S \times N \times BW \tag{1.1}$$

where $S \times$ BW is seen to be the number of samples per second and N is the number of bits per sample. We can estimate the bandwidth B of a channel that carries a data rate of DR as

$$B = DR/2. \tag{1.2}$$

(It is important to separate in your mind the bandwidth (BW) of the information signal from the bandwidth B of the carrier required for the digitized version of the signal.)

Hence, we see that the data rate will always be a multiple of the bandwidth of the information signal. The size of the multiplier is $S \times N$, whose values can range from a typical low of $2 \times 8 = 16$ up to a typical high of $10 \times 16 = 160$. From this multiplier factor we understand that increased accuracy in the data requires a significant increase in the data rate—the perfect justification for fiber optics.

1.2 Fiber Advantages

Among the potential advantages offered by optical fiber communications are:

- **Wide bandwidth.** The bandwidth of a medium depends directly on the carrier frequency. Optical carriers are superior to rf and microwave carriers because of their higher frequencies. Optical fibers offer the possibility of several thousands of GHz (i.e., several THz) of bandwidth.

- **Light weight and small size.** Due to their small volume and lower density, optical fiber cables enjoy considerable weight advantages over typical coaxial cables. (As a measure of the size of a reel of fiber-optic cable, a rule of thumb is that one can achieve "fifty miles per gallon," i.e., a spool containing fifty miles of fiber occupies the volume of a one-gallon gasoline can [approximately 30 cm $\times$ 15 cm $\times$ 18 cm].)

- **Immunity to Electromagnetic Interference.** Since optical fibers are nonconducting, they will neither generate nor receive *electromagnetic interference*. This feature allows the use of fibers in regions of high electric fields, as with power electronics, radar feed horns and antennas, nuclear explosions, and other such sources of intense electromagnetic fields. Indeed one of the thriving applications of fiber optics is sending control signals into power stations, where switching transients could obliterate them. Telemetry links for bringing information out of a system exposed to high electromagnetic signals, such as EMP (electromagnetic pulse) or lightning-strike testing of military aircraft and missiles, also use fiber optics. Finally, optical fibers are used to telemeter information out of underground atomic-bomb test caverns, where the EMP from the blast would contaminate the data from the experiment.

- **Lack of EMI Crosstalk Between Channels.** One form of EMI occurs when two conducting lines lie near enough to each other to allow the signal from one to leak into the other (called *crosstalk*) because of overlapping electromagnetic fields. Traditional solutions have included further separation of the cables or increased shielding (i.e., including a grounded wire mesh around the individual cables to short out the fields) thereby increasing the size, weight, and cost of the coaxial cable. However, the optical fields extending from an optical fiber are negligible, eliminating optical pickup between adjacent cables.

- **Lack of Sparking.** For special purpose applications that require transmission of information through hazardous cargo areas (e.g., areas with explosive or flammable fuel), fibers offer the potential advantage of not sparking if there is a break in the transmission line. For example, circuitous routes are followed in electrically transmitting information from an aircraft fuselage to the wing stations to avoid the fuel tanks in the wings. Use of fiber optics allows the line to be routed by the most direct path.

- **Compatibility with Solid-State Sources.** The physical dimensions of the fiber-optic sources, detectors, and connectors, as well as the fiber itself, are compatible with modern miniaturized electronics. Most components are available in dual in-line packaging (DIP packs), making mounting on a printed circuit board extremely easy. This compact size of components is of prime importance in making this new technology acceptable to today's electronic designer.

- **Low Cost.** Copper is a critical commodity on the world market; as such, it is subject to rapid upward and downward fluctuations in price. The primary ingredient of silica-based glass fibers, on the other hand, is widely available and is not a critical commodity. The price tradeoffs, comparing a fiber-optic link with alternative technology, are usually dependent on data rate since low data rates can use more economically coaxial cable. All economic analyses show that, at some value of data rate, a fiber link becomes cost effective. The exact location of the price crossing point is sensitive to many variables, but the trend is clear—fiber-optic technology is cost effective only for wideband signals (with data rates typically in excess of tens of megabits per second). Of course, if one is using another of the special properties of fiber optics (e.g., immunity to EMI), then this benefit alone might justify its use (as frequently occurs in military applications). This requirement for wideband signals to justify an economic advantage for fiber optics works to a disadvantage in the American market, where regulation inhibits a telephone carrier from providing other services such as cable TV. While other nations combine their telephone, postal and broadcasting services and are investigating the use of a single-fiber drop to provide broadband services, in the United States only cable television companies currently have the potential bandwidth demand to justify fiber-optic home hookups (unless new services, such as electronic-catalog shopping, are offered to telephone subscribers).

- **No Emission Licenses.** Since fiber optics is a nonradiating means of information transfer, no government licenses are required to implement a link. This offers potential advantages to companies that desire a point-to-point communication link between two locations for temporary use and that wish to avoid the time-consuming licensing process. Although rights-of-way must be negotiated for the fiber-cable route, this is frequently preferable to the licensing process.

1.3 Fiber Disadvantages

Along with the advantages, of course, come some potential disadvantages. These include:

- **Lack of Bandwidth Demand.** Although fiber-optic systems have been demonstrated with multigigabit data rates (and even terabit/s rates), few users have requirements for such data rates today. (The total traffic in North America was estimated to be about one-third of a terabit per second (i.e., 333 Gb/s) in 1995.) Since at lower data rates, fiber links cost more than conventional links, they are not yet suited for these applications. Many installed fiber links have excess capacity beyond their present usage. Historic trends predict, however, that society will require high data rates as we become more information dependent. Even at present, applications appear on the horizon that require the real-time transmission of high-resolution pictorial information, or the multiplexing of many channels into a data superhighway.

  In addition, lower-cost components are being devised for low-data-rate links. The cost of low-loss fibers, plastic connectors, long-wavelength light-emitting diode sources, and other economic components is steadily reducing the cost of the lower-performance links.

  Users of fiber optics find it advantageous, in some cases, to anticipate future demands for bandwidth, since installation of a fiber optic link is typically a capital investment with an anticipated lifetime of over 20 years. It is possible in some links to upgrade the link performance by exchanging the sources and receivers, while keeping the same installed fiber.

- **Lack of Standards.** Because fiber optics is a relatively new technology, standards are just evolving.

  Classically, the decision time to set standards has been difficult to establish, since one freezes technology in order to establish the standards. In a rapidly developing field such as fiber optics, committees are hesitant to freeze the technology prematurely lest they stifle a major breakthrough. The penalty for this behavior is that the design components are in a constant state of flux, with each manufacturer touting product superiority. Until the marketplace makes a decision on the worth of the various products, the casual user is unable to make the necessary technical design decisions and consequently sits on the sidelines. Coupled with the capital investment required for point-to-point communications links, conservatism is highly rewarded.

  Fiber-optic standards are being set by manufacturer trade groups, telecommunications groups (both national and international), and governmental agencies (e.g., military standards for defense systems). Currently standards exist for fiber dimensions, some measurement techniques, and two communication networks (FDDI and SONET, discussed in Chapter 8). Committees exist to standardize some connector designs and electro-optic components, but few decisions have been made.

The consequence of this lack of standards is seen, for example, in the incompatibility of connectors. Connectors of varying design are available from a variety of manufacturers, and most are incompatible with each other. Consequently, a major portion of the design process is an evaluation of connector designs for the link under consideration.

Standardization will eventually come as the technology matures and a consensus is reached regarding the best techniques.

- **Radiation Darkening.** Optical glass darkens when exposed to nuclear radiation [2, 3]. Although the specifics of the interaction depend on the dose rate and time history of the dose, as well as the type of radiation and the material and dopants of the glass [4–6], optical fibers are generally susceptible to interruption by nuclear radiation. While research continues to study the interaction and to identify glasses that minimize the effect [3, 7, 8], most optical fibers cannot be used in a nuclear environment, such as reactor spaces or platforms exposed to nuclear effects.

1.4   History

The principle of total internal reflection was studied by Tyndall in the 1850s. Development of optical fibers occurred in the 1950s driven by imaging applications in the medical and non-destructive testing fields. Plastic fibers were also used for lighting effects. During the late 1960s [9, 10], the communications aspects of fiber optics came into prominence when various schemes for atmospheric transmission or elaborate guided-wave schemes proved to be impractical for the transmission of light. The telecommunications industry has a rule of thumb that a guided-wave transmission system would have to achieve transmission losses of 20 dB/km or less to be economically competitive with current repeater spacings. Fibers at that time were highly lossy, but it became evident that purification techniques similar to those developed for silicon in the semiconductor industry could be used to reduce those losses (which were determined by impurities in the glass). Developments in the late 1960s and early 1970s brought losses steadily lower, until current loss levels of a few tenths of a dB/km were achieved.

Now approaching maturity, the technology has been through three distinct stages. The first stage used visible and near-IR (from 600–920 nm) light sources combined with fiber bundles. The fibers carried redundant optical signals to the receiver. The second stage used single, multimode fibers as the channel, with the sources still in the visible and near IR (called *short-wavelength sources*). Many current short-distance commercial installations are of this type. The third stage, widely used in long-haul communications, uses so-called *long-wavelength sources* operating between 1300 and 1600 nm in concert with fibers of smaller diameters (called *single-mode* fibers). The data rate and distance between repeaters of this latter combination are superior to the second stage systems, but the lower cost of the second-stage systems will ensure their continued use in applications requiring modest bandwidths and/or short transmission distances. Current efforts [1] in the development of fiber-optic communication systems are proceeding in two directions. The first is high-data-rate systems for long-distance or high throughput applications using long-wavelength sources and single-mode fibers, combined with optical amplifiers. These amplifiers replace the electronic repeaters in long-distance links. The second thrust is toward high-density, short-distance, moderate-data-rate applications such as home data services (e.g., video on demand or database applications such as airline-ticket booking) and computer local-area networks (LANs). Such applications are focusing on reducing the cost of the components used, defining services and providers, and standardizing the technology to allow a multivendor environment, where the components and systems are treated as commodities rather than pieces of custom-designed systems.

1.5   Textbooks and Other Sources

The following is a list of some textbooks on fiber-optic systems (listed alphabetically by author). The reader is encouraged to consult these sources for alternative or more complete viewpoints.

G. P. Agrawal, *Nonlinear Fiber Optics, Second Edition*. San Diego: Academic Press, 1995.

G. P. Agrawal, *Fiber-Optic Communication Systems*. New York: John Wiley & Sons, 1992.

F. C. Allard, ed., *Fiber Optics Handbook for Engineers and Scientists*. New York: McGraw-Hill, 1990.

B. Bendbow and S. Mitra, *Fiber Optics Advances in Research and Development*. New York: Plenum Press, 1979.

John A. Buck, *Fundamentals of Fiber Optics*. New York: Wiley, 1995.

A. Bruce Buckman, *Guided-Wave Photonics*. Orlando, FL: Harcourt Brace Jovanovich, 1992.

C. D. Chaffee, *The Rewiring of America: The Fiber Optics Revolution*. New York: Academic Press, 1988.

A. Cherin, *Introduction to Optical Fibers*. New York: McGraw-Hill, 1983.

Staff of CSELT, *Optical Fibre Communications*. New York: McGraw-Hill, 1981.

R. Gagliardi and S. Karp, *Optical Communications*. New York: Wiley, 1976.

R. Gallawa, *A User's Manual for Optical Waveguide Communications*. Pub. No. OTR76-83: US Dept. of Commerce, 1976.

D. Gloge, ed., *Optical Fiber Technology*. New York: IEEE Press, 1976.

P. E. Green, Jr., *Fiber Optic Networks*. Englewood Cliffs, NJ: Prentice Hall, 1993.

M. Howes and D. Morgan, *Optical Fibre Communications*. New York: Wiley, 1980.

W. B. Jones, Jr., ed., *Introduction to Optical Fiber Communications Systems*. New York: Holt, Rinehart and Winston, 1988.

C. Kao, ed., *Optical Fiber Technology*. New York: IEEE Press, 1981.

C. Kao, *Optical Fiber Systems*. New York: McGraw-Hill, 1982.

G. Keiser, *Optical Fiber Communications*. New York: McGraw-Hill, 1983.

G. Keiser, *Optical Fiber Communications, Second Edition*. New York: McGraw-Hill, 1991.

H. Kressel, ed., *Semiconductor Devices for Optical Communications*. New York: Springer, 1980.

A. Kumar, *Antenna Design with Fiber Optics*. Boston: Artech House, 1996.

C. Lin, ed., *Review of Optoelectronic Technology and Lightwave Communication System*. New York: Van Nostrand Reinhold, 1989.

D. Marcuse, *Principles of Optical Fiber Measurement*. New York: Academic Press, 1981.

J. Midwinter, *Optical Fibers for Transmission*. New York: Wiley, 1979.

S. E. Miller and A. G. Chynoweth, eds., *Optical Fiber Telecommunications*. New York: Academic Press, 1979.

S. E. Miller and I. P. Kaminow, eds., *Optical Fiber Telecommunications II*. New York: Academic Press, 1988.

J. G. Nellist, *Understanding Telecommunications and Lightwave Systems: An Entry-Level Guide, Second Edition*, New York: IEEE Press, 1996.

J. C. Palais, *Fiber Optic Communications,* Third Edition. Englewood Cliffs, NJ: Prentice Hall, 1992.

S. Personick, *Optical Fiber Transmission Systems.* New York: Plenum Publishing, 1981.

S. Personick, *Fiber Optics.* New York: Plenum Press, 1985.

C. R. Pollock, *Fundamentals of Optoelectronics.* Chicago: Richard D. Irwin, Inc., 1995.

C. Sandbank, ed., *Optical Fibre Communication Systems.* New York: Wiley, 1980.

J. Senior, *Optical Fiber Communications: Principles and Practice.* Englewood Cliffs, NJ: Prentice Hall, 1985.

Research Journals: Through 1983, fiber optic research results in the United States were published primarily in the *IEEE Journal on Quantum Electronics, Applied Optics,* the *Bell System Technical Journal,* and other journals. Since then, US research results are typically published in the *Journal of Lightwave Technology,* published jointly by the IEEE and the Optical Society of America since 1983. In 1990, the IEEE Communications Society initiated publication of *IEEE LCS (Lightwave Communications Systems) Magazine* (later changed to *IEEE LTS [Lightwave Telecommunications Systems] Magazine*); this magazine ceased publication in 1992. Device research results are presented in *IEEE Photonics Technology Letters* and in *Electronic Letters.*

1.6   Problems

1.   (a) How many US standard voice channels could a 10 Gb/s channel accommodate?

(b) What is the bit period T_b of each bit in a 10 Gb/s signal?

2.   (a) Calculate the data rate required to transmit a 20,000 Hz audio signal at a sampling rate of four times the Nyquist rate with a digitization of 8 bits per sample. Calculate the channel bandwidth.

(b) ... for a 5 MHz television signal?

3. Calculate the data rate required to transmit a high-definition television (HDTV) signal if the image is 1000×1000 pixels, each pixel is tri-color with 12 bits of definition per color, and the frame rate is 70 frames per second.

References

1. T. Ikegami, "Survey of telecommunications applications of quantum electronics—Progress with optical fiber communications," *Proc. IEEE*, vol. 80, no. 3, pp. 411–419, 1992.

2. George H. Sigel, Jr., "Fiber transmission losses in high radiation fields," *Proc. IEEE*, vol. 68, no. 10, pp. 1236–1240, 1980.

3. R. A. Greenwell, "Reliable fiber optics for the adverse nuclear environment," *Optical Engineering*, vol. 30, no. 6, pp. 802–807, 1991.

4. E. J. Friebele, P. B. Lyons, J. Blackburn, H. Henschel, A. Johan, J. A. Krinsky, A. Robinson, W. Schneider, D. Smith, E. W. Taylor, G. Y. Turquet de Beauregard, R. H. West, and P. Zagarino, "Interlaboratory comparison of radiation–induced attenuation in optical fibers. Part III: Transient exposures," *J. Lightwave Technology*, vol. 8, no. 6, pp. 977–989, 1990.

5. E. J. Friebele, E. W. Taylor, G. Turquet, J. A. Wall, and C. E. Barnes, "Interlaboratory comparison of radiation-induced attenuation in optical fibers. Part I: Steady-state exposures," *J. Lightwave Technology*, vol. 6, no. 2, pp. 165–171, 1988.

6. E. Taylor, E. J. Friebele, H. Henschel, R. H. West, J. Krinsky, and C. Barnes, "Interlaboratory comparison of radiation-induced attenuation in optical fibers. Part II: Steady-state exposures," *J. Lightwave Technology*, vol. 8, no. 6, pp. 967–976, 1990.

7. T. Kakuta, N. Wakayama, K. Sanada, O. Fukuda, K. Inada, T. Suematsu, and M. Yatsuhashi, "Radiation resistance characteristics of optical fibers," *J. Lightwave Technology*, vol. LT–4, no. 8, pp. 1139–1143, 1986.

8. A. Iino and J. Tamura, "Radiation resistivity in silica optical fibers," *J. Lightwave Technology*, vol. 6, no. 2, pp. 145–149, 1988.

9. K. Kao and G. Hockham, "Dielectric-fiber surface waveguides for optical frequencies," *Proc. IEE*, vol. 133, pp. 1151–1158, 1966.

10. A. Werts, "Propagation de la lumiere coherente dans les fibres optique," *L'One Electrique*, vol. 46, pp. 967–980, 1966.

Chapter 2

The Optical Fiber

2.1 Introduction

In this chapter we will describe the geometry of the optical fiber and define some of the fiber parameters that are used to characterize its physical shape. We begin by describing the confinement process that makes optical fibers useful, defining several parameters, and introducing several terms that are associated with optical fibers. We will see that fibers can propagate various electromagnetic modes. Some fibers are designed to propagate a single mode; others will propagate multiple modes. Each type of fiber has its own advantages.

References [1–11] represent early writings and other descriptions can be found in most fiber optics texts.

2.2 Optical Confinement

Optical fibers work by confining the light within a long strand of glass. In their simplest form, they are cylindrical dielectric waveguides made up of central cylinder of glass with one index of refraction, surrounded by an annulus with a slightly different index of refraction.

One confinement process that traps the light inside the fiber and allows it to propagate down the length of the fiber is based on the principle of *total internal reflection* at the interface of two dielectric media. Consider an optical-plane-wave incident on an infinite planar interface between two dielectric media, as shown in Fig. 2.1 on the next page. The propagation direction of the plane wave makesan *angle of incidence* θ_i with the normal to the interface as shown in the figure. The *index of refraction* (or *refractive index*) n of a medium is given by the equation

$$n = \frac{c}{v} \tag{2.1}$$

where c is the velocity of light in a vacuum (3×10^8 m/sec) and v is the velocity of light in the medium. (Note that n is always greater than 1; for glass, n is between 1.4 and 1.5.) For total internal reflection to occur, the index of refraction of the medium containing the incident plane wave is required to be larger than the index of refraction of the other medium (i.e., the wave is traveling from the higher-index medium into the lower-index medium).

Snell's law governs the transmission of the plane wave through the interface and is given by

$$n_1 \sin \theta_i = n_2 \sin \theta_t \,, \tag{2.2}$$

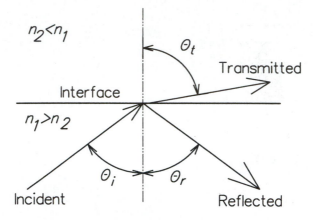

Figure 2.1 Planar interface geometry.

where θ_i is the angle of incidence and θ_t is the angle of transmission. From this relation, one can see that θ_t reaches 90 degrees when θ_i reaches the value of

$$\theta_i = \theta_c \equiv \sin^{-1}\left(\frac{n_2}{n_1}\right) \tag{2.3}$$

where θ_c is called the *critical angle* of incidence and is given by the latter part of the equation. For angles of incidence equal to or exceeding the critical angle, the energy of the incident wave is totally reflected back into medium 1. It is this total internal reflection that allows the light to propagate with no loss. Light that is incident at an angle below the critical angle is partially transmitted and partially reflected, losing a significant fraction of the power into the transmitted beam.

———————————————— ∞ ————————————————

Example: (a) Assuming that n_2 is 1% smaller than n_1, find n_2 if $n_1 = 1.45$.

<u>Solution:</u> Since n_2 is 1% smaller than n_1, it is 99% of n_1, so we have

$$n_2 \;=\; 0.99n_1 = 0.99 \times 1.45 = 1.435\,. \tag{2.4}$$

(b) Find the value of the critical angle.

<u>Solution:</u> The critical angle is found from

$$\theta_c \;=\; \sin^{-1}\left(\frac{n_2}{n_1}\right) = \sin^{-1}\left(\frac{0.99n_1}{n_1}\right) = \sin^{-1}(0.99) = 81.9^\circ\,. \tag{2.5}$$

———————————————— ∞ ————————————————

A typical optical fiber looks as shown in Fig. 2.2 on the facing page. The central portion, called the *core*, is cylindrically shaped and is surrounded by an annular shaped outer region called the *cladding*. The index of refraction radial variation $n(r)$ for a *step-index fiber* is plotted in Fig. 2.3 on the next page. The lower index of refraction of the cladding is related to the index of the core by

$$n_2 \approx n_1(1 - \Delta) \tag{2.6}$$

where Δ is the *fractional change in the index of refraction*, given by

$$\Delta = \frac{n_1^2 - n_2^2}{2n_1^2} \approx \frac{n_1 - n_2}{n_1} . \tag{2.7}$$

(We have used the approximation for Δ (good for $\Delta \ll 1$, called a *weakly guided* waveguide design) in Eq. 2.6 on the facing page.) Typical values of Δ can range between 0.001 and 0.02 (i.e., 0.1% to 2%) with a nominal core refractive index of 1.47. Typically, the cladding is also surrounded by a plastic protective *jacket* for handling purposes and to protect the fiber surface from forming flaws that will weaken its strength.

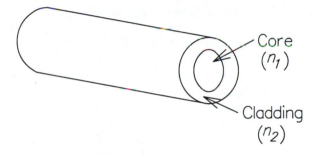

Figure 2.2 Typical step-index fiber showing the core and cladding.

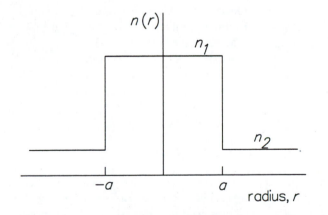

Figure 2.3 Refractive index profile of a step-index fiber.

Table 2.1 on the next page summarizes some of the typical parameters associated with standard fibers. (While the sizes are standardized (except for the core diameter of single-mode fiber), the other values are representative values only.)

Type	core diam. μm	cladding diam. μm	Δ	Application
8/125 single-mode	8	125	0.1% to 0.2%	Long distance, high data rate
50/125 multimode	50	125	1% to 2%	Short distance, moderate data rate
62.5/125 multimode	62.5	125	1% to 2%	Local area networks
100/140 multimode	100	140	1% to 2%	Local area networks, short distance

Table 2.1 Representative parameters for standard fibers.

2.3   Waveguide Modes in Step-Index Fibers

The fibers just described are called *step-index* fibers due to the step discontinuity in the index of refraction at the core-cladding interface, as seen in Fig. 2.3 on the preceding page. Such cylindrically symmetric dielectric waveguides can be modeled and solved for the electromagnetic fields that propagate in the waveguide. The solutions involve Bessel functions and will not be displayed here, but can be found in Refs. [12–15], as well as many other fiber-optic texts.

Waveguides present certain boundary conditions to the electromagnetic fields that must be met at the interfaces between the waveguide regions. As a result of these boundary conditions, only certain combinations of waves (called *modes*) will meet the conditions and be allowed to propagate. All other combinations that cannot meet the boundary conditions will not be supported and cannot propagate without extreme losses. Like other waveguides, this glass dielectric waveguide of a fiber results in characteristic electromagnetic modes that can propagate. The simplest modes are those with an electric field transverse to the propagation direction, called TE (transverse electric) modes, with radial symmetry and those with a magnetic field transverse to the propagation direction, called TM (transverse magnetic) modes. (These modes also exist in metallic waveguides.) In addition, hybrid modes (i.e., combinations of TE and TM modes) also exist and are called HE_{mn} and EH_{mn} modes. (These modes have electric and magnetic fields along the propagation direction.) Some modes are linearly polarized and are called LP modes.

As a mode $U(x, y, z, t)$ propagates down the fiber, it has a longitudinal propagation term that can be separated out. We can represent the wave as phasor,

$$U(x, y, z, t) = U'(x, y, z)e^{j(\omega t - \beta z)} . \tag{2.8}$$

We find the actual wave by taking the real part of the phasor. The exponential term represent the propagating optical carrier wave. The temporal radian frequency is ω [radians/s] and β is the longitudinal propagation coefficient [radians/m]. It should be noted that the propagation coefficient is also sometimes called the propagation "constant." While it is true that a particular mode in a particular waveguide will propagate with a constant-valued propagation coefficient, in general the propagation coefficient differs from mode to mode and differs with waveguide geometry and physical parameters. Hence, we will avoid describing β as the propagation "constant" and will prefer to call it the propagation coefficient.

Each mode not only has its own geometric electric and magnetic fields, but also will have a unique value of propagation coefficient. Much of the electromagnetic analyses of optical fibers is the description of these modes and their propagation coefficients. We will selectively extract the results of these analyses for our use.

A key parameter that describes the mode structure is the *V-parameter V*, defined by

$$V = \frac{2\pi a}{\lambda}\sqrt{n_1^2 - n_2^2} = \frac{2\pi a}{\lambda}n_1\sqrt{2\Delta}, \tag{2.9}$$

where a is the radius of the core of the fiber, λ is the nominal free-space wavelength of the light, and n_1 and n_2 are the indices of the core and cladding, respectively. The V-parameter is important because it determines the number of electromagnetic modes in the fiber. Many other properties of the fiber (e.g., the ability to couple light into the fiber or the losses in a fiber splice) rely, in turn, on these modes.

Figure 2.4 shows the "relative" modal propagation coefficient for the lowest-order modes of a step-index fiber as a function of V. (Similar plots can obtained for fibers with different refractive-index profiles.) Each mode shown has two polarizations associated with it and is described as *doubly degenerate*. The *relative modal propagation coefficient* (shown on the vertical axis) is defined as

$$b = \frac{(\beta/k)^2 - n_2^2}{n_1^2 - n_2^2} \approx \frac{(\beta/k) - n_2}{n_1 - n_2}, \tag{2.10}$$

where β is the mode's propagation coefficient and $k = 2\pi/\lambda$.

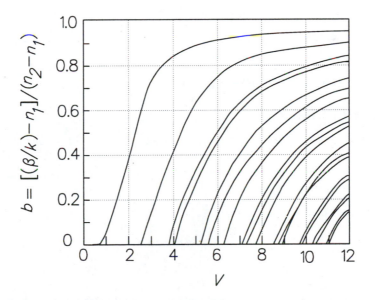

Figure 2.4 Relative modal propagation coefficient vs. V-parameter for a step-index fiber. (After D. Gloge, *Applied Optics*, vol. 10, no. 10, pp. 2252–2258, 1971.)

We note from Fig. 2.4 that, for any given value of V, there are, in general, several modes that can propagate in a step-index fiber. The number of modes present in a fiber is dependent on the value of V (i.e., the number of modes present increases as V increases). The value of the propagation coefficient of any particular mode can vary from a minimum of $n_2 k$ (i.e., the propagation coefficient of the cladding) to a maximum of $n_1 k$ (i.e., the propagation coefficient of the core). The actual value of the propagation coefficient (and, hence, the mode's velocity) is somewhere between these extremes; each mode travels at its own velocity.

---------------- ∞ ----------------

Example: (a) Consider a fiber with $a = 25$ μm, $n_1 = 1.45$ and $\Delta = 0.9\%$. If this fiber is operated at 1550 nm, how many modes will it have?

Solution: The value of V is found as

$$V \;=\; \frac{2\pi a}{\lambda}\sqrt{2\Delta} = \frac{2\pi(4 \times 10^{-6})}{1550 \times 10^{-9}}\sqrt{2(0.009)} = 2.17. \tag{2.11}$$

From Fig. 2.4 on the page before, we see that at $V = 2.17$, there is only one mode; it has a normalized propagation coefficient value, $b \approx 0.35$.

(b) Repeat if Δ is increased to 3.5%.

Solution: The new value of V is found as

$$V \;=\; \frac{2\pi a}{\lambda}\sqrt{2\Delta} = \frac{2\pi(4 \times 10^{-6})}{1550 \times 10^{-9}}\sqrt{2(0.035)} = 4.29. \tag{2.12}$$

From Fig. 2.4 on the preceding page, we see that at $V = 4.29$, there are four modes; they have approximate normalized propagation coefficient values of 0.14, 0.28. 0.65, and 0.88. (Note: fibers are not usually designed to have values of V in this range.)

---------------- ∞ ----------------

2.4   Multimode Step-Index Fibers

2.4.1   Number of Modes

For the step-index profile curves of Fig. 2.4 on the page before, we observe that, for any value of V larger than 2.405, more than one mode will exist. For large values of V, many electromagnetic modes can be supported by the fiber waveguide structure. An estimate of the *total number of modes* N supported at any given large value of V ($V \gg 2.405$) is [16]

$$N \;\approx\; \frac{V^2}{2} \quad \text{(for } V \gg 2.405\text{)} \tag{2.13}$$

$$\approx\; (kan_1)^2 \, \Delta = \left(\frac{2\pi a n_1}{\lambda}\right)^2 \Delta.$$

Such a fiber, operated in this region of many modes, is called a *multimode* fiber.

In a fiber with a large number of modes, we can use ray theory to describe the modes (otherwise, an electromagnetic field approach must be used, as is done with single-mode fibers). As seen in Fig. 2.5 on the facing page, the highest-order modes correspond to the rays that are at the steepest angles (i.e., closest to the critical angle); the lowest-order modes correspond to those rays that strike the interface at low, grazing angles. This simple picture is useful for envisioning the behavior of the modes and in performing simple modeling of connectors, splices, and coupling of light from sources into fiber. (For single-mode fibers, we emphasize that the ray model of the fiber-optic modes is *not* valid. The electromagnetic wave is no longer confined primarily to the core and can have appreciable energy in the cladding. Hence, the propagation properties of the wave are a combination of the core propagation properties and the cladding propagation properties. This leads us to use the mode field diameter description of the single-mode fiber, described in a later section.)

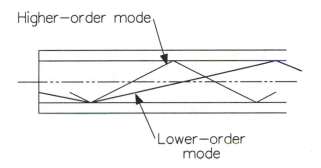

Figure 2.5 Ray diagram illustrating high-order modes and low-order modes as rays in a step-index multimode fiber.

Deviations from the ideal waveguide (because of geometry fluctuations, impurities, bends, etc.) cause energy to be coupled out of the lower-order modes into the higher-order modes and from the higher-order modes into the cladding modes where the energy is lost. (The process of transferring energy between modes is called *mode mixing*.) Using the ray approach, a localized bend in the core-cladding interface, for example, will change the angle of incidence of the ray that would have been at the critical angle in an unperturbed fiber. Because of this new angle of incidence, some of the light will be transmitted into the cladding and lost. The total amount lost depends on the severity and the frequency of such perturbations.

The excitation of the modes is a function of the beam pattern of the exciting source at the input end of the fiber, the geometry of the fiber, the optical properties of the fiber, and the alignment of the source and fiber. Phenomena involving multimode fibers are usually modeled assuming that the modes are all equally excited. Measurements of multimode fiber parameters must incorporate features to ensure excitation of all of the modes in the fiber.

2.4.2 Power Distribution Between Core and Cladding

The electromagnetic fields in the fiber must meet the boundary conditions across the core-cladding interface. Unlike the metallic waveguide problem where the fields are zero in the walls of the waveguide, the field in the fiber cladding is *not* zero. Fields exist in both the core and the cladding. Therefore, some of the power of the mode is carried in the core and some in the cladding. It is instructive to consider the power distribution in the core and the cladding for the different modes in a step-index multimode fiber, as shown in Fig. 2.6 on the next page [1]. Using the left axis, we can estimate, for a given mode, the fraction of the total power in the cladding; with the right axis, we can find the fraction of the power in the core. We also note that, as the V parameter approaches cutoff for any particular mode, more of the power of the mode is in the cladding. At cutoff, all of the power transfers to the cladding, the mode becomes radiative, and it ceases to exist as a propagating mode. For large values of V, the modes that are close to cutoff can be ignored, compared to the propagating modes, and the fraction of the total power that is found in the cladding can be estimated by [1]

$$\frac{P_{\text{cladding}}}{P_{\text{core}}} \approx \frac{4}{3\sqrt{N}} \qquad (V \text{ large}), \qquad (2.14)$$

where N is the total number of modes in the fiber, as given by Eq. 2.13 on page 16. Efforts to reduce the number of modes by reducing V usually cause more of the total power to be carried in the cladding where it is susceptible to effects from the outside environment, which can lead to undesirable high losses.

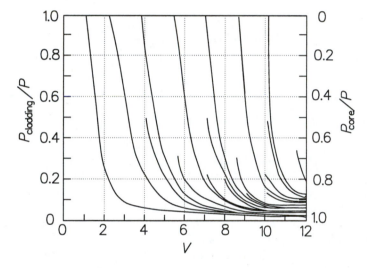

Figure 2.6 Power distribution in cladding (left axis) or core (right axis) vs. V-parameter for a step-index fiber. (After D. Gloge, *Applied Optics*, vol. 10, no. 10, pp. 2252–2258, 1971.)

2.5 Numerical Aperture

One of the primary parameters of a fiber is its *numerical aperture*. This parameter is used in equations describing the coupling losses of the source light entering the fiber, the mode excitation, connector and splice losses, and other system performance equations. We will now use our ray theory that is applicable to multimode fibers to derive an expression for the numerical aperture.

Figure 2.7 on the next page shows a step-index fiber. Consider a ray, internal to the fiber, that is incident on the core-cladding interface at exactly the critical angle. If we extend that ray back through the fiber-air interface at the input by applying Snell's law, it will have an angle of incidence given by θ_{max}. Some thought should convince you that all incident rays with angles *less than* θ_{max} will have a core-cladding angle of incidence greater than the critical angle and will be guided down the (ideal) fiber. All input rays with an angle *greater than* θ_{max} will intercept the core-cladding interface at an angle less than the critical angle, will be partially transmitted through the interface, and, hence, will suffer some loss at each reflection. They will soon be attenuated into insignificance if the fiber is relatively long. (Low-loss fibers of short length [i.e., a few meters] can carry significant power to the receiver in these cladding modes.) It can be shown (see the problems at the end of chapter) that the expression for $\theta_{\max}$ is given by

$$\theta_{\max} = \sin^{-1}\left(\sqrt{n_1^2 - n_2^2}\right) \approx \sin^{-1}\left(n_1\sqrt{2\Delta}\right). \tag{2.15}$$

Hence, this angle depends only on the indices of refraction of the core and cladding; it is

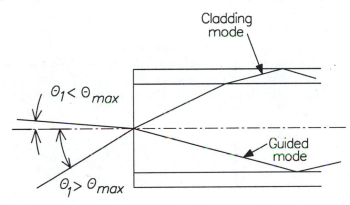

Figure 2.7 Entry conditions for a step-index fiber showing maximum acceptance angle.

independent of the diameter of the fiber. The *numerical aperture* NA of the fiber is the sine of this maximum input angle and is given by

$$NA = \sin\theta_{\max} = \sqrt{n_1^2 - n_2^2} \approx n_1\sqrt{2\Delta}. \tag{2.16}$$

(It is worth noting that the numerical aperture of a fiber is also independent of the fiber-core radius.)

———————————— ∞ ————————————

Example: Consider a 50/125 fiber with a core index of 1.49 and $\Delta = 1.5\%$.

(a) Calculate the maximum acceptance angle,

Solution: Using Eq. 2.15,

$$\theta_{\max} = \sin^{-1}(n_1\sqrt{2\Delta}) = \sin^{-1}\left(1.49\sqrt{2(0.015)}\right) = 14.96°. \tag{2.17}$$

(b) Find the numerical aperture of the fiber.

Solution: Using Eq. 2.16,

$$NA = n_1\sqrt{2\Delta} = 1.49\sqrt{2(0.015)} = 0.258. \tag{2.18}$$

(c) Calculate the maximum acceptance angle and NA for the same fiber, except with a 62.5/125 diameter.

Solution: Since $\theta_{\max}$ and NA are independent of the fiber size, the same answers apply (i.e., $\theta_{\max} = 14.96°$ and NA = 0.258).

———————————— ∞ ————————————

2.5.1   Cladding Modes and Leaky Modes

As noted in Fig. 2.7 on the preceding page, certain rays in the incident radiation are not captured by the core of the fiber, but pass through the core-cladding interface into the cladding region. Because of the finite radius of curvature of the outer cladding surface, some of this light at this boundary will be reflected back into the cladding, where it can be trapped and propagated. This light forms the *cladding modes* of the fiber, and appreciable coupling can occur with the higher-order modes of the core, resulting in increased loss of the core power. Cladding modes are suppressed by placing a high-loss material outside of the cladding surface that will absorb the light as it strikes the interface, by increasing the scattering at the cladding interface to extract the cladding modes, or by surrounding a portion of the fiber with a material whose index of refraction matches that of the cladding, causing the cladding light to transmit into the index-matching material (see Fig. 2.8). This latter technique is called *mode stripping*.

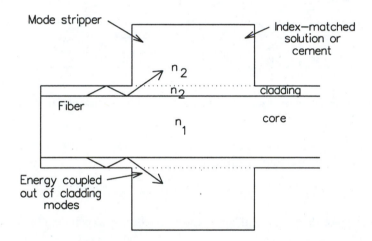

Figure 2.8 Removal of cladding power by a mode stripper.

A third type of mode is the *leaky mode*, which is a nonpropagating mode with significant power shared between the core and the cladding. They are predicted by theory and occur near the cutoff conditions for propagating modes. They are attenuated in long fibers but can carry significant power in short fibers. These modes also play a role at connectors and splices, occurring in fibers where the connector or splice can cause conversion of energy from a propagating mode to a leaky mode.

Having considered the properties of multimode fibers, we now turn our attention to fibers that support a single electromagnetic mode.

2.6   Step-Index Single-Mode Fibers

Single-mode (or *monomode*) fibers are characterized as having only one propagating mode. In the next chapter, we will find that these fibers offer superior performance to multimode fibers for carrying high-data-rate signals.

2.6.1 Cutoff Wavelength

From the mode diagram of Fig. 2.4 on page 15, we observe that, for small values of V ($V < 2.405$), only one mode, called the LP_{01} mode, will propagate. Although this mode is doubly degenerate (i.e., two orthogonal polarizations of this mode can simultaneously exist), this region of operation is called *single-mode operation*. We note that, generally, one wants a small core radius (typically several wavelengths) and a small index difference (typically less than 1%) between the core and cladding to achieve this relatively low value of V. For a given fiber geometry, the value of λ that makes V equal to 2.405 is the (theoretical) *cutoff wavelength* of the fiber. (The actual cutoff wavelength is very susceptible to variations in the fiber parameters and, so, is usually a measured fiber parameter.)

From the power distribution curve (Fig. 2.6 on page 18), we note that, at $V = 2.405$, approximately 84% of the mode's power is in the core; while at $V = 1$, only 30% is in the core. We note, also, that about 75% of the power is in the core when $V = 2.0$. The power in the cladding is at risk to being removed by jacket losses, splices, and other loss mechanisms. For this reason single-mode fibers keep V in the range $2.0 < V < 2.405$. Some compromise is required here. If V is made too close to 2.405, the cutoff wavelength of the fiber will be too close to the designed operating wavelength; if V is made too small, significant cladding power will be removed, raising the optical losses too much.

$$\text{———————} \infty \text{———————}$$

Example: (a) Calculate the required Δ if a fiber with an 8 μm core and a 125 μm cladding is to be single-mode at 1300 nm. Assume that the core index is 1.46.

Solution: For single-mode operation we want V to be in the range, $2.0 < V < 2.405$. We will arbitrarily choose $V = 2.1$.

$$V = \frac{2\pi a}{\lambda} n_1 \sqrt{2\Delta} \tag{2.19}$$

$$\sqrt{2\Delta} = \frac{V\lambda}{2\pi a n_1}$$

$$\sqrt{2\Delta} = \frac{2.1(1300 \times 10^{-9})}{2\pi(4 \times 10^{-6})(1.46)} = 7.44 \times 10^{-3}$$

$$\Delta = 0.277\%.$$

Note: Tight control on the difference in refractive index is required, as well as the ability to make a small core.

(b) Calculate the cutoff wavelength for this single-mode fiber.

Solution: The cutoff wavelength λ_c is the value that will make $V = 2.405$,

$$V = 2.405 = \frac{2\pi a n_1 \sqrt{2\Delta}}{\lambda_c} \tag{2.20}$$

$$\lambda_c = \frac{2\pi a n_1 \sqrt{2\Delta}}{2.405} \tag{2.21}$$

$$= \frac{2\pi(4 \times 10^{-6})(1.46)(\sqrt{2(2.77 \times 10^{-3})})}{2.405} = 1136 \text{ nm}.$$

This fiber is no longer single-mode for wavelengths below 1136 nm.

———————————————— ∞ ————————————————

We now want to turn our discussion to the parameters used to describe the behavior of single-mode fibers. Since the ray theory portrayal of the electromagnetic modes does not apply to single-mode fiber, we need an alternative representation. Also, since the size of the core a and the fractional index difference Δ are smaller than multimode fibers, some of the measurement techniques applied to single-mode are more difficult, due to the reduced tolerances.

2.6.2   Mode Field Diameter

In single-mode fibers we have seen that an appreciable part of the wave is contained in the cladding outside of the core. For such waves the use of the core diameter to express the "width" of the wave is no longer possible, since the wave "width" is now larger (or smaller) than the core. We still want to know the "width" of the wave because this quantity is important in predicting the cabling losses, the losses due to bends, and the joining losses when the cables are connected or spliced together. (These losses are discussed in detail in Chapter 3 on fiber properties and Chapter 4 on splices and connectors.)

For single-mode fibers a useful measure of the field "width" has been the *mode field diameter* (usually abbreviated "MFD"). This measure has, unfortunately, been mathematically defined in three different ways, as we shall describe. If the fiber produces a Gaussian-shaped field, then the definitions all reduce to the same value; if the field is non-Gaussian (as occurs in varying degrees in dispersion-adjusted fibers) or the fiber is non-symmetric (as might intentionally be the case in a polarization-maintaining fiber), then the definitions diverge slightly. Originally, simple step-index single-mode fiber designs produced Gaussian-shaped beams; recent advances in designing dispersion-shifted fibers and dispersion-flattened fibers have produced non-Gaussian beams, resulting in revised definitions of the MFD.

Usually, circular fiber geometries are assumed. (New definitions are also required to accommodate noncircular geometries [e.g., elliptical-core polarization-maintaining fibers]. Reference [17] suggests methods that can be applied to these nonsymmetric fibers.) We can measure or observe the near-field spatial optical amplitude distribution [18], $e(r)$ (where r is the radial coordinate), and the far-field angular amplitude distribution [18], $E(\rho)$ (where ρ is an angular variable defined below). These distributions are related by

$$E(R,\rho) = \frac{k\cos\theta}{iR}\exp\left(ikR\right)\int_0^\infty e(r)J_0(r\rho)r\,dr\,, \qquad (2.22)$$

where R is the observation distance from the end of the fiber and

$$\rho = k\sin\theta\,. \qquad (2.23)$$

Here r, ϕ, and θ are the spherical coordinates with the origin located at the center of the fiber end. To ensure that the far-field observation is truly in the far-field requires that the observation distance R from the end of the fiber obey $R >> r_{\max}^2/\lambda$, where $r_{\max}$ is the maximum extent of the near-field (typically estimated as being a little larger than the radius of the fiber core). Usually the $\cos\theta$ term is neglected for the small-valued range of θ expected. Then

$$E(\rho) \quad = \quad \int_0^\infty e(r)J_0(r\rho)\,r\,dr = \frac{1}{\sqrt{2\pi}}\mathcal{H}\left\{e(r)\right\}\,. \qquad (2.24)$$

where $\mathcal{H}$ is the Hankel-transform operator. This means that the far-field angular distribution $E(\rho)$ is the zero-order Hankel transform of the near-field distribution.

The light *intensity* distributions are given by $|e(r)|^2$ and $|E(\rho)|^2$. Since $e(r)$ and $E(\rho)$ are real functions [17], we can use $e^2(r)$ and $E^2(\rho)$ to represent the intensity distributions. The "width" of the wave, or MFD, can be defined from the measurement of the widths of these intensity patterns, either in the fiber near-field or far-field.

Three definitions of the mode field diameter have been made [17], depending on how the extent of the field is measured.

- MFD I: Near-field mode field diameter—The definition for the *near-field mode field diameter* d_n is

$$d_n = 2\sqrt{2}\sqrt{\frac{\int_0^\infty e^2(r)\,r^3\,dr}{\int_0^\infty e^2(r)\,r\,dr}}.$$ (2.25)

  This definition is the square root of the ratio of the third moment of the near-field intensity pattern to the first moment of the pattern. This MFD is also called the *Petermann I MFD* [19].

- MFD II: Far-field mode field diameter—Since the angular width of the far-field pattern (to first-order approximation of the diffraction phenomenon) is inversely proportional to the width of the near-field intensity pattern, we can also consider the *inverse rms (root-mean-square) width* w_{ff} of the far-field intensity pattern (with units of inverse length) as

$$w_{ff} = \sqrt{\frac{\int_0^\infty E^2(\rho)\rho^3\,d\rho}{\int_0^\infty E^2(\rho)\rho\,d\rho}}.$$ (2.26)

Note that this definition is the square root of the ratio of the third moment of the far-field intensity pattern to the first moment of that pattern. It is similar in form to the MFD-I definition. The *far-field MFD* or the *Petermann II MFD* [19] is

$$d_f = \frac{2\sqrt{2}}{w_{ff}} = 2\sqrt{2}\sqrt{\frac{\int_0^\infty E^2(\rho)\rho\,d\rho}{\int_0^\infty E^2(\rho)\rho^3\,d\rho}}.$$ (2.27)

This definition of the mode field diameter has been adopted by various international standards-setting organizations.

A few observations [17] are in order about the near-field and far-field MFDs.

- We note that the upper limit on the ρ integrals in Eq. 2.27 should actually be $\rho = k$ (corresponding to $\theta = \pi/2$); we use ∞ to be able to use Hankel transforms, with the understanding that $E^2(\rho)$ is zero for $\rho \geq k$. For small values of θ, e and E are, then, a Hankel transform pair. If so, then d_n can be expressed in terms of $E(\rho)$, and d_f can be found from $e(r)$ as

$$d_n = 2\sqrt{2}\sqrt{\frac{\int_0^\infty |E'(\rho)|^2\,\rho\,d\rho}{\int_0^\infty E^2(\rho)\,\rho\,d\rho}}$$ (2.28)

  and

$$d_f = 2\sqrt{2}\sqrt{\frac{\int_0^\infty e^2(r)\,r\,dr}{\int_0^\infty |e'(r)|^2\,r\,dr}},$$ (2.29)

where the prime indicates differentiation. Hence, we are able to find the near-field MFD from a far-field measurement and the far-field MFD from a near-field measurement, if we so desire.

– It can be shown [17] that, as defined, $d_f \leq d_n$.

– The near-field and far-field MFD definitions are equal if and only if [17] $re(r)$ and $e'(r)$ are proportional to each other. This is true for a Gaussian-shaped wave. If

$$e(r) = e_g(r) = A \exp \left(-\frac{r^2}{2w^2} \right), \qquad (2.30)$$

where w is the Gaussian beam radius, then $d_f = d_n = 2\sqrt{2}\,w$. (From this diameter, we can define $\sqrt{2}\,w$ as the *field beam radius*.)

How good is an assumption that the field in a single-mode fiber is Gaussian-shaped? A comparison of the measured near-field MFD and the far-field MFD would tell us. Reference [17] demonstrates the variation of d_n and d_f with wavelength for two single-mode fibers, a step-index fiber and a special fiber called a "dispersion-flattened" fiber. The two measured MFDs were found to be within 5% of each other for the conventional step-index single-mode fiber, indicating that the Gaussian assumption is pretty good for this fiber. However, the MFDs were found to vary significantly (20% to 45%) for the dispersion-shifted fiber, showing that a Gaussian assumption should not be made for this fiber.

• MFD III: Transverse-offset mode field diameter—Yet another definition of MFD is related to the transverse offset technique [18] that can be used to measure the MFD. In this measurement, two identical single-mode fibers are butt-coupled (i.e., aligned with a tiny longitudinal separation) and laterally aligned for maximum transmission. One fiber is then laterally displaced with a micrometer and the transmitted power is recorded. As the translation increases, less of the power is collected by the receiving fiber. The MFD d_a is defined with this technique as the $1/e$ width of the transmission curve.

It can be shown [17] that all three definitions of mode field diameter are the same *if* the field is Gaussian (as in step-index fibers); otherwise, the alternative definition d_a has little utility (other than being fairly easy to measure) and has grown out of favor with the increased use of dispersion-shifted and dispersion-flattened fibers.

2.6.3   Multi-Step Single-Mode Fibers

The expression for V that we have been using is valid for the simple step-index profile. The reader should be aware that other profiles are being increasingly used in single-mode fibers in an effort to improve the information-carrying capacity of the fiber. Some of these profiles are illustrated in Fig. 2.9 on the next page. The top curve is the simple step-index fiber as described earlier. Other profiles shown include the *depressed cladding*, the *W profile*, the *triangular profile*, and the *segmented profile*. Each of these profiles has its own electromagnetic mode distributions and the value of V at cutoff will be different. These profiles are under research investigation by various manufacturers, and it is not yet obvious which ones offer superior properties over the others. These more elaborate profiles are discussed further in the section on dispersion shifting in Chapter 3.

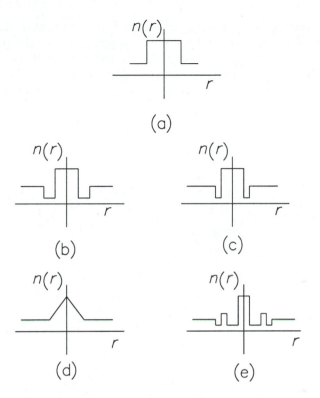

Figure 2.9 Variations of the step-index profile: (a) Regular profile, (b) Depressed cladding profile, W profile, (d) Triangular profile, (e) Segmented profile.

2.7 Graded-Index Multimode Fibers

So far, we have described fibers with a step shape to their index profile $n(r)$. A second type of fiber profile does not have the abrupt jump in the index of refraction nor does it use total internal reflection at the core-cladding interface to guide the wave. It has a parabolic (or almost-parabolic) variation in the index, as shown in Fig. 2.10 on the following page. A fiber with this index profile is a *graded-index fiber*. Notice that the index of refraction is a maximum at the center (with value n_1) and tapers off to a minimum value (n_2) at the edge of the core. The confinement process in such a fiber is a consequence of the solution of the electromagnetic propagation problem in a dielectric medium with such a refractive index profile. These waves follow a sinusoidal path, as shown in Fig. 2.11 on the next page. We see from the figure that the light path undulates from one side to the other without any reflection from the interface. One mathematical model for the radial index profile is

$$n(r) = \begin{cases} n_1\sqrt{1 - 2\Delta \left(\dfrac{r}{a}\right)^g} & \text{for } r < a \\ n_1\sqrt{1 - 2\Delta} \approx n_1(1 - \Delta) = n_2 & \text{for } r > a, \end{cases} \tag{2.31}$$

where n_1 is the index of refraction at the center, Δ is the fractional change in the index of refraction from the center to the edge of the core, r is the radial distance, and a is the radius of the core. The quantity g is the *profile parameter* or the *gradient* of the graded-index fiber. Note that $g = 2$ is a parabolic profile, $g = 1$ is a triangular profile, and $g = \infty$ is a step-index profile. In these fibers the cladding $(r > a)$ plays the role of a mechanical support and buffer from the outside world; it has no role in guiding the light. While a parabolic gradient is close to the optimum profile, other profiles also have been analyzed and evaluated in various laboratories.

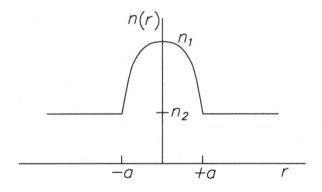

Figure 2.10 Index of refraction vs. radial position for a graded-index fiber.

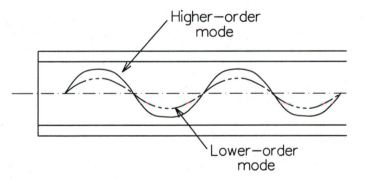

Figure 2.11 Ray path of modes in a graded-index fiber.

The approximation in the above relation follows from the definition of Δ for the graded-index fiber as

$$\Delta = \frac{n_1^2 - n_2^2}{2n_1^2} = \frac{(n_1 + n_2)(n_1 - n_2)}{2n_1^2} \approx \frac{n_1 - n_2}{n_1}, \tag{2.32}$$

which is the same expression as in the step-index fiber.

A typical graded-index multimode fiber has an outside diameter of 125 or 140 μm with a fractional change of the index of refraction of 1–2% induced by varying the doping level of impurities radially.

2.7.1   Number of Modes: GI Fiber

The V-parameter for graded-index fibers (for typically small values of Δ) is defined [16] in the same way as step-index fibers,

$$V = n_1 k a \sqrt{2\Delta} = \frac{2\pi a n_1 \sqrt{2\Delta}}{\lambda}. \tag{2.33}$$

The *number of modes* N in a multi-mode graded-index fiber can be estimated [16] by the approximation

$$N \approx \left(\frac{g}{g+2}\right)\left(\frac{4a^2\pi^2 n_1^2 \Delta}{\lambda^2}\right) \approx \left(\frac{g}{g+2}\right)\left(\frac{V^2}{2}\right). \tag{2.34}$$

────────────────── ∞ ──────────────────

Example: Consider a 50/125 graded-index fiber with $g = 2$. If $n_1 = 1.48$ and $n_2 = 1.46$,

(a) Calculate Δ from the exact expression of Eq. 2.32 on the facing page.

Solution: The fractional difference in refractive index, Δ, is

$$\Delta = \frac{n_1^2 - n_2^2}{2n_1^2} = \frac{(1.48)^2 - (1.46)^2}{2(1.48)^2} = 1.342\%. \tag{2.35}$$

(b) Calculate Δ from the approximation of Eq. 2.32 on the preceding page.

Solution: Using the approximation, we find

$$\Delta \approx \frac{n_1 - n_2}{n_1} \approx \frac{1.48 - 1.46}{1.48} \approx 1.351\%. \tag{2.36}$$

Hence, we see that the approximation is quite good.

(c) Calculate the number of modes in the graded-index fiber at an operating wavelength of 850 nm.

Solution: Using Eq. 2.34,

$$N \approx \left(\frac{g}{g+2}\right)\left(\frac{4a^2\Delta\pi^2 n_1^2}{\lambda^2}\right) \tag{2.37}$$

$$\approx \left(\frac{2}{2+2}\right)\left(\frac{4(25\times10^{-6})^2(0.0135)\pi^2(1.48)^2}{(850\times10^{-9})^2}\right) \approx 504 \text{ modes}.$$

────────────────── ∞ ──────────────────

2.7.2   Numerical Aperture

The *numerical aperture of a graded-index fiber* is more difficult to define than a step-index fiber since, unlike the step-index fiber, the maximum acceptance angle $\theta_{\max}$ of a ray is a function of the radial position of the entry location. A local NA can be defined by

$$\mathrm{NA}(r) = \begin{cases} \mathrm{NA}(0)\sqrt{1 - \left(\frac{r}{a}\right)^g} & r < a \\ 0 & r > a \end{cases} \tag{2.38}$$

where $\mathrm{NA}(r)$ is the *local numerical aperture* and $\mathrm{NA}(0)$ is the numerical aperture at the center of the fiber core (as computed by Eq. 2.16 on page 19). This equation predicts that the local NA decreases from the on-axis value as one moves away from the center of the fiber.

2.7.3   Single-Mode Graded-Index Fibers

It is possible to make single-mode graded-index fibers. The cutoff value of V for single-mode operation in a fiber with a graded refractive index profile that follows Eq. 2.31 on page 25 is different from the 2.405 value used for a step-index fiber. For a parabolic-index fiber ($g = 2$) the cutoff value is 3.53; for a triangular-index profile ($g = 1$), it is 4.38 [20]. An estimate of the cutoff value of V for single-mode propagation in a graded-index fiber can be shown to be [7]

$$V_{\text{co}} \approx 2.405 \sqrt{1 + (2/g)}. \tag{2.39}$$

If other parameters are the same, the core diameter of the graded-index single-mode fiber can be a factor of $\sqrt{1 + (2/g)}$ larger than the equivalent step-index fiber. This provides an increased ease of coupling light, an increased ease in splicing, and a reduction in the susceptibility of the fiber to microbend-induced losses (discussed in Chapter 3). A parabolic-index fiber ($g = 2$) provides an improvement by a factor of $\sqrt{2}$; the triangular-profile fiber ($g = 1$) provides a $\sqrt{3}$ improvement.

2.8   Summary

In this chapter we have introduced the fiber and some of the parameters used to describe the fiber's performance. The basic optical fiber consists of a core and a cladding; more advanced designs can add more cladding layers. The refractive index profile is generally modeled as either a step profile or a power-law profile described by Eq. 2.31 on page 25. We have seen that a fiber can be designed to be either single-mode or multimode. In a multimode fiber, the fiber V-parameter, diameter $2a$, and the fiber numerical aperture are three of the prime fiber parameters. In a single-mode fiber the mode field diameter and cutoff wavelength describe its characteristics.

Multimode fibers (especially graded-index fibers) are used for applications calling for moderate distances or data rates or both. The larger core size makes coupling light into them relatively easy. Their primary disadvantage is a lack of bandwidth capacity (compared to single-mode fibers). Also, because of the mode mixing effects described, it is difficult to model and predict the loss mechanisms in the fiber and at connections and splices.

Single-mode fibers are currently the fiber of choice for applications combining long-distance and high data-rate. Their advantages in high data-capacity and low attenuation have overcome the disadvantages of more difficult fabrication tolerances and more difficult coupling of light from the source. Currently, single-mode fibers are price-competitive with most multimode fibers. Single-mode fibers are more susceptible to increases in losses (excess losses) due to bends in the fiber as it turns corners and to losses induced by core-diameter fluctuations because of spooling and handling.

Both multimode and single-mode fibers are readily available commercially. The choice of the fiber type is dictated by the system design trade-offs described in Chapter 7.

2.9   Problems

1. If $n = 1.5$, calculate the velocity of light in glass.
2. (a) Find n_2 if $\Delta = 1\%$ and $n_1 = 1.48$.
   (b) ...if $\Delta = 2\%$?

(c) ...if $\Delta = 0.1\%$?

3. Light traveling in air strikes a glass plate with an angle of incidence of 57°. If the reflected and refracted beams make an angle of 90° with each other,...

   (a) ...calculate the refractive index of the glass.

   (b) What is the critical angle for this material if the light travels from glass into air?

4. A point source of light is located 1 m below a water-air interface. Find the radius of the light circle seen by an observer positioned over the source (outside of the water). The refractive index of water is 1.333.

5. (a) Consider a fiber with a 100 μm core diameter and a 140 μm cladding diameter. If $n_1 = 1.48$ and $\Delta = 1\%$, calculate the V-parameter if the operating wavelength is 850 nm.

   (b) ...if the wavelength is 1300 nm?

   (c) Find the value of V at a wavelength of 850 nm if the diameter of the core is 50 μm?

6. (a) Calculate the number of modes for each case described in the previous problem.

   (b) Calculate the percentage of the optical power that is carried in the cladding for each case described in the prior problem.

7. (a) Calculate the numerical aperture of a step-index fiber having a core index of 1.47 and a cladding index of 1.45.

   (b) Find the value of the largest angle made by a ray that is accepted by the fiber if the outer medium is air.

   (c) ...if the outer medium is water? $(n = 1.33)$

8. (a) Determine the mode parameter V at 820 nm for a step-index fiber with a 50 μm core diameter, $n_1 = 1.47$ and $n_2 = 1.45$.

   (b) How many modes will propagate in this fiber at 820 nm?

   (c) ...at 1300 nm?

   (d) What percentage of the optical power is flowing in the cladding for each operating wavelength?

9. (a) Find the core diameter required to ensure single-mode operation of a step-index fiber with $n_1 = 1.485$ and $n_2 = 1.480$ at a wavelength of 820 nm. At 1300 nm?

   (b) What are the NA and θ_{max} for this fiber?

10. (a) Design a single-mode step-index fiber for operation at 1300 nm with a fused silica core ($n_1 = 1.458$). Find n_2 and the diameter of the core.

    (b) Is the fiber still single-mode at 820 nm? If not, how many modes are there?

    (c) Calculate the cutoff wavelength λ_c.

11. (a) Consider a step-index fiber with a core diameter of 50 μm, a core index of 1.450, and a fractional index difference of 1.3%. Find the values of the V-parameter and NA of the fiber if the operating wavelength is 820 nm.

    (b) Does the number of modes in the fiber increase or decrease if n_1 increases?

    (c) ...if λ increases?

12. A step-index fiber has $n_{core} = 1.450$, $n_{cladding} = 1.440$, and will operate at 820 nm.

    (a) Find the diameter of the core that will ensure single-mode operation.

    (b) If the core diameter is 50 μm, how many modes will the fiber have?

    (c) Calculate the numerical aperture of the fibers in parts (a) and (b).

13. Calculate V and the numerical aperture of a step-index multimode fiber if $n_1 = 1.450$, $\Delta = 1.3\%$, $\lambda_0 = 0.82$ μm, and $a = 25$ μm.

14. Design a single-mode fiber (with $V = 2.3$) for operation at 1300 nm with a fused silica core ($n_1 = 1.458$). The numerical aperture of the fiber is to be 0.10.

    (a) Find the cladding index n_2 and the radius a of the fiber.

    (b) Calculate the approximate number of modes in the fiber for operation at 820 nm.

15. Using a computer, plot the power-law refractive index profile from n_1 to n_2 vs. radial position for $g = 1, 2, 4, 8$ and ∞. Assume a core diameter of 50 μm, $n_1 = 1.480$, and $\Delta = 1.00\%$.

16. (a) Calculate the number of modes in a 50/125 graded-index fiber having a parabolic index (i.e., $g = 2.0$), $n_1 = 1.485$ and $n_2 = 1.460$ at an operating wavelength of 820 nm.

    (b) ...at a wavelength of 1300 nm?

    (c) Calculate the number of modes in an equivalent step-index fiber at both wavelengths.

17. Prove Eq. 2.15 on page 19 by applying Snell's law at the fiber input face for the ray that meets the critical angle condition at the core-cladding interface. For generality, assume that the medium outside the fiber has an index n_0 and then check the equation for $n_0 = 1$.

References

1. D. Gloge, "Weakly guiding fibers," *Applied Optics*, vol. 10, no. 10, pp. 2252–2258, 1971.

2. D. Gloge, "Dispersion in weakly guiding fibers," *Applied Optics*, vol. 10, no. 11, pp. 2442–2445, 1971.

3. L. Felsen, "Rays and modes in optical fibers," *Electronics Letters*, vol. 10, pp. 95–96, 1974.

4. D. Marcuse, "Theory of dielectric optical waveguides," in *Theory of Optical Waveguides*, New York: Academic Press, 1974.

5. J. Conradi, F. Kapron, and J. Dyment, "Fiber optical transmission between 0.8 and 1.4 micrometers," *IEEE Trans. on Electron Devices*, vol. ED–25, pp. 180–193, 1978.

6. D. Gloge, "The optical fiber as a transmission medium," *Rep. Prog. Phys.*, vol. 42, pp. 1777–1824, 1979.

7. D. Marcuse, D. Gloge, and E. A. Marcatili, "Guiding properties of fibers," in *Optical Fiber Telecommunications* (S. E. Miller and A. G. Chynoweth, eds.), pp. 37–100, New York: Academic Press, 1979.

8. R. Olshansky, "Propagation in glass optical waveguides," *Review of Modern Physics*, vol. 51, pp. 341–367, 1979.

9. D. Payne, A. Barlow, and J. R. Hansen, "Development of low and high birefringence optical fibres," *IEEE J. on Quantum Electronics*, vol. QE–18, no. 4, pp. 477–487, 1982.

10. M. Ramsay and G. Hockham, "Propagation in optical fibre waveguides," in *Optical Fibre Communication Systems* (C. Sandbank, ed.), pp. 25–41, New York: Wiley, 1980.

11. D. B. Keck, "Optical fiber waveguides," in *Fundamentals of Optical Fiber Communications* (M. F. Barnoski, ed.), pp. 1–107, New York: Academic Press, 1981.

12. C. Yeh, "Guided-wave modes in cylindrical optical fibers," *IEEE Trans. on Education*, vol. E–30, no. 1, pp. 43–51, 1987.

13. A. B. Buckman, *Guided-Wave Photonics*. Orlando, FL: Harcourt Brace Jovanovich, 1992.

14. G. P. Agrawal, *Fiber Optic Communication Systems*. New York: John Wiley & Sons, 1992.

15. J. A. Buck, *Fundamentals of Fiber Optics*. New York: Wiley, 1995.

16. A. H. Cherin, *Introduction to Optical Fibers*. New York: McGraw-Hill, 1983.

17. M. Artiglia, G. Coppa, P. DiVita, M. Potenza, and A. Sharma, "Mode field diameter measurements in single–mode optical fibers," *J. Lightwave Technology*, vol. 7, no. 8, pp. 1139–1152, 1989.

18. F. P. Kapron, "Fiber-optic test methods," in *Fiber Optics Handbook for Engineers and Scientists* (F. C. Allard, ed.), pp. 4.1–4.54, New York: McGraw-Hill, 1990.

19. K. Petermann, "Constraints for fundamental-mode spot size for broadband dispersion-compensated single-mode fibres," *Electronics Letters*, vol. 19, no. 18, pp. 712–714, 1983.

20. B. J. Ainslie and C. R. Day, "A review of single-mode fibers with modified dispersion characteristics," *J. Lightwave Technology*, vol. LT–4, no. 8, pp. 967–979, 1986.

Chapter 3

Fiber Properties

3.1 Introduction

In this chapter, we continue our discussion of the important optical fiber properties by considering optical attenuation, fiber dispersion, and an introduction to some of the effects of fiber nonlinearities.

As the signal propagates through the fiber, the optical power loss will eventually attenuate the signal until it is lost in the noise at the receiver. We want to understand the mechanisms that cause this loss and to show that there is a minimum loss in silica optical fibers at an operating wavelength of 1550 nm.

A narrow pulse that originates at the optical source will spread out as it propagates down the fiber. This spread in the pulse width (called pulse dispersion) can limit the data rate since we want to avoid overlapping pulses at the receiver. Multimode fibers are especially susceptible to this pulse spread; single-mode fibers have inherent advantages over multimode fibers in reducing the dispersion. In fact, we will find that a dispersion-free wavelength that eliminates the total dispersion in the fiber can be found near 1300 nm.

Now that we have fairly powerful laser sources (i.e., >1 mW) and small fiber cores, we are able to generate high power densities (mW/μm^2) in the fiber. Nonlinear effects that are negligible at lower power densities or at shorter propagation lengths can become important. These nonlinearities can limit the power that we can put into the fiber and/or limit the data rate (or link distance) that we can achieve. This chapter will also introduce these nonlinearities and their effects.

3.2 Fiber Losses

The power in the optical signal in a fiber decreases exponentially with distance. We can write

$$P(z) = P(0)e^{-\alpha_p z}, \tag{3.1}$$

or

$$\alpha_p = -\left(\frac{1}{z}\right)\ln\left(\frac{P(z)}{P(0)}\right) \tag{3.2}$$

where $P(z)$ is the power at a position z from the origin, $P(0)$ is the power in the fiber at the origin, and α_p is the fiber *attenuation coefficient* (in units of m^{-1}, cm^{-1}, or km^{-1}). The attenuation coefficient α_p is a function of several variables to be described in this section.

To avoid having to compute exponentials repeatedly, the optical losses of a fiber are given by an *attenuation coefficient*, α, expressed in *decibels per kilometer (dB/km)*. The expression for α is

$$\alpha = -\frac{10}{z[\text{km}]} \log\left(\frac{P(z)}{P(0)}\right). \tag{3.3}$$

For a given fiber, these losses are wavelength-dependent, and either a spectral distribution should be given (as in Fig. 3.1) or the losses must be specified at the operating wavelength. The value of the attenuation factor depends greatly on the fiber material and the manufacturing tolerances, but the general spectral shape of the loss curve shown in Fig. 3.1 is representative. In particular, it is important to note that there is an optimum operating wavelength (1550 nm for silica fibers) that reduces fiber loss to a minimum.

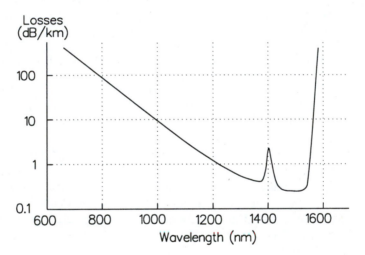

Figure 3.1 Spectral distribution of losses for a typical multimode silica fiber.

Example: An optical fiber has losses of 0.6 dB/km at 1300 nm. If 100 μW of power is injected into the fiber at the transmitter, how much will the power be at a distance of 22 km down the fiber?

Solution: We will use a dB method of calculating the output power. We begin by choosing an arbitrary power reference; useful references in fiber optics are 1 mW and 1 μW. We calculate the input power relative to 1 mW; the result will be expressed as *dBm* (i.e., in dB relative to 1 mW).

$$P(\text{dBm}) = 10 \log\left(\frac{P_{\text{in}}(\text{watts})}{1 \times 10^{-3}}\right). \tag{3.4}$$

(An alternative is to calculate the power in dB relative to 1 μW; the result is expressed as $dB\mu$. The two results can be shown to be related by the expression, $P(dB\mu) = P(dBm)+30$.)

$$P(\text{dBm}) = 10\log\left(\frac{P_{\text{in}}(\text{watts})}{1\times 10^{-3}}\right) \tag{3.5}$$

$$= 10\log\left(\frac{100\times 10^{-6}}{1\times 10^{-3}}\right) = 10\log(10^{-1}) = -10.0 \text{ dBm}.$$

The output power is reduced by 0.6 dB/km times the distance of 22 km ($= 0.6 \times 22 = 13.2$ dB). Subtracting the losses, we have

$$P_{\text{out}}(\text{dBm}) = P_{\text{in}}(\text{dBm}) - \text{losses(dB)} = -10 - 13.2 = -23.2 \text{ dBm}. \tag{3.6}$$

Finding the power, we compute

$$P_{\text{out}} = 10^{-23.2/10} = 4.78 \times 10^{-3} \text{ mW} = 4.78 \ \mu\text{W}. \tag{3.7}$$

Note that the result of a conversion from dBm has the units of milliwatts.

Alternative solution I: Working this problem in $dB\mu$, we convert the input power to units of $dB\mu$,

$$P_{\text{in}}(\text{dB}\mu) = 10\log\left(\frac{P_{\text{in}}(\text{watts})}{1\times 10^{-6}}\right) \tag{3.8}$$

$$= 10\log\left(\frac{100\times 10^{-6}}{1\times 10^{-6}}\right) = 20 \text{ dB}\mu.$$

The output power (in $dB\mu$) is the input power (in $dB\mu$) minus the losses (in dB),

$$P_{\text{out}}(\text{dB}\mu) = P_{\text{in}}(\text{dB}\mu) - \text{losses(dB)} = +20 - 13.2 = 6.8 \text{ dB}\mu. \tag{3.9}$$

Finding the power we have

$$P_{\text{out}} = 10^{6.8/10} = 4.78 \ \mu\text{W}. \tag{3.10}$$

Note that the result of a conversion from $dB\mu$ has the units of microwatts. Also you should note that we subtract the same number of dB for the losses in both cases, regardless of our choice of reference power.

Alternative solution II: The -13.2 dB loss in the fiber can be expressed as an equivalent transmission factor of

$$T = 10^{-13.2/10} = 47.9 \times 10^{-3}. \tag{3.11}$$

The output power, then, is

$$P_{\text{out}}(W) = T P_{\text{in}}(W) = (47.9 \times 10^{-3})(100 \times 10^{-6}) \tag{3.12}$$

$$= 47.9 \times 10^{-7} \text{ W} = 4.79 \ \mu\text{W}.$$

∞

Fiber losses are due to several effects; among the more important are

- material absorptions,

- impurity absorptions (particularly metallic ions and the hydroxyl ion from water vapor),

- scattering effects,

- interface inhomogeneities (both impurities and geometry imperfections), and

- radiation from bends.

We describe the effects of each by considering it in the absence of the other effects. The combined losses are just the sum of the losses (in dB) due to the individual effects.

3.2.1   Material Absorptions

The *material absorptions* are those due to the molecules of the basic fiber material, either glass or plastic. These losses represent a fundamental minimum to the attainable loss (in the absence of the other loss mechanisms) and can be overcome only by changing the fiber material. Indeed, the search continues for materials with ultra-low losses with several candidates (especially halide glasses and metal halides) being identified with extremely low losses in the middle-infrared region from 2 μm to 4 μm.

Impurity Ions

The primary source of *material impurities* in a glass fiber are metallic ions (primarily iron, cobalt, copper, and chromium) and the OH^- ion from water. Losses due to the metallic ions can be reduced to contributions below 1 dB/km by refining the glass mixture to an impurity level below 1 part per billion. The effects of water vapor in the glass are primarily located at wavelengths of 2.7 μm (a fundamental absorption) and of 950 and 725 nm (characteristic of second and third overtones of the fundamental). Peaks of attenuation located at these wavelengths are sometimes observed in the loss spectral distribution curve, similar to Fig. 3.1 on page 34. Minimizing this loss also requires hydroxyl-ion concentrations of one part per billion (which can be achieved by drying the glass in chlorine gas to leach out the water vapor).

Hydrogen Effects

An increased loss can occur when the fiber is exposed to hydrogen gas (which can be produced by corrosion of steel-cable strength members or by certain bacteria [1]). The hydrogen either can interact with the glass to produce hydroxyl ions and their losses or it can infiltrate the fiber and produce its own losses near 1.2 μm and 1.6 μm [2]. The solution is to eliminate the hydrogen-producing sources or to add a coating to the fiber that is impermeable to hydrogen.

Ultra-Low-Loss Fiber

Research [3–6] is being conducted to fabricate fibers from materials with lower intrinsic losses than the silica that is used in current fibers. Most of these predicted loss minima occur in the mid-IR portion (3–6 μm) of the optical spectrum. Of course, low loss does not, by itself, mean that a material is suitable for a fiber; other properties, such as mechanical strength, ability to fabricate the fiber, and the index of refraction and its control, help to determine the candidates for ultra-low-loss fibers.

Low-loss IR materials have already been studied, in general, for possible use in military systems and in mid-IR laser systems. They can be used to make optical windows, IR missile domes, and IR lenses, as well as optical fibers. Present candidates for materials suitable for ultra-low-loss fibers include *halide glasses* (e.g., $BeFl_2$ and $ZnCl_2$), single-component *chalcogenide glasses* (e.g., As_2S_3, GeS_2) or multicomponent chalcogenide glasses (with mixtures of As, Ge, P, S, Se, and Te), and *heavy-metal oxide glasses*, such as GeO_2, Sb_2O_3, TeO_2, and La_2O_3. Work is also being done at exploring *polycrystalline materials* [4] such as some of the *alkali halides* (e.g., AgCl, AgBr, TlBrI, TlBr, and an IR material known as KRS-5), *alkaline earth fluorides*, and some oxides (such as MgO and Al_2O_3). Some *single-crystal materials* such as AgBr, CsI, and KCl offer a theoretical minimum of 0.001 dB/km, but have proved hard to fabricate into fiber form (i.e., it is difficult to grow a crystal that is over 1 km long!).

Whether any of these materials are amenable to the growing of fibers and have the mechanical properties of glass fibers remains to be seen, but research is progressing rapidly.

3.2.2 Scattering Losses

Scattering losses occur when a wave interacts with a particle in a way that removes energy in the directional propagating wave and transfers it to other directions.

Linear Scattering

For *linear scattering*, the amount of light power that is transferred from a wave is proportional to the power in the wave. It is characterized by having no change in frequency in the scattered wave (unlike nonlinear scattering, where the frequency is changed upon scattering).

- *Rayleigh scattering* results from light interacting with inhomogeneities in the medium that are much smaller than the wavelength of the light. Such inhomogeneities can be minute changes in the refractive index of the glass at some locations, caused by changes in the composition of the glass (controllable with improving quality control), or by fluctuations in the glass density (fundamental and not improvable). This scattering strength is proportional to $1/\lambda^4$.

  Because the fundamental limits for a silica-based glass are the Rayleigh scattering at short wavelengths and the material absorption properties of silica at long wavelengths, a theoretical *attenuation minimum* for silica fibers can be predicted at a wavelength of 1550 nm where the two curves cross, as shown in Fig. 3.2 on the following page. This predicted minimum of attenuation has been one reason for the implementation of *long-wavelength* sources and receivers that work in this portion of the spectrum. Attenuations below 0.2 dB/km have been measured at these wavelengths.

- *Mie scattering* occurs at inhomogeneities that are comparable in size to a wavelength. They can be core-cladding refractive index variations over the length of the fiber, impurities at the core-cladding interface, strains or bubbles in the fiber, or diameter fluctuations. This scattering can have a large angular dependence (especially when the size of the scatterer is bigger than $\lambda/10$). This scattering can be reduced by carefully removing imperfections from the glass material, carefully controlling the quality and cleanliness of the manufacturing process, and keeping Δ relatively large.

Nonlinear Scattering

High values of electric field within the fiber (associated with fairly modest amounts of optical power) leads to the presence of nonlinear scattering interactions [7–10]. Such scattering causes

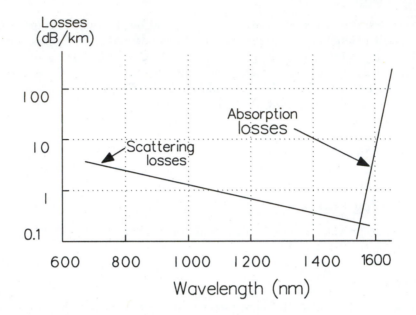

Figure 3.2 Attenuation valley between scattering and absorption in silica fibers.

significant power to be scattered in the forward, backward, or sideways directions, depending on the nature of the interaction. The scattering is accompanied by a frequency shift of the scattered light.

The major nonlinear scattering mechanisms are Brillouin scattering and Raman scattering. At normal power densities (i.e., the power per unit area), the interactions are negligible; at high power densities, such as might be encountered with more powerful diode lasers or smaller fiber cores, significant power can be removed from the light wave.

- *Brillouin scattering* is modeled as a modulation of the light by the thermal energy in the material. The incident photon of light undergoes the nonlinear interaction to produce vibrational energy (or "phonons") in the glass as well as the scattered light (as "photons"). The scattered light is found to be frequency modulated by the thermal energy, and both upward and downward frequency displacement is observed. The size of the frequency shift and the strength of the scattering vary as functions of the scattering angle, with the maximum occurring in the backward direction and the minimum of zero being observed in the forward direction. Hence, Brillouin scattering is mainly in the backward direction which directs the power toward the source and away from the receiver, thus reducing the power at the receiver. The optical power level at which Brillouin scattering becomes significant in a single-mode fiber is given by the empirical formula [7, 8],

$$P_B = (17.6 \times 10^{-3}) \, a'^2 \lambda'^2 \alpha \Delta \nu' , \tag{3.13}$$

where

- P_B is the power level (in watts) required for the onset of Brillouin scattering,
- a' is the radius of the fiber (in μm),

    — λ' is the wavelength of the source (in μm),

    — α is the fiber loss (in dB/km), and

    — $\Delta\nu'$ is the linewidth of the source (in GHz).

Power levels above this threshold will lose significant amounts of the signal power to the scattering.

- In *Raman scattering*, the nonlinear interaction produces a high-frequency phonon and a scattered photon, instead of the low-frequency phonon of Brillouin scattering. The scattering is predominately in the *forward* direction, hence the power is not lost to the receiver. The power level threshold for significant Raman scattering to occur is given by [7, 8]

$$P_R = (23.6 \times 10^{-2}) a'^2 \lambda' \alpha, \tag{3.14}$$

where a', λ', and α are the quantities that were defined above with the units noted.

$$\rule{4cm}{0.4pt}\ \infty\ \rule{4cm}{0.4pt}$$

Example: Consider an 8/125 single-mode fiber operating at 1300 nm with a loss of 0.8 dB/km. The linewidth of the source is 0.013 nm.

(a) Calculate the threshold power level for Brillouin scattering in the fiber.

<u>Solution:</u> We begin by finding the linewidth in Hertz.

$$\begin{aligned}
\Delta\nu &= \frac{c}{\lambda^2}\Delta\lambda = \frac{3 \times 10^8}{(1300 \times 10^{-9})^2}(0.013 \times 10^{-9}) \tag{3.15}\\
&= 2.31 \times 10^9 \ \text{Hz} = 2.31 \ \ \text{GHz}.
\end{aligned}$$

The threshold power for Brillouin scattering is then found as

$$\begin{aligned}
P_B &= (17.6 \times 10^{-3}) a'^2 \lambda'^2 \alpha \Delta\nu' \tag{3.16}\\
&= (17.6 \times 10^{-3})(4)^2(1.3)^2(0.8)(2.31) = 0.879 \ \text{watts}.
\end{aligned}$$

(b) Calculate the threshold power for the beginning of Raman scattering in the same fiber.

<u>Solution:</u> The threshold power for Raman scattering is given by

$$\begin{aligned}
P_R &= (23.6 \times 10^{-2}) a'^2 \lambda' \alpha \tag{3.17}\\
&= (23.6 \times 10^{-2})(4)^2(1.3)(0.8) = 3.93 \ \text{watts}.
\end{aligned}$$

(c) Find the ratio of the Brillouin scattering threshold to the Raman scattering threshold.

<u>Solution:</u> The ratio of the threshold power levels is given by

$$\begin{aligned}
\frac{P_B}{P_R} &= \frac{17.6 \times 10^{-3} \, a^2 \lambda^2 \alpha \Delta\nu}{23.6 \times 10^{-2} a^2 \lambda \alpha} = 0.0746\lambda\Delta\nu \tag{3.18}\\
&= 0.0746(1.3)(2.31) = 0.224.
\end{aligned}$$

So, we find that the threshold for Brillouin scattering is about 22% of that for Raman scattering, indicating that Brillouin scattering will be the limiting factor in nonlinear scattering.

$$\rule{4cm}{0.4pt}\ \infty\ \rule{4cm}{0.4pt}$$

When we have more than one signal present in the fiber (as in a wavelength-division multiplexed [WDM] system), more complicated interactions can occur because of nonlinear interactions between the signals. Chraplyvy [9] describes both single-signal effects and multiple-signal effects that occur when several discrete optical wavelengths are simultaneously present. He finds that Brillouin scattering will affect single-signal links but will not affect multiple-signal channels. Four-photon mixing, which requires at least two signals to occur, will have the greatest impact on multi-signal links. These and other effects of nonlinearities will be discussed in detail in a later section of this chapter.

3.2.3   Interface Inhomogeneities

Interface inhomogeneities have the effect of converting high-order modes into lossy modes extending into the cladding where they are removed by the jacket losses. Such inhomogeneities may be due to impurities trapped at the core-cladding interface or impurities in the fiber buffering. The inhomogeneities may also be geometric changes in the shape and/or size of the core due to manufacturing tolerance allowances. Generally, single-mode fibers are more susceptible to losses from geometric irregularities or defects in the jacket material, because a defect of a given size will represent a larger fractional portion of the single-mode fiber diameter than a multimode fiber of larger diameter.

These losses are minimized by reducing the source of the problem. Manufacturing improvements have reduced the losses due to geometric variation in core diameter and coating imperfections.

3.2.4   Macrobending and Microbending Losses

Fibers show increased losses due to bending effects [11–14]. Large bends of the cable and fiber are *macrobends* (see Fig. 3.3a); small-scale bends in the core-cladding interface are *microbends* (see Fig. 3.3b). These latter localized bends can develop during deployment of the fiber, or can be due to local mechanical stresses placed on the fiber (e.g., stresses induced by cabling the fiber or wrapping the fiber on a spool or bobbin). These latter losses are called the *cabling loss* and *spooling loss*, respectively. Typical additional losses caused by microbends added during cabling can be 1 to 2 dB/km.

Bend Losses: Multimode and Single-Mode Fibers

Radiation losses occur at bends in the fiber path. At a bend, the geometry of the core-cladding interface changes and some of the guided light is transmitted from the core into the cladding

Most power is lost from the higher-order modes; less power is lost from the lower-order modes. The radius of curvature of fiber bend is critical to the amount of power lost.

The loss coefficient associated with a fiber bend is given by [15]

$$\frac{P_{\text{out}}}{P_{\text{in}}} = e^{-\alpha_{\text{bends}} z} , \tag{3.19}$$

and the attenuation coefficient α_{bends} is given by [15]

$$\alpha_{\text{bends}} = c_1 e^{-c_2 r} , \tag{3.20}$$

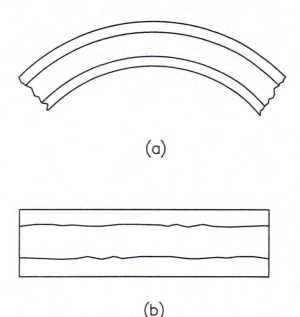

(a)

(b)

Figure 3.3 Example of (a) a macrobend and (b) microbends. (The microbends are greatly exaggerated in size in this figure.)

where r is the radius of curvature of the fiber bend and c_1 and c_2 are constants. The losses are negligible until the radius reaches a critical size given by [15]

$$r_{\text{critical}} \approx \frac{3n_2\lambda}{4\pi(\text{NA})^3}. \tag{3.21}$$

From this relation we see that, to minimize these losses, we want a fiber with a large NA and we want to operate at a short wavelength.

Fortunately, macrobending does not cause appreciable losses until the radius of curvature of the bend is below (approximately) 1 cm. This requirement does not present much problem in the practical utilization of fiber cables, but does present a minimum curvature to the fiber. Frequently the fiber jacket is stiffened to prevent an attempt to loop the fiber into too small a curvature.

∞

Example: (a) Calculate the critical radius of curvature for a multimode 50/125 fiber with an NA of 0.2 operating at 850 nm.

Solution: Lacking information about the value of n_2, we have to assume a reasonable value. (We note that the critical radius is linearly proportional to the index n_2, so any error in n_2 will be directly reflected in r_{critical}.) We will assume a value of $n_2 = 1.48$.

The critical radius of curvature is

$$r_{\text{critical}} \approx \frac{3n_2\lambda}{4\pi(\text{NA})^3} \approx \frac{3(1.48)(850 \times 10^{-9})}{4\pi(0.2)^3} \approx 37.5 \ \mu\text{m}. \tag{3.22}$$

(b) For a 9/125 single-mode fiber with an NA of 0.08 operating at 1300 nm?

<u>Solution:</u> The critical radius is found from

$$r_{\text{critical}} \approx \frac{3n_2\lambda}{4\pi\text{NA}^3} \approx \frac{3(1.48)(1300 \times 10^{-9})}{4\pi(0.08)^3} \approx 897 \ \mu\text{m}. \tag{3.23}$$

We note that most fiber bends (typically greater than a few centimeters) will well exceed these values. We note, however, that the calculated values differed by more than an order of magnitude. We also note that some single-mode fiber designs carry more power in the cladding, making them more susceptible to this loss mechanism than designs that carry more power within the core. This observation helps to explain the increased sensitivity of single-mode long-wavelength fibers to microbends.

─────────────────── ∞ ───────────────────

Bend Losses: Single-Mode Fiber

Bend losses are particularly important in single-mode fiber. In these fibers, the bend losses show a dramatic increase above a critical wavelength when the fiber is bent or perturbed. In particular, it has been observed that the bend losses can be appreciably high at 1550 nm in fibers designed for operation at 1300 nm [16]. The susceptibility of a fiber to these losses depends on the mode-field diameter and the cutoff wavelength [12, 17–19]. The analysis is beyond the level of this text, but, in general, the worst-case condition is in a fiber with a large mode-field diameter and a low cutoff wavelength. Bending losses are minimized in single-mode fibers by avoiding this combination of features.

Losses due to mechanical stresses can be reduced by the fiber design (e.g., choosing a small ratio of core diameter to fiber diameter and having a large Δ value), or by jacketing the fiber with a compressible material that distributes external mechanical stresses more uniformly along a portion of the fiber, or both. Further resistance to microbending losses can be achieved by encasing the fiber and its compliant jacket in a highly rigid sheath that resists bending.

3.2.5   Attenuation Measurements

The typical user is interested primarily in the attenuation per kilometer of a fiber at the required wavelength. The following techniques can be used to measure the fiber attenuation losses [11, 20–22] .

Insertion Loss Method

The fiber losses can be determined by the *insertion loss method*. Here, the output powers transmitted by two different lengths of fiber are measured, with the input and output coupling losses held constant. The power lost in the longer fiber (in dBm or dBμ) is subtracted from the power lost in the shorter fiber, and this lost power is attributed to the extra length. Assuming that the additional losses are uniformly distributed, dividing this dB figure by the increased length of the longer fiber will give the losses in dB/km. The spectral distribution of the fiber losses is obtained by using a tunable source such as a dye laser or a filtered white light source to vary the exciting wavelength.

When used with multimode fibers, one key implicit assumption of the insertion loss technique (and other loss measurement techniques) is that all modes are equally excited, either by using special purpose devices (called *mode mixers* or *mode scramblers*) or ensuring enough length of fiber to allow sufficient mode mixing to occur. A physical test to examine whether all modes are excited is to look at the far-field radiation pattern of the light exiting at two different lengths of the fiber. (The far-field pattern can be observed conveniently by using a lens positioned one focal length from the end of the fiber to collect the light. The far-field pattern will then be imaged at a distance of one focal length behind the lens.) If the half-cone angles of the observed far-field patterns are equal, then sufficient equality in mode excitation exists. Unequal angles imply insufficient mixing. (If measurement of the loss is made with unequal angles, then the results are valid only for the particular length of fiber between the points that were used and for the particular excitation condition of the test. While this information is of some use for characterizing an existing system that is in place, it is otherwise limited, because of its lack of generality.)

The Cutback Method

In the *cutback method,* the power out of a long length of fiber is measured. The fiber is then cut to a short length and the measurement is repeated. The ratio is expressed in dB and divided by the length of the piece that was removed to compute the average loss per length. This method is accurate but cannot be used to characterize the fiber loss in a working link. The following technique does not require any destructive cuts to the fiber.

Optical Time-Domain Reflectometry

The third technique for measuring losses in fibers is *optical time-domain reflectometry (OTDR)* [11, 23–25]. This technique has the primary advantage of not requiring any cuts in the fiber as it works by measuring backscattered light rather than transmitted light. It also requires access at only one end of the fiber.

Figure 3.4 on the next page illustrates an OTDR setup. The pulsed excitation from the transmitter is coupled into the fiber (while the detector is blanked to avoid saturation). The pulse is transmitted down the length of the fiber. In an ideal fiber, Rayleigh backscattering will occur uniformly along the length of the fiber. In a fiber without losses, the returned radiation would be constant in time as the excitation pulse traverses the length of the fiber. For fibers with a uniform loss, the returned radiation declines with a constant slope (when the measured power is expressed on a logarithmic or dB axis). The loss per length can be readily measured by determination of the slope.

Figure 3.5 on page 45 illustrates representative results of the experiment of Fig. 3.4 on the next page. The trace begins with the detection of the light reflected from the connector where the fiber under test connects to the OTDR. (In some systems, the detector is saturated by the partial reflections of the strong outgoing pulse from the coupler. During this saturation time the detector cannot detect the backscatter light or any reflected pulses. The length of fiber corresponding to the time of detector saturation is called the OTDR's *dead zone.* A length of fiber is frequently included within the OTDR between the coupler and the OTDR connector to allow sufficient time for the detector to come out of saturation. However, the losses of this piece of fiber decrease the power available to the test fiber.) The backscattered light is measured from fiber #1 and diminishes with time (and distance). The slope of segment #1 is the loss factor for that fiber (in dB/km). At the splice the discontinuity will cause a pulse in reflected light. The difference between the optical power level (in dBm or dBμ) just before the splice and the

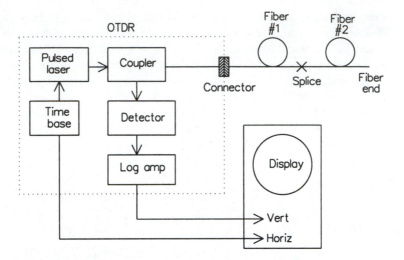

Figure 3.4 Diagram of an optical time-domain reflectometer.

power level just after the splice is a measure of the loss introduced by the splice. The light continues to be backscattered in segment #2 of the fiber in a fashion similar to segment #1. A different fiber loss will be reflected as a different slope. The discontinuity at the end of the fiber produces a large reflected pulse. Other fiber segments can be added. The limit on the total losses that can be measured is determined by the dynamic range of the detector (i.e., the difference between the maximum power (in dBm or dBμ) that can be detected reliably without saturation and the minimum power that can be detected as different from the receiver noise). One problem with current OTDR techniques is the limited dynamic range available (approximately 20 dB in commercially available devices), thereby limiting the length of fiber that can be characterized. Advanced techniques for increasing the OTDR sensitivity are discussed in Refs. [25–27].

Note that accurate position information is also available from this trace. The technique also can be used to localize breaks, abrupt dislocations, and other localized sources of loss. Obviously, the position resolution is determined by the pulse width of the optical source. Typically, pulsed semiconductor lasers with pulse widths on the order of 10 ns are used (providing a resolution of 6 m). For increased accuracy, pulse-compression techniques can be applied to increase the resolution of the system.

3.3   Dispersion

As a pulse of light proceeds through the fiber, it widens in the time domain (see Fig. 3.6 on page 46). This spreading is caused by *dispersion*. The amount of pulse spreading determines how close (in time) two adjacent output pulses are. For any given receiver, there is a minimum spacing required between output pulses, since the receiver must be able to resolve the two separate pulses. Hence, the amount of pulse spreading in the fiber limits the maximum rate at which data can be sent (or, the spreading determines the maximum length of the fiber, if the data rate is fixed). Low dispersion means a high data rate, since pulses can be transmitted closer together with less overlap at the output. There are three primary sources of dispersion in fibers:

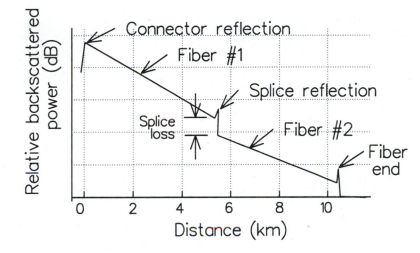

Figure 3.5 Typical result from an OTDR experiment.

- material dispersion (also called chromatic dispersion),

- waveguide dispersion, and

- modal delay (or group delay).

In actuality, material dispersion and waveguide dispersion are caused by the same physical effect (the dependence of the index of refraction of glass on wavelength), but they are accurately modeled as separate effects with additive results [28].

We now want to describe material dispersion and waveguide dispersion and to consider their effects on the performance of a fiber.

3.3.1   Material Dispersion

Material dispersion is caused by the velocity of light (or, equivalently, the index of refraction) being a function of wavelength. Figure 3.7 on page 47 shows the refractive index $n(\lambda)$ in silica. In single-mode fibers operating at shorter wavelengths (i.e., 800–900 nm), this will be the dominant source of dispersion. All sources of light have some degree of spectral width to their output. (A laser source has a considerably narrower spectral width than an LED.) This nonzero spectral width implies that, even with single-mode propagation, the longer wavelengths with their faster velocities will arrive at the receiver before the shorter wavelengths, thereby stretching the pulse.

As shown in the following example, the pulse spreading due to material dispersion is given by

$$\Delta\tau_{\text{mat}} = -\frac{L}{c}\frac{\Delta\lambda}{\lambda}\left(\lambda^2\frac{d^2 n_1}{d\lambda^2}\right), \tag{3.24}$$

where $\Delta\lambda$ is the spectral width of the source, λ is the nominal wavelength of the source, and $d^2 n_1/d\lambda^2$ is the second derivative of the core index of refraction with respect to wavelength. The product of λ^2 and the second derivative characterizes the material dispersion of the fiber and is so grouped in writing the equation. A plot of $\lambda^2(d^2 n/d\lambda^2)$ for the data of Fig. 3.7 is shown in

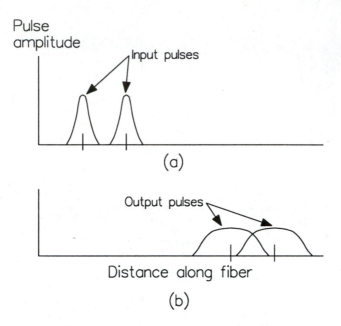

(a)

(b)

Figure 3.6 Effects of pulse spreading on data rate: (a) Well-resolved pulses at input, (b) Unresolved (overlapping) pulses at output.

Fig. 3.8 on page 48. The interesting thing about the material dispersion of silica is that it changes sign at a wavelength of 1.27 μm (see Fig. 3.8) and, hence, has a zero value at that wavelength, offering the possibility of a zero-dispersion fiber. (We shall see that the other sources of dispersion move the wavelength for zero total dispersion up to 1300 nm.)

———————————————— ∞ ————————————————

Example: Derive the expression for the material dispersion in a fiber.

Solution: The arrival time τ of light after traversing a length L of fiber is

$$\tau = L/v_g, \tag{3.25}$$

where v_g is the group velocity of the fiber, given by

$$v_g = \frac{1}{\frac{d\beta}{d\omega}}. \tag{3.26}$$

We have, then,

$$\tau = L\frac{d\beta}{d\omega} = L\frac{d\beta}{d\lambda}\frac{d\lambda}{d\omega}. \tag{3.27}$$

Since $\lambda = c/\nu = 2\pi c/\omega$, we find

$$\frac{d\lambda}{d\omega} = -\frac{2\pi c}{\omega^2} = -\frac{1}{\omega}\frac{2\pi c}{\omega} = -\frac{\lambda}{\omega}. \tag{3.28}$$

Substituting Eq. 3.28 into Eq. 3.27, we obtain

$$\tau = L\frac{d\beta}{d\lambda}\left(-\frac{\lambda}{\omega}\right) = -\frac{L\lambda}{\omega}\frac{d\beta}{d\lambda} = -\frac{L\lambda^2}{2\pi c}\frac{d\beta}{d\lambda}. \tag{3.29}$$

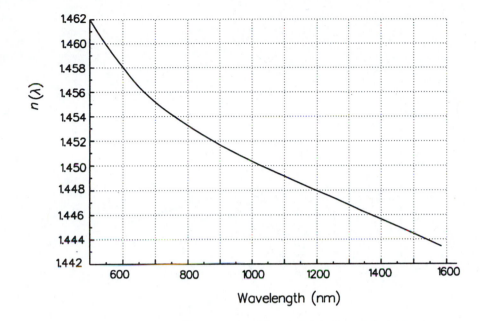

Figure 3.7 Index of refraction vs. wavelength for typical glass.

We know that $\beta = 2\pi n(\lambda)/\lambda$, so

$$
\begin{aligned}
\tau &= -\frac{L\lambda^2}{2\pi c}\frac{d\beta}{d\lambda} \\
&= -\frac{L\lambda^2}{2\pi c}\left[-\frac{2\pi n}{\lambda^2} + \frac{2\pi n'}{\lambda}\right] \\
&= -\frac{L}{c}\left[-n + \lambda n'\right] = +\frac{L}{c}\left[n(\lambda) - \lambda\frac{dn(\lambda)}{d\lambda}\right].
\end{aligned}
\tag{3.30}
$$

The pulse spread $\Delta\tau$ due to a source linewidth of $\Delta\lambda$ is

$$
\frac{\Delta\tau}{\Delta\lambda} = \frac{d\tau}{d\lambda} = \frac{L}{c}\left[\frac{dn(\lambda)}{d\lambda} - \lambda\frac{d^2n}{d\lambda^2} - \frac{dn}{d\lambda}\right] = -\frac{L\lambda}{c}\frac{d^2n}{d\lambda^2}.
\tag{3.31}
$$

Multiplying by $\Delta\lambda$, we find the desired expression for the material dispersion,

$$
\Delta\tau = -\frac{L\lambda\,\Delta\lambda}{c}\frac{d^2n}{d\lambda^2} = -\frac{L}{c}\frac{\Delta\lambda}{\lambda}\left(\lambda^2\frac{d^2n}{d\lambda^2}\right).
\tag{3.32}
$$

∞

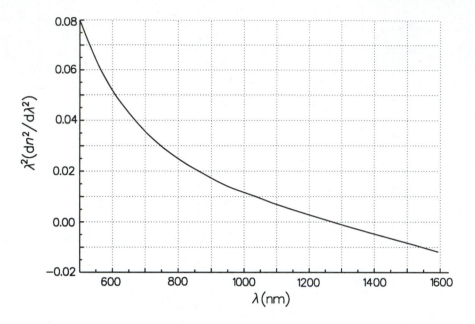

Figure 3.8 Plot of $\lambda^2(d^2n/d\lambda^2)$ vs. wavelength for typical silica glass.

———————————————— ∞ ————————————————

Example: Consider the material dispersion in a 62.5/125 fiber with $n_1 = 1.48$ and $\Delta = 1.5\%$.

(a) Calculate the material dispersion in normalized units of ps·km^{-1}·nm^{-1} at 850 nm.

<u>Solution:</u> The pulse spreading is

$$\Delta\tau_{\text{mat}} = -\frac{L}{c}\frac{\Delta\lambda}{\lambda}\left(\lambda^2\frac{d^2n_1}{d\lambda^2}\right). \tag{3.33}$$

The normalized delay is

$$\frac{\Delta\tau_{\text{mat}}}{L\,\Delta\lambda} = -\frac{1}{c\lambda}\left(\lambda^2\frac{d^2n}{d\lambda^2}\right). \tag{3.34}$$

From Fig. 3.8, we see that $\lambda^2 d^2n_1/d\lambda^2$ is approximately 0.022 at $\lambda = 850$ nm; hence,

$$\frac{\Delta\tau_{\text{mat}}}{L\,\Delta\lambda} = -\frac{1}{c\lambda}\left(\lambda^2\frac{d^2n}{d\lambda^2}\right) = -\frac{1}{(3.0\times10^8)(850\times10^{-9})}(0.022) \tag{3.35}$$

$$= -8.63\times10^{-5}\ \text{s}\cdot\text{m}^{-1}\cdot\text{m}^{-1} = -86.3\ \text{ps}\cdot\text{km}^{-1}\cdot\text{nm}^{-1}.$$

(b) ...at 1500 nm?

<u>Solution:</u> From Fig. 3.8, we estimate that $\lambda^2 d^2n_1/d\lambda^2 \approx -0.007$, so

$$\frac{\Delta\tau_{\text{mat}}}{L\,\Delta\lambda} = -\frac{1}{c\lambda}\left(\lambda^2\frac{d^2n}{d\lambda^2}\right) = -\frac{1}{(3.0\times10^8)(1500\times10^{-9})}(-0.007) \tag{3.36}$$

$$= +1.55\times10^{-5}\ \text{s}\cdot\text{m}^{-1}\cdot\text{m}^{-1} = +15.5\ \text{ps}\cdot\text{km}^{-1}\cdot\text{nm}^{-1}.$$

———————————————— ∞ ————————————————

3.3.2 Waveguide Dispersion

For the low material-dispersion region near 1.27 μm, *waveguide dispersion* becomes important. Usually this source of dispersion is negligible in multimode fibers and in single-mode fibers operated at wavelengths below 1 μm, but it is *not* negligible for single-mode fibers operated in the vicinity of 1.27 μm. This waveguide dispersion results from the propagation constant of a mode (and, hence, its velocity) being a function of a/λ. In particular, the net delay due to waveguide dispersion is expressed by [28]

$$\tau_{wg} = \frac{L}{c}\frac{d\beta}{dk} .$$

(3.37)

We again define the normalized propagation constant b as

$$b = \frac{(\beta^2/k^2) - n_2^2}{n_1^2 - n_2^2} .$$

(3.38)

An approximation for b is

$$b \approx \frac{(\beta/k) - n_2}{n_1 - n_2} ,$$

(3.39)

thereby giving

$$\beta \approx n_2 k\,(b\Delta + 1) .$$

(3.40)

Here b is a function of V (and of k). Substitution of Eq. 3.40 into Eq. 3.37 gives

$$\tau_{wg} \approx \frac{L}{c}\left(n_2 + n_2\Delta\frac{d(kb)}{dk} \right) .$$

(3.41)

Using the approximation,

$$V \approx kan_2\sqrt{2\Delta} ,$$

(3.42)

we can write

$$\tau_{wg} \approx \frac{L}{c}\left(n_2 + n_2\Delta\frac{d(Vb)}{dV} \right) .$$

(3.43)

Neglecting the constant term in the delay, we have a wavelength-dependent time delay of

$$\tau_{wg}(\lambda) \approx \frac{n_2\Delta\,L}{c}\frac{d(Vb)}{dV} .$$

(3.44)

Figure 3.9 on the next page shows a plot of $d(Vb)/dV$ vs. V for the lowest-order mode in the fiber. Similar curves [29] can be obtained for higher-order modes, if desired. (For large values of V [i.e., multimode fibers], however, the value of the delay due to waveguide effects is small and negligible compared to the delay due to modal or material dispersion.) In single-mode fibers, however, modal dispersion is not present and, in the region near 1270 nm, the waveguide dispersion is comparable with the material dispersion. Therefore, we want to concentrate the waveguide dispersion in that region of the spectrum.

Continuing the development, we now want to include the effects of source linewidth. We know that the difference in propagation time $\Delta\tau_{wg}$ is

$$\Delta\tau_{wg} = \Delta\lambda\frac{d\tau_{wg}}{d\lambda} = \Delta\lambda\frac{dV}{d\lambda}\frac{d\tau_{wg}}{dV} .$$

(3.45)

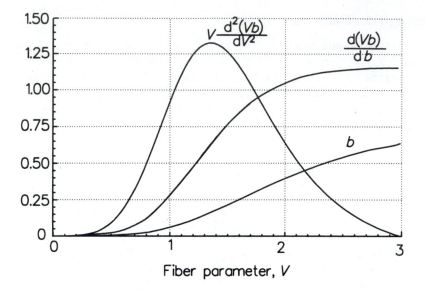

Figure 3.9 Plot of b, $d(Vb)/dV$, and $V\,d^2(Vb)/dV^2$ vs. V for the lowest-order fiber mode.

Since $V = 2\pi a n_1 \sqrt{2\Delta}$, we can show that $dV/d\lambda = -V/\lambda$. We can use Eq. 3.44 on the preceding page to find $d\tau_{wg}/dV$ and, hence, can show that

$$\Delta\tau_{wg} = -\frac{V}{\lambda}\,\Delta\lambda\,\frac{d\tau_{wg}}{dV} \approx -\frac{n_2 L\Delta}{c}\,\frac{\Delta\lambda}{\lambda}\left(V\frac{d^2(Vb)}{dV^2}\right). \tag{3.46}$$

This is our desired expression for the waveguide dispersion.

In Ref. [30], Gloge defines the *normalized propagation constant, b,* as

$$b(V) = 1 - \frac{u^2}{V^2} = \frac{\left(\dfrac{\beta^2}{k^2}\right) - n_2^2}{n_1^2 - n_2^2}. \tag{3.47}$$

Solving for β in terms of b, we find

$$\beta = k\sqrt{n_2^2 + (n_1^2 - n_2^2)b} \tag{3.48}$$

so we can express b in terms of β as,

$$b = \frac{\dfrac{\beta^2}{k^2} - n_2}{n_1^2 - n_2^2} \approx \frac{\dfrac{\beta}{k} - n_2}{n_1 - n_2}. \tag{3.49}$$

Gloge [30] also shows that, for the lowest-order mode (i.e., the HE_{11} mode), the expression for u is

$$u(V) = \frac{(1 + \sqrt{2})V}{1 + (4 + V^4)^{0.25}}. \tag{3.50}$$

By substituting Eq. 3.50 into the middle expression of Eq. 3.47 on the preceding page, we can find $b(V)$,

$$b(V) \;=\; 1 - \left(\frac{(1+\sqrt{2})^2}{\sqrt{1+(4+V^4)}} \right) \tag{3.51}$$

Figure 3.9 on the facing page shows a plot of $b(V)$. From this expression, we can also obtain plots of $d(Vb)/dV$ and $V\,d^2(Vb)/dV^2$ as functions of V, as also shown in the figure (see problem set at end of chapter). The single-mode values of interest are from $V = 2.0$ to 2.4, as shown in Fig. 3.10. The value of $V\,d^2(Vb)/dV^2$ decreases monotonically from 0.64 down to 0.25. Hence, the waveguide dispersion is, in general, a small negative value.

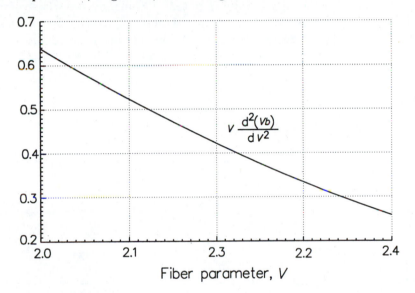

Figure 3.10 Plot of $V\,d^2(Vb)/dV^2$ vs. V for the lowest-order fiber mode in the region of practical single-mode fiber design.

$$\infty$$

Example: Calculate the waveguide dispersion in units of ps·km^{-1}·nm^{-1} for a 9/125 single-mode fiber with $n_1 = 1.48$ and $\Delta = 0.22\%$ operating at 1300 nm.

<u>Solution:</u> We begin by calculating V from Eq. 2.9 on page 15,

$$V = \frac{2\pi a}{\lambda}\, n_1 \sqrt{2\Delta} = \frac{2\pi(4.5 \times 10^{-6})}{1300 \times 10^{-9}}(1.48)\sqrt{2(0.0022)} = 2.14\,. \tag{3.52}$$

(We note that V falls within the expected range of $2.0 < V < 2.405$ for single-mode fiber.) From Fig. 3.10, we find $V\,d^2(Vb)/dV^2 \approx 0.480$ at $V = 2.14$.

We also have $n_2 = n_1(1 - \Delta) = 1.48(1 - 0.0022) = 1.477$, so

$$\frac{\Delta\tau_{wg}}{L\,\Delta\lambda} \;=\; -\left(\frac{n_2\Delta}{c}\right)\left(\frac{1}{\lambda}\right)\left(V\frac{d^2(Vb)}{dV^2}\right) \tag{3.53}$$

$$= -\left(\frac{(1.477)(0.0022)}{3 \times 10^8}\right)\left(\frac{1}{1300 \times 10^{-9}}\right)(0.48)$$

$$= -4.00 \times 10^{-6} = -4.00 \text{ ps} \cdot \text{km}^{-1} \cdot \text{nm}^{-1}.$$

$$\underline{\hspace{5cm}} \infty \underline{\hspace{5cm}}$$

3.3.3 Total Dispersion: Single-Mode Fiber

To minimize the total dispersion of a single-mode fiber, it is necessary to operate at a wavelength longer than 1.27 μm to allow the small positive material dispersion to cancel the small negative waveguide dispersion, causing a net dispersion of zero, as illustrated in Fig. 3.11. This zero-dispersion point occurs near 1300 nm, a wavelength that, fortunately, has a fairly low attenuation (although not as low as the attenuation minimum at 1550 nm), allowing operation of high-data-rate links at this wavelength. This low dispersion at 1300 nm has provided the impetus for the development of a family of sources and receivers that operate at this wavelength.

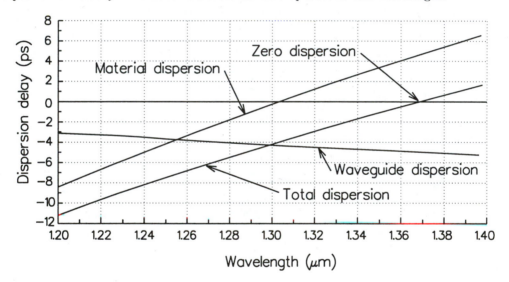

Figure 3.11 Addition of material dispersion and waveguide dispersion to achieve zero dispersion at a value just below 1.32 μm. (This plot is the solution of a problem at the end of this chapter.)

The size of the waveguide dispersion has been found to be sensitive to the doping levels as well as the values of Δ and a. For various combinations of Δ and a, and for triangular and other doping profiles (as in Fig. 2.9 on page 25), it has proved possible to achieve zero dispersion at other wavelengths between 1300 and 1700 nm, allowing the development of fibers that combine minimum losses at 1550 nm with zero dispersion at this wavelength. This process is called *dispersion shifting* and results in so-called *dispersion-shifted fiber* .

3.3.4 Dispersion-Adjusted Single-Mode Fibers

We have found that the lowest losses occur at a 1500 nm wavelength, while the lowest total dispersion occurs (in a step-index single-mode fiber) at 1300 nm wavelength. Since these wavelengths are near each other, it has become a goal to combine the two features so that optimum performance in terms of bandwidth and loss occurs at the same wavelength. Of the two parameters, dispersion is the most adjustable and is the one that is manipulated. This manipulation has led to two adjustments; one is to move the zero-dispersion wavelength (*dispersion shifting*) to higher values in the vicinity of 1550 nm (as in Fig. 3.12 and the other is to minimize or flatten the dispersion over a range of wavelength values (*dispersion flattening*).

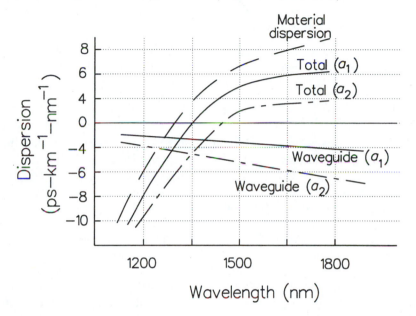

Figure 3.12 Example of a dispersion-shifted fiber. The waveguide dispersion is modified by changing a (while also changing Δ to keep V fixed) to increase the zero-dispersion wavelength.

Dispersion-Shifted Step-Index Fibers

The *material dispersion* of silica can be adjusted in small amounts by doping the core of the fiber with GeO_2 or other dopants. (Doping with GeO_2 has the greatest effect on the material dispersion.)

The *waveguide dispersion* depends on the fiber-core radius, the fractional change in the fiber index of refraction Δ, and on the shape of the fiber profile g. A large number of fiber profile adjustments can be made to tune the zero dispersion wavelength. For the step-index profile the core radius a is reduced while increasing the fractional index change Δ. (We still want to keep the cutoff wavelength for the second-lowest mode, the LP_{11} mode, between 1000 to 1300 nm to keep the power distribution of the lowest-order mode reasonably close to the center of the fiber. This requires that the value of V be kept between 1.5 and 2.4.) The equation for V indicates that a linear decrease in a requires a quadratic increase in Δ in order to keep V constant. It has been found that the waveguide dispersion is very sensitive to changes in the fiber parameters. For example, reducing a core radius from 5.5 μm to 1.8 μm changes the zero-dispersion wavelength from 1300 to 1750 nm [31]. Experimental results quoted in Ref. [31] indicate that dispersion-shifted step-index fibers are easily made and achieve the desired dispersion shift. Unfortunately, these fibers also exhibited an increased loss (0.30 dB/km or greater at 1550 nm) that seems to be due to the strong concentration of germanium doping required in the core. While various models have been proposed [31] for various single-cladding designs, none have enabled the production of dispersion-shifted step-index fibers with losses comparable to the unshifted fibers. An alternative is to consider dispersion shifting with a non-step-index profile, either a graded-index profile or a multi-layer profile.

Dispersion-Shifted Graded-Index Fibers

Graded-index fibers offer more parameters for the fiber designer to manipulate. One additional parameter is the profile g of the core from the refractive index equation

$$n(r) = n_1 \sqrt{1 - 2\Delta \left(\frac{r}{a}\right)^g}. \tag{3.54}$$

Studies [32–37] of the propagation of modes in these structures show that the single-mode behavior exists for $V \leq 3.53$ for a parabolic profile ($g = 2$) and for $V \leq 4.38$ for a triangular profile ($g = 1$).

The triangular profile (Fig. 3.13 on the next page) was demonstrated [31] to shift the zero-dispersion wavelength successfully without any penalty in fiber losses with a Δ of 0.115 and a core radius a of 3.2 μm. The measured losses were 0.25 dB/km at 1.56 μm. (Other experiments have achieved losses as low as 0.21 dB/km [31].) The triangular profile has the advantage of the mode-field diameter staying quite small. The cutoff wavelength for the LP_{11} mode is usually fairly low (i.e., in the vicinity of 850 nm), a potential disadvantage in that it raises the susceptibility to microbending losses. A remedy to this low value of cutoff wavelength is to use a depressed-cladding triangle profile, as in Fig. 3.13b. The extra cladding tends to guide the LP_{01} mode better. In a design cited in Ref. [31], the ratio of a_2/a_1 was > 8.5.

Dispersion-Shifted Multi-Index Fibers

More complicated profiles give the designer ample parameters to manipulate in an attempt to optimize the fiber performance. The *double-clad fiber* (or *"W" fiber* of Fig. 3.14a has been used widely for dispersion-flattening, but also works for dispersion shifting. The dispersion-shifted fibers were sensitive to microbending [31], however. Other designs incorporating a triangular center profile with a raised outer cladding have produced dispersion-shifted fibers with losses as low as 0.17 dB/km at 1550 nm [31].

Dispersion-Flattened Fibers

An alternative approach to minimizing dispersion effects with less loss penalty than the step-index dispersion-shifted fibers is to attempt to reduce the dispersion to a nonzero minimum between 1300 and 1500 nm, as shown in Fig. 3.15 on page 57 for the quadruple-clad profile of Fig. 3.14b. This technique is called *dispersion flattening*. (Zero dispersion can occur toward the ends of the range in some designs.) These dispersion-flattened fibers allow the use of multiple wavelengths with reasonable loss and dispersion performance; the advantages of this are discussed in Chapter 10 on wavelength-division multiplexing. As noted above, multi-layer profiles have been used successfully to flatten the dispersion characteristics of fibers. Step-index designs can achieve dispersion flattening only with unacceptably high losses.

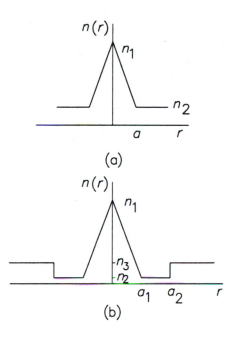

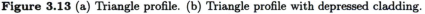

Figure 3.13 (a) Triangle profile. (b) Triangle profile with depressed cladding.

3.3.5 Modal Dispersion

Group delay (or *modal dispersion*) is important in multimode fibers. It is caused by the different path lengths associated with each of the modes of a fiber, as well as the differing propagation coefficients associated with each mode, as was discussed in the last chapter when we considered Fig. 2.4 on page 15.

Modal Dispersion I: Step-Index Fiber

Assuming a multimode step-index fiber and considering each mode as a ray of light at a slightly different angle, we can see from Fig. 3.16 on page 57 that each mode will have a slightly different path length and velocity in traversing the fiber. For an ideal, straight step-index fiber, the fastest mode is the one that travels straight down the fiber (i.e., at an angle of incidence of 90° to the core-cladding interface). The slowest mode is that which is incident at the critical angle of the fiber.

The time delay between the fastest and slowest pulse is the *modal pulse delay distortion* $\Delta\tau_{\text{modal}}$ and, for a step-index fiber, is given by [38]

$$\Delta\tau_{\text{modal}} = \frac{L(n_1 - n_2)}{c}\left(1 - \frac{\pi}{V}\right) \tag{3.55}$$

where L is the length of the fiber and V is the parameter given by Eq. 2.9 on page 15. (This expression ignores the constant delay common to all modes and represents only the *difference* in propagation times.) We note that for most multimode fibers, V is larger than 10, so a useful

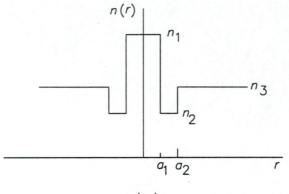

(a)

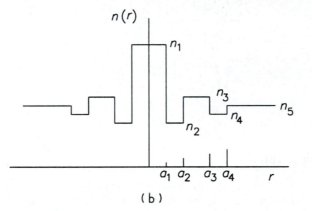

(b)

Figure 3.14 Refractive index profile of (a) a double-clad or "W" profile fiber and (b) a quadruple-clad fiber.

———————————————— ∞ ————————————————

approximation is

$$\Delta\tau_{\text{modal}} \approx \frac{L\,\Delta n_1}{c}. \tag{3.56}$$

Since modal dispersion is due to the existence of various modes, it is *not* present in single-mode fibers.

Example: Consider a 50/125 step-index fiber with $n_1 = 1.47$ and $\Delta = 1.5\%$. Calculate the group delay (or modal dispersion) in units of ns·km^{-1} for this fiber at an operating wavelength of 850 nm.

<u>Solution:</u> We find the normalized dispersion as

$$\frac{\Delta\tau_{\text{modal}}}{L} \approx \frac{\Delta n_1}{c} = \frac{(0.015)(1.47)}{3 \times 10^8} \tag{3.57}$$

$$= 7.35 \times 10^{-11} \text{ s} \cdot \text{m}^{-1} = 73.5 \text{ ns} \cdot \text{km}^{-1}.$$

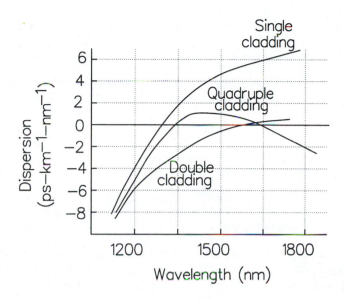

Figure 3.15 Example of a dispersion-flattened fiber. The quadruple-clad fiber provides enough design variables to produce a flattened small-dispersion region between 1300 nm and 1600 nm.

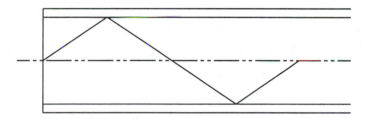

Figure 3.16 Step-index fiber showing high-order mode path.

<u>Alternative solution:</u> We can use the more exact formula,

$$\frac{\Delta\tau_{\text{modal}}}{L} = \frac{(n_1 - n_2)}{c}\left(1 - \frac{\pi}{V}\right). \tag{3.58}$$

Finding $(n_1 - n_2)$, we have

$$n_1 - n_2 \approx \Delta n_1 \approx (0.015)(1.47) \approx 2.21 \times 10^{-2}. \tag{3.59}$$

To find V, we use

$$V = \frac{2\pi a}{\lambda}n_1\sqrt{2\Delta} \tag{3.60}$$

$$= \left(\frac{2\pi(25 \times 10^{-6})}{850 \times 10^{-9}}\right)(1.47)\left(\sqrt{2(0.015)}\right) \approx 47.1.$$

Hence,

$$\frac{\Delta\tau_{\text{modal}}}{L} \approx \frac{(n_1 - n_2)}{c}\left(1 - \frac{\pi}{V}\right) = \frac{(2.21 \times 10^{-2})}{3 \times 10^8}\left(1 - \frac{\pi}{47.1}\right) \tag{3.61}$$

$$= \quad 6.87 \times 10^{-11} \ \text{s} \cdot \text{m}^{-1} = 68.7 \ \text{ns} \cdot \text{km}^{-1}.$$

--------------------------------------- ∞ ---------------------------------------

Once we find the pulse spread $\Delta\tau$ caused by modal dispersion, we need a method of calculating the maximum data rate that the fiber can support. This maximum bit rate $B_{R\text{max}}$ can be found from

$$B_{R\text{max}} = \frac{1}{4\Delta\tau}. \tag{3.62}$$

If the limiting pulse spread is caused by modal dispersion, for example, we would have

$$B_{R\text{max}} = \frac{1}{4\Delta\tau_{\text{modal}}}. \tag{3.63}$$

Modal Dispersion II: Graded-Index Fiber

For the graded-index fiber, we must account for the inhomogeneous velocity of the light in the fiber as well as the sinusoidal paths as shown in Fig. 2.11 on page 26. In particular, we note that, although the higher-order modes have longer path lengths because of their further excursion from the axis, their average velocity will be higher because of the increased velocity as the ray moves away from the center axis. To a first-order approximation, the effect of the longer path length is canceled by the higher velocity, and modes that leave the axis at the same time will arrive at the next axis crossing at the same time. More exact analysis of the graded index fiber predicts [28] that the delay time incurred by the m-th mode of the fiber will be given by

$$\begin{aligned}
\tau_{\text{modal}} \quad = \quad & \frac{LN_{g1}}{c} \\
& \times \left(1 + \frac{g-2-\epsilon}{g+2} \Delta \left(\frac{m}{N}\right)^{g/(g+2)} \right. \\
& \qquad + \frac{\Delta^2}{2} \frac{3g-2-2\epsilon}{g+2} \left(\frac{m}{N}\right)^{2g/(g+2)} \\
& \qquad \left. + \text{other terms with } \Delta^3, \ \Delta^4, \text{etc.}\right),
\end{aligned} \tag{3.64}$$

where

$$\epsilon \quad = \quad -\frac{2n_1}{N_{g1}} \frac{\lambda}{\Delta} \frac{d\Delta}{d\lambda}, \tag{3.65}$$

$$N_{g1} \quad = \quad n_1 - \lambda \frac{dn_1}{d\lambda}, \quad \text{and} \tag{3.66}$$

$$N \quad = \quad a^2 \Delta k^2 n_1^2 \left(\frac{g}{g+2}\right). \tag{3.67}$$

The information on $d\Delta/d\lambda$ and $dn_1/d\lambda$ is material-dependent and wavelength-dependent and would have to be provided from a study of the materials used in fabrication of the fibers.

From Eq. 3.64, we note that the term that is linear in Δ can be eliminated if the profile index is set to an optimum value, g_{opt}, given by

$$g = g_{\text{opt}} = 2 + \epsilon = 2 - \frac{2n_1}{N_{g1}} \frac{\lambda}{\Delta} \frac{d\Delta}{d\lambda}. \tag{3.68}$$

A more useful approximation for g_{opt} that is cited frequently [39, 40] is

$$g_{\text{opt}} \approx 2 - \frac{12\Delta}{5}. \tag{3.69}$$

This approximation is accurate only if the modal dispersion is considered and the material dispersion is ignored. References [40] and [41] give more involved expressions for the optimum profile (beyond the scope of this text) that include material and waveguide dispersion.

From this discussion, we see that graded-index fibers have an *optimum index profile* g_{opt} that minimizes the time delay of the modes. The net delay $\Delta\tau_{\text{modal}}$ from the lowest-order modes to the highest order modes is given by [38, 42, 43]

$$\Delta\tau_{\text{modal}} \approx \begin{cases} n_1 \Delta \dfrac{(g - g_{\text{opt}})L}{(g+2)c} & g \neq g_{\text{opt}} \\[4mm] \dfrac{n_1 \Delta^2 L}{2c} & g = g_{\text{opt}}. \end{cases} \tag{3.70}$$

We observe that $\Delta\tau_{\text{modal}}$ can be positive or negative depending on the size of g relative to g_{opt}. For negative $\Delta\tau_{\text{modal}}$, the interpretation is that the higher-order modes are arriving before the lower-order modes.

––––––––––––––––––––––––– ∞ –––––––––––––––––––––––––

Example: Consider a graded-index fiber with $\Delta = 2\%$ and $g_{\text{opt}} = 2.0$. If $g = 95\%$ of g_{opt}, calculate the ratio of $\Delta\tau_{\text{modal}}|_{g=g_{\text{opt}}}$ to $\Delta\tau_{\text{modal}}|_{g\neq g_{\text{opt}}}$.

Solution: We have $g = 0.95 g_{\text{opt}} = 0.95(2.0) = 1.90$, so

$$\frac{\Delta\tau_{\text{modal}}|_{g\neq g_{\text{opt}}}}{\Delta\tau_{\text{modal}}|_{g=g_{\text{opt}}}} = \frac{n_1 \Delta \dfrac{g - g_{\text{opt}}}{(g+2)c}L}{\dfrac{n_1 \Delta^2 L}{2c}} = \frac{2(g - g_{\text{opt}})}{\Delta(g+2)} \tag{3.71}$$

$$= \frac{(2)(1.90 - 2)}{(0.02)(1.90 + 2)} = -2.56 = -256\%.$$

We note that small differences between g_{opt} and the achieved g cause a major change in the value of this ratio. This high sensitivity to values of g leads to tight tolerances on its value.

––––––––––––––––––––––––– ∞ –––––––––––––––––––––––––

We now want to compare the modal dispersion of a step-index multimode fiber with that of a graded-index fiber of the same size. A comparison of Eq. 3.56 on page 56 and Eq. 3.70 shows a graded-index-fiber dispersion that is a factor of Δ smaller than the equivalent step-index fiber (see the following example). This improved modal-dispersion performance is one of the primary advantages of the graded-index fiber over a multimode step-index fiber of the same diameter.

––––––––––––––––––––––––– ∞ –––––––––––––––––––––––––

Example: (a) Calculate the ratio of the modal delay per km in a 50/125 graded-index fiber with $n_1 = 1.46$, $\Delta = 1.5\%$, and $g = g_{\text{opt}} = 2$ to the modal delay in a step-index fiber of the same size with the same n_1 and Δ.

Solution: The time delays are given by

$$\frac{n_1 \Delta\tau(\text{GI})|_{g=g_{\text{opt}}}}{L} \approx \frac{\Delta^2}{2c} \tag{3.72}$$

$$\frac{\Delta\tau(\text{SI})}{L} \approx \frac{n_1 \Delta}{c}. \tag{3.73}$$

Taking the ratio,

$$\frac{\Delta\tau(\text{GI})|_{g=g_{\text{opt}}}}{\Delta\tau(\text{SI})} = \frac{\dfrac{n_1 \Delta^2}{2c}}{\dfrac{n_1 \Delta}{c}} = \frac{\Delta}{2} = \frac{0.015}{2} = 0.00750. \tag{3.74}$$

Hence we see that, when optimized, the graded-index fiber can have one to two orders of magnitude less dispersion than the equivalent step-index fiber.

(b) Consider the same question if the graded-index fiber is not optimized. Let $g = 2.1$ and $g_{\text{opt}} = 2.0$.

Solution: We find that

$$\frac{\Delta\tau(\text{GI})|_{g \neq g_{\text{opt}}}}{\Delta\tau(\text{SI})} = \frac{\dfrac{n_1 \Delta(g - g_{\text{opt}})}{(g + 2)c}}{\dfrac{n_1 \Delta}{c}} = \frac{g - g_{\text{opt}}}{(g + 2)} \tag{3.75}$$

$$= \frac{2.1 - 2.0}{4.1} = 0.0244.$$

The dispersion of the graded-index fiber is still less than that of the step-index fiber, but the ratio has been reduced by more than one order of magnitude due to the non-optimum value of profile.

———————————————— ∞ ————————————————

Current practical values of the total dispersion in multimode graded-index fibers are on the order of 0.2 ns·km^{-1} for laser sources and 1.0 ns·km^{-1} for LED sources. (The LED dispersion is higher because the material-dispersion effects for the wide-spectral-width LED are not negligible.) Practical data bandwidth-distance products for a laser source and a multimode graded-index fiber with an optimized profile are on the order of 0.5 to 2.5 GHz·km.

3.3.6 Dispersion Units

Modal dispersion is independent of the source linewidth and is usually specified for a fiber or cable by the characteristic pulse spread per kilometer length (i.e., in units of ns·km^{-1}). For multimode fibers dominated by modal dispersion, the specification can be alternatively given as an analog bandwidth-distance product (in units of GHz·km).

Both material dispersion and waveguide dispersion are dependent on the source linewidth. To remove the dependence, the specification for these quantities can also be given per nanometer of linewidth. Hence, the units are ns·km^{-1}·nm^{-1}.

3.3.7 Pulse-Spreading Approach

We have been using the time delay of waves to characterize the dispersion of a fiber. An alternative approach is to consider quantitatively the pulse spreading that is induced by the dispersion. In this approach we can imagine measuring the pulse width at the input and at the output and attributing the increase in pulse width to the dispersion.

We begin by defining the *RMS pulse width*. The RMS pulse width σ_s is defined as the standard deviation of the pulse width (or, equivalently, the square root of the variance) and is related to the first moment of the pulse M_1 and the second moment of the pulse M_2 by

$$\sigma_s^2 = M_2 - M_1^2 \tag{3.76}$$

where

$$M_1 = \int_{-\infty}^{\infty} tp(t)\, dt \tag{3.77}$$

and

$$M_2 = \int_{-\infty}^{\infty} t^2 p(t)\, dt. \tag{3.78}$$

Many optical pulses are symmetric (or are assumed to be). The mean value M_1 of a symmetric input pulse is zero; so, for symmetric pulses, the *mean-square pulse width* σ_s^2 is

$$\sigma_s^2 = \int_{-\infty}^{\infty} t^2 p(t)\, dt. \tag{3.79}$$

The pulse width at the output of the fiber is a combination of the pulse width of the source and the pulse spreading of the fiber caused by dispersion. For the mean-square pulse width, the effects are combined according to

$$\sigma_{\text{out}}^2 = \sigma_{\text{in}}^2 + \sigma_{\text{fiber}}^2. \tag{3.80}$$

We usually determine σ_{fiber} by measuring the input and output pulses, calculating σ_{in} and σ_{out} from the measured waveforms, and computing σ_{fiber} from Eq. 3.80 as

$$\sigma_{\text{fiber}} = \sqrt{\sigma_{\text{out}}^2 - \sigma_{\text{in}}^2}. \tag{3.81}$$

Similar results are also available in terms of the RMS pulse widths of the dispersion terms, where, again, the results are combined as

$$\sigma_{\text{fiber}}^2 = \sigma_{\text{modal}}^2 + \sigma_{\text{material}}^2. \tag{3.82}$$

We will now give some expressions for the dispersion-induced pulse spreading.

Material Dispersion

The RMS pulse spread for material dispersion is given [28] by

$$\sigma_{\text{mat}} = \frac{L}{c}\frac{\sigma_\lambda}{\lambda}\left(\lambda^2 \frac{d^2 n_1}{d\lambda^2}\right)^2, \tag{3.83}$$

where σ_λ is the RMS spectral width of the source.

Modal Dispersion: Step-Index Fiber

For a step-index fiber, Senior [15] gives an estimated spreading caused by modal dispersion as

$$\sigma_{\text{modal SI}} \approx \frac{L n_1 \Delta}{2c\sqrt{3}} \approx \frac{L\,(\text{NA})^2}{4\sqrt{3}n_1 c}. \tag{3.84}$$

The maximum bit rate that can be achieved in a fiber (assuming a negligible input pulse width) can be estimated as [44]

$$B_{\max} \approx \frac{0.2}{\sigma_s}. \tag{3.85}$$

For the step-index fiber this gives

$$B_{R\max} = \frac{0.8\sqrt{3}\,n_1 c}{L\,(\text{NA})^3}. \tag{3.86}$$

Modal Dispersion: Graded-Index Fiber

Keiser [28] gives a complete expression for the pulse spreading due to the modal dispersion σ_{modal} in a graded-index fiber as

$$\sigma_{\text{modal GI}} = \left(\frac{L N_{g1}\Delta}{2c}\right)\left(\frac{g}{g+1}\right)\sqrt{\frac{g+2}{3g+2}}$$
$$\times \sqrt{c_1^2 + \frac{4c_1 c_2(g+1)\Delta}{2g+1} + \frac{16\Delta^2 c_2^2(g+1)^2}{(5g+2)(3g+2)}} \tag{3.87}$$

where

$$N_{g1} = n_1 - \lambda\frac{dn_1}{d\lambda}, \tag{3.88}$$

$$c_1 = \frac{g-2-\epsilon}{g+2}, \tag{3.89}$$

$$c_2 = \frac{3g-2-2\epsilon}{2(g+2)}, \text{ and} \tag{3.90}$$

$$\epsilon = -\frac{2n_1}{N_g 1}\frac{\lambda}{\Delta}\frac{d\Delta}{d\lambda}. \tag{3.91}$$

While this expression is a complete solution, we would like to have a simpler estimate of the RMS pulse spread in the graded-index fiber.

To find the optimum value of g, we can use the estimate [39, 40]

$$g_{\text{opt}} = 2 - \frac{12\Delta}{5}. \tag{3.92}$$

Reference [42] gives the following expression for the pulse spread in a graded-index fiber.

$$\sigma_{\text{modal}} = \begin{cases} \dfrac{0.246 L N_{g1}\Delta|g-g_{\text{opt}}|}{c(g+2)} & 1 > |g-g_{\text{opt}}| \gg \Delta \\[2ex] \dfrac{0.150 L N_{g1}\Delta^2}{c} & g = g_{\text{opt}}. \end{cases} \tag{3.93}$$

The dependence of the pulse spread is very sensitive to small changes in g, as seen in Fig. 3.17. An alternative expression for the RMS pulse spread at $g = g_{\text{opt}}$ is [15]

$$\sigma_{\text{modal}} = \frac{n_1 \Delta L}{2c\sqrt{3}} \qquad (\text{for } g = g_{\text{opt}}). \tag{3.94}$$

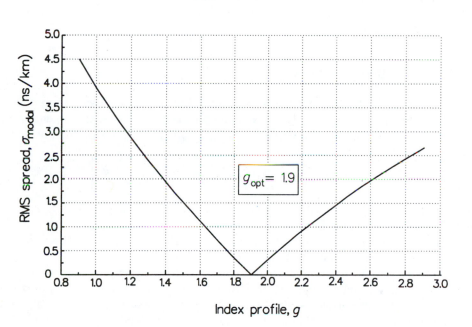

Figure 3.17 Typical modal pulse spread, σ_{modal} for a graded-index fiber vs. the refractive-index profile, g.

Estimates of expected RMS pulse spread can be made by calculating the pulse spread expected when $g = g_{\text{opt}}$ and then multiplying that value by 10 to allow for expected variations.

3.3.8 Data Rate-Distance Product

From the discussion of dispersion characteristics, we see that, for a specified dispersion, longer distances require an increase in pulse spacing (or, equivalently, a reduction in the data rate) to avoid intersymbol interference that would increase the error-rate to unacceptable levels. Since the dispersions that we have considered are all linearly dependent on distance, the product of the data rate and the distance is (theoretically) a constant. This tradeoff between distance and data rate is fundamental to fiber-optic system designs. Individual fibers are specified by their data rate-distance product.

A typical single-mode fiber will have a product of several Gb·s^{-1}·km, a typical graded-index fiber product will be several 100s of Mb·s^{-1}·km, and a typical multimode step-index fiber will have a product of a few 10s of Mb·s^{-1}·km. Thus we see that, based on data rate alone, the single-mode fiber provides the best performance (especially when operated near 1300 nm), followed by the graded-index fiber, with the multimode fiber step-index fiber offering the least data handling capacity. Note, however, that other considerations (especially the cost of the fiber and system components) can dominate the fiber selection decision in any system where data rate is not the prime consideration.

Our discussion of information-handling capacity has been couched in terms of digital transmission since that is the primary format that is used in fiber-optic systems. Fibers can also carry analog information, and the same discussion can be formulated in terms of bandwidth rather than data rate without changing the conclusions on the order of information-handling capacity among the different types of fibers.

3.4   Fiber Nonlinearities

We have seen earlier that nonlinear effects can limit the power that we can put into a fiber (especially when the core is small). We now want to consider some of the other aspects of the nonlinear behavior of the fiber. The *amplitude* of light as it propagates in glass behaves as

$$E(z + dz) = E(z) \exp\left[\left(-\frac{\alpha_p}{2} + ik + \frac{\gamma P(z)}{2A_{\text{eff}}}\right) dz\right] \tag{3.95}$$

where $E(z)$ and $E(z + dz)$ are the amplitudes of the light at two positions in space that are separated by a distance dz, α_p is the *power* loss coefficient, k is the propagation constant in the fiber ($k = 2\pi n/\lambda$), γ is the coefficient of nonlinearity, $P(z)$ is the power in the fiber at position z, and A_{eff} is the *effective area* of the fiber core (to be defined).

The value of γ is small and its effects are unobserved for most power levels, but in a fiber $P(z)/2A_{\text{eff}}$ can be made quite large and the lengths can be quite long, so the effects of γ can become noticeable. For stimulated scattering, γ is real and it can lead to either a gain or a loss in the fiber. For nonlinear index effects, γ is imaginary and can lead to phase modulation effects. The following are some of the effects that we wish to discuss.

- Stimulated scattering

    - Raman scattering
    - Brillouin scattering

- Nonlinear index effects

- Single-signal

    - Self-phase modulation

- Multi-signal effects

    - Cross-phase modulation
    - Intermodulation (mixing)

These effects have deleterious impact of the performance of some fiber-optic systems [9, 45]. The single-signal effects will occur in the presence of one signal. The multi-signal effects require at least two signals at different wavelengths to be present. The former will affect all fiber transmissions; the latter will affect wavelength-division multiplexed (WDM) transmissions (discussed in some detail in Chapter 9) [46–48]. Scattering can limit the amount of power that we are able to introduce into a fiber, thereby affecting the achievable distance between repeaters or lowering the bit-error rate of a link. Self-phase modulation broadens the spectral width of a pulse resulting in increased material and waveguide dispersion effects. Cross-phase modulation causes signals in a multiwavelength system to interfere with each other, lowering their performance.

Even the single-channel problems can affect the multichannel systems, however [46]. For example, we will find out that stimulated Brillouin scattering generates a wave at a frequency that is different than the original wave. As the scattered light grows in amplitude, it will decrease the strength of the original signal. In a single-channel system this decreased strength is a weakening of the signal beam. In a multichannel system the new wave frequency might fall near to one of the other channels in the system, causing degradation.

3.4.1   Effective Area and Effective Length

We wish to define the *effective area* of the fiber and the *effective length* of the fiber. We have seen that the field in a single-mode fiber extends out of the core into the cladding and is somewhat Gaussian shaped. The effective area concept assumes that the field can be modeled as a constant value over a certain area of the fiber as shown in Fig. 3.18. The effective area A_{eff} is defined by

$$A_{\text{eff}} = \frac{\left[\int \int I(r,\theta)\, r\, dr\, d\theta\right]^2}{\int \int I^2(r,\theta)\, r\, dr\, d\theta}.$$
(3.96)

A typical value of the effective area is on the order of 50 μm^2. Frequently, the effective area is approximated as $\pi d^2/4$ where d is the mode-field diameter (MFD) of the field distribution.

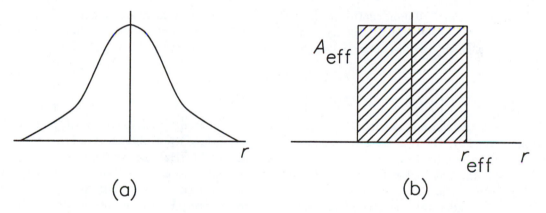

(a)   (b)

Figure 3.18 Field distribution in fiber (left) and assumed distribution over effective area (right).

The *effective length* of the fiber is an equivalent length of the fiber with constant power down the length as shown in Fig. 3.19 on the next page. The *effective length* is defined by

$$L_{\text{eff}} = \frac{1 - e^{-\alpha_p L}}{\alpha_p}$$
(3.97)

where α_p is the power attenuation coefficient (discussed in Chapter 3) and L is the actual length of the fiber. For fibers of long length (i.e., $L \gg 1/\alpha_p$), we have $L_{\text{eff}} \approx 1/\alpha_p$. We also need a definition of the effective length that applies when we have several segments of fiber separated by optical amplifiers. In this case, we will assume that the effective length of the total chain is the sum of the effective lengths of each segment.

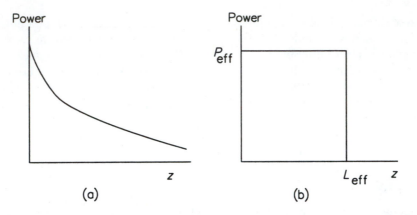

Figure 3.19 Power vs. length in actual fiber (left) and assumed power behavior in fiber of effective length (right).

3.4.2   Nonlinear Scattering

Linear scattering is characterized by the frequency of the scattered light being the same as the incident light. In *nonlinear scattering*, the scattered light frequency is different than the incident light.

In Raman and Brillouin scattering [10], the light in a medium has a small probability of interacting with the medium to produce some vibrational energy (a *phonon*) and a lower frequency (longer wavelength) photon. Energy must be conserved, so

$$h\nu_{in} = h\nu_{out} + h\nu_{phonon} \tag{3.98}$$
$$\nu_{out} = \nu_{in} - \nu_{phonon}.$$

The process can happen spontaneously or it can be stimulated by the presence of light at the output frequency. For Raman scattering, the frequency of the output phonon is at lower frequencies (called "acoustic" frequencies but being on the order of 30 GHz); for Brillouin scattering, the output photon is at higher frequencies ("optical" phonons). Stimulated Raman scattering can occur in all directions; stimulated Brillouin scattering will occur preferentially in the backward direction. The spectral tolerance of the stimulating wave is very narrow for stimulated Brillouin scattering. (It is ~135 GHz for silica fiber.) The spectral tolerance for Raman scattering is considerably broader; it is 6000 GHz (or 6 THz) for silica.

Stimulated Raman Scattering

If the molecule is in the ground state when the incident light passes, then the situation of Fig. 3.20a might occur (with a small probability). Here, the phonon is excited from its ground state into its excited state and $\nu_{out} = h\nu_{in} - h\nu_{phonon}$. The scattered photon is at a lower energy and frequency and is called the *Stokes photon*. In the case shown in Fig. 3.20b, the molecule is already in the excited vibration state and can give up energy when interacting with the incident photon. We have a higher-energy (and higher-frequency) photon exiting the process (where $\nu_{out} = \nu_{in} + \nu_{phonon}$). This is an *anti-Stokes photon*. At thermal equilibrium there are many more molecules in the ground state rather than the excited state so, normally, the Stokes emission will

Parameter	Value
Frequency shift (peak)	14 THz
Linewidth ($\Delta\nu$)	6 THz
Raman gain (m/W)	0.9×10^{-13}

Table 3.1 Raman frequencies and gain coefficients in silica at $\lambda = 0.694$ μm. From Ref. [49].

dominate the anti-Stokes emission. Note that Raman process *always* occurs when light passes through materials. Usually, however, the effect is too weak to be noticed. In long fibers, though, the effect may be more pronounced due to the possibility of high power densities in small-core fibers and the long lengths of glass that can be involved.

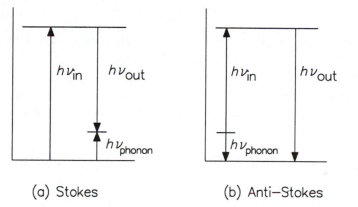

(a) Stokes   (b) Anti-Stokes

Figure 3.20 (a) Raman process for molecule in the ground state and (b) Raman process for molecule in the excited vibrational state.

If the interaction occurs spontaneously, the emitted light is incoherent and is emitted in random directions. In the presence of a strong field, the emissions are coherent and offer the possibility of making a Raman amplifier for optical fibers. (Raman amplifiers [10] are still of laboratory interest, although the efficiency of the erbium-doped fiber amplifiers has made these latter amplifiers dominant in commercial use.)

Once the stimulated field begins, it will grow exponentially until the stimulating source gain is saturated. Thus, it can deplete energy from the original input wavelength by generating increasing fields at other wavelengths. The generation of these other wavelengths can also have the effect of increasing the effective $\Delta\lambda$ of the source (and thereby increase the material and waveguide dispersion). The growth of the intensity of the Stokes wave, I_2, with distance is given by

$$dI_2 = G_r I_1 I_2 \, dz \qquad (3.99)$$

where G_r is the Raman gain ($G_r = 0.9 \times 10^{-13}$ m/W in silica at $\lambda = 0.694$ μm), and I_1 is the intensity of the stimulating wave (called the "pump beam"). (The value of the Raman gain scales as the inverse of the wavelength squared.) Other Raman scattering information is given in Table 3.1. The solution to the equation for a weak I_2 (i.e., $I_2 \ll I_1$) is

$$I_2(z) = I_2(0)e^{G_r I_1 z} \qquad \text{for } I_1 \ll I_1. \qquad (3.100)$$

---------------------------------- ∞ ----------------------------------

Example: Consider a fused silica fiber with an effective area of 1×10^{-6} cm^2 (MFD ≈ 11 μm). A laser diode couples 100 mW of light into the fiber at a wavelength of 1 μm. If the fiber is 1 km long, find (a) the peak wavelength where the gain will appear.

Solution: (a) We know from Table 3.1 that the peak of the gain will be 14 THz below the stimulating wavelength. The stimulating frequency is

$$\nu_{\text{pump}} = \frac{c}{\lambda} = \frac{3.0 \times 10^8}{1 \times 10^{-6}} = 14 \times 10^{12} \text{ Hz}. \tag{3.101}$$

The gain frequency and gain wavelengths, then, are

$$\nu_{\text{gain}} = \nu_{\text{pump}} - \nu_{\text{Raman}} = 300 \times 10^{12} - 14 \times 10^{12} = 286 \times 10^{12} \tag{3.102}$$

$$\lambda_{\text{gain}} = \frac{c}{\nu_{\text{gain}}} = 1.048 \ \mu\text{m}. \tag{3.103}$$

Hence, the stimulated Raman emission will be centered at 1.088 μm (286 THz) and will have a linewidth of 6 THz. (How many nanometers is this?)

Solution: (b) Find the magnitude of the Raman gain.

In order to find the overall gain, we first need to find the Raman gain at the peak wavelength. We know from the text that this gain coefficient scales as the square of the inverse wavelength ratios, so

$$G_r[1\mu\text{m}] = G_r[0.694 \ \mu\text{m}] \left(\frac{0.694}{1}\right)^2 = \left(0.9 \times 10^{-13}\right) \left(\frac{0.694}{1}\right)^2 \tag{3.104}$$

$$= 4.33 \times 10^{-14} \text{ m} \cdot \text{W}.$$

The overall gain is found from

$$\frac{I_2(z)}{I_2(0)} = e^{G_r I_1 z} = e^{(4.33 \times 10^{-14})(1 \times 10^9)(1000)} = 1.044. \tag{3.105}$$

Hence, we see that the Raman gain is about 4% per kilometer at 1.049 μm. This gain is spectrally quite broad. (The linewidth of the gain curve is 6 THz.) If the stimulating pump is not attenuated too much, added length can provide more amplification.

---------------------------------- ∞ ----------------------------------

Some issues that can arise in the stimulated Raman amplifier include the following.

• Frequently the pump beam and amplified wave are pulses. Because they are at different wavelengths, material dispersion will cause them to travel at different velocities. Eventually the pulses will no longer overlap and the amplification process will stop. This problem becomes more severe as the pulses become shorter. The difference in propagation time for two pulses at a nominal frequency ν, separated by $\delta\nu$, is the same as that found for our material dispersion discussion,

$$\Delta\tau = \frac{L}{c} \frac{\delta\nu}{\nu} \left(\lambda^2 \frac{dn^2}{d\lambda^2}\right). \tag{3.106}$$

From this, we see that a set of 1-ps pulses that are initially launched together will separate in about 30 cm (before much amplification can be accomplished).

- The pump power decreases as it traverses the fiber. This implies that Eq. 3.99 on page 67 should be written as

$$dI_2 = G_r I_2 I_1 e^{-\alpha z}\, dz.$$ (3.107)

The integral solution will be in terms of L_{eff} instead of L [49].

- In a fiber link carrying a single wavelength, the presence of the signal will offer the chance for Raman amplification, using the signal as the pump. The amplification will begin from a spontaneous Raman scattered light wave that is propagating along the fiber at the proper wavelength. The stimulated process will begin and will rob power from the signal and divert it into other wavelengths. (The situation is even more complicated in fibers carrying multiple wavelength signals.) The power lost to the Stokes signals depends on the fiber loss, the Raman gain, and the number of modes carried by the fiber. The upper limit on the power that can be injected into the fiber is determined by the power that causes the Stokes power P_r to equal the signal power. This power is [49]

$$P \approx \frac{16 A_{\text{eff}}}{G_r L_{\text{eff}}}.$$ (3.108)

(Note that we are no longer are using the small-signal assumption that was used to obtain Eq. 3.100 on page 67.)

Stimulated Brillouin Scattering

Brillouin scattering [10] is caused by the interaction of the input optical photon (the "pump" photon) and moving index variations caused by vibrations within the material. The scattered light behaves as if it were reflected from the moving wave and results in a Doppler-like frequency shift of

$$\delta\nu = \frac{2n V_s}{\lambda}$$ (3.109)

where V_s is the acoustic velocity of the vibration. For the case where the input light frequency linewidth, $\Delta\nu_{\text{pump}}$, is narrow compared to the frequency linewidth of the Brillouin scatter, the unsaturated *Brillouin gain coefficient*, G_{B0}, is given by

$$G_{B0} = \frac{2\pi n^7 p_{12}^2}{c\lambda^2 \rho V_s \Delta\lambda_B}$$ (3.110)

where p_{12} is the elasto-optic coefficient of the material, ρ is the material density, and $\Delta\nu_B$ is the Brillouin frequency linewidth ($\sim$135 MHz in silica). At $\lambda = 1$ μm, $G_{B0} = 4.5 \times 10^{-9}$ cm/W. If the pump linewidth $\Delta\nu_{\text{pump}}$ is larger than the gain linewidth, then

$$G_B \approx G_{B0} \frac{\Delta\nu_B}{\Delta\nu_{\text{pump}}}.$$ (3.111)

Since the Brillouin scattering is primarily in the reverse direction, light is removed from the signal beam and put into the Brillouin scattered light. We note from Eq. 3.111 that one way to reduce the scatter is to broaden $\Delta\nu_p$ (i.e., use a broadband source). For short pulses (i.e., higher data rates), $\Delta\nu_p$ is also broadened and reduces the Brillouin scatter. The critical power level to avoid much impact from Brillouin scatter is

$$P = \frac{21 A_{\text{eff}}}{G_B L_{\text{eff}}}.$$ (3.112)

We note from this equation that reducing G_B causes the threshold to increase. For data rates exceeding a few hundred Mb/s, the power level is fairly high and Brillouin scattering is not a major problem. The effect of *stimulated Brillouin scattering* begins at the smallest value of power of any of the nonlinear effects. In a fiber without amplifiers, the threshold can be as low as 12 mW. For a fiber chain with N cascaded amplifiers, the threshold becomes as low as $12/N$ mW.

3.4.3   Index of Refraction Effects

The index of refraction in silica can be written as

$$n(I) = n_0 + n_2 I \tag{3.113}$$

where n_0 is the usual index of refraction, n_2 is the nonlinear refractive component ($n_2 = 3.2 \times 10^{-16}$ cm^2/W for silica), and I is the irradiance of the light ($I = P/A_{\text{eff}}$). Because of the small value of n_2 most of its effects are negligible under ordinary circumstances.

Self-Phase Modulation

If we consider a plane propagating through a nonlinear medium, then the field can be written as

$$E(z) = E_0 e^{i(\omega_0 t - k_0 n z)} \tag{3.114}$$

where $k_0 = \omega_0/c$. The instantaneous phase of the wave is

$$\phi = \omega_0 t - \frac{\omega_0 n z}{c} . \tag{3.115}$$

For a time-varying wave $n(t) = n_0 + n_2 I(t)$ and

$$\phi(t) = \omega_0 t - \left(\frac{\omega_0 z}{c}\right)(n_0 + n_2 I(t)) . \tag{3.116}$$

The instantaneous frequency of the wave is

$$\omega(t) = \frac{d\phi}{dt} = \omega_0 - \frac{\omega_0 z n_2}{c}\frac{dI(t)}{dt} . \tag{3.117}$$

We observe that the instantaneous frequency is now a function of the time-rate-of-change of the intensity of the wave. Due to the minus sign, the frequency decreases when the wave power increases (and vice versa). For example, as a light pulse builds, the instantaneous frequency decreases. As a light pulse fades in amplitude, the frequency increases. This time-varying frequency change of the light is called a *frequency chirp*. Figure 3.21 illustrates the effects of a light pulse on the frequency content of the light. We see from the figure that the power variation of the wave is modulating the phase of the wave (and, hence, the name *self-phase modulation* [10]). As a result of the frequency chirp, the spectral width of the source $\Delta\nu$ is broader than it otherwise would be and the effects of material and waveguide dispersion are increased.

The magnitude of the frequency chirp is

$$\delta\omega = \Delta k L = \frac{2\pi L n_2}{\lambda}\frac{dI}{dt} . \tag{3.118}$$

The total accumulated phase is

$$\Delta\phi = \frac{2\pi L}{\lambda}\delta n_{\text{max}} \tag{3.119}$$

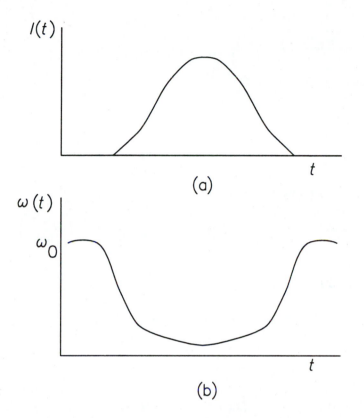

Figure 3.21 Effects of (a) time-varying light pulse on (b) light frequency, causing a frequency chirp.

where

$$\delta n_{\max} = n_2 I_{\max}. \tag{3.120}$$

We can define the *critical length* as the propagation length that causes the temporal pulse length to double, i.e.,

$$L_{\text{crit}} = \frac{cT}{n}. \tag{3.121}$$

Assuming that the pulse shape has exponential tails (i.e., that the rate of change of the pulse intensity is $dI/dt = I/\tau$ where I is the peak intensity and τ is the pulse width), then the increase in frequency width from travel of L_{crit} is

$$\Delta\omega = \frac{2\pi L_{\text{crit}} n_2}{\lambda} \frac{I}{\tau}. \tag{3.122}$$

If we define the group velocities of two different wavelengths as v_1 and v_2, then in a time T the waves travel $z_1 = v_1 T$ and $z_2 = v_2 T$, respectively. The time T required for the pulses to separate by a distance one pulse width (i.e., to separate by a distance of $\tau c/n$) is

$$T = \frac{\tau c}{n(v_2 - v_1)}. \tag{3.123}$$

Hence, we have

$$L_{\text{crit}} = \frac{\tau c^2}{(v_2 - v_1)n^2} \ .$$ (3.124)

We also note that

$$v_2 - v_1 = \frac{dv_g}{d\omega} \Delta\omega$$ (3.125)

and find

$$v_g = \frac{c}{n_0 + \omega \dfrac{dn}{d\omega}} \ .$$ (3.126)

Taking the derivative, we can show that

$$\frac{dv_g}{d\omega} = -\frac{1}{2\pi} \frac{1}{\left(n_0 + \omega \dfrac{dn}{d\omega}\right)} \left(\lambda^2 \frac{d^2n}{d\lambda^2}\right) \approx -\frac{1}{2\pi} \frac{1}{n_0} \left(\lambda^2 \frac{d^2n}{d\lambda^2}\right) \ .$$ (3.127)

Substitution of Eqs. 3.127 and 3.122 into Eq. 3.124, gives

$$L_{\text{crit}} = \frac{c\tau}{\sqrt{\delta n \, \lambda^2 \dfrac{d^2n}{d\lambda^2}}} \ .$$ (3.128)

———————————————— ∞ ————————————————

Example: Consider a fiber with a 10 μm mode-field diameter that is excited by a 10-ps pulse at 1550 nm. Find the critical length of this fiber if the peak power of the pulse is 10 W.

Solution: The peak change in index is found from Eq. 3.120 on the preceding page where $n_2 = 3.2 \times 10^{-16}$ cm^2/W,

$$\delta n_{\text{max}} = n_2 I_{\text{max}} = (3.2 \times 10^{-16}) \left(\frac{(4)(10)}{\pi(10 \times 10^{-4})^2}\right) = 4 \times 10^{-9} \ .$$ (3.129)

Note that we have kept the units in terms of cm and that the resulting index change is very small. The critical length is

$$L_{\text{crit}} = \frac{c\tau}{\sqrt{\delta n \lambda^2 \dfrac{d^2n}{d\lambda^2}}} = \frac{(3.0 \times 10^8)(10 \times 10^{-12})}{\sqrt{(4 \times 10^{-9})(0.009)}} = 500 \text{ m} \ .$$ (3.130)

The pulse in this problem will double its width after propagating only 500 m due to the self-phase modulation. This is a considerable pulse spread.

———————————————— ∞ ————————————————

In single-channel links operating at the wavelength of zero dispersion, the broadening effects of the self-phase modulation can be negligible. In systems operating away from the zero-dispersion wavelength, the broadening effect can be noticeable and, without increased spacing between pulses (i.e., lower data rates) can lead to intersymbol interference due to the pulse spread.

Cross-Phase Modulation

In *cross-phase modulation* [10], which occurs in multichannel links, variations in the intensity in one channel will change the phase in the other channels. The cause of the effect is the same as self-phase modulation; changes in intensity leads to changes in the index, changes in the phase, and changes in the frequency of the channel. The cross-phase modulation is complicated by the "collisions" that occur between the pulses in different channels. Since the wavelengths differ, the pulses at the shorter wavelengths will be traveling slower than the long-wavelength pulses. The long-wavelength pulse will, generally, overtake and pass through the slower short-wavelength pulses. During the first half of the pulse collision, the leading edge of passing pulse will cause a wavelength chirp that lengthens the instantaneous wavelength (a *"red" shift*). In the second half of the collision, as the overtaking pulse leaves, the wavelength of the overtaken pulse shortens (a *"blue" shift*). The shifts cancel if the intensities of the waves have not changed during the interaction.

Four-Photon Mixing

Four-photon mixing or *four-wave mixing* [10, 50] occurs on multichannel links. The intensity of the combined waves is proportional to the square of the sum of the separate field amplitudes, i.e.,

$$I = k \left| \sum_i (E_1 + E_2 + \cdots + E_i) \right|^2 .$$

(3.131)

If there are N channels, there will be $N^2(N-1)/2$ mixing terms in the expansion. It is predicted that this effect becomes noticeable at power levels of about 1 mW per channel. This level of power is quite low and is the limitation to power levels in systems with ten or fewer channels [9]. The first effect of this mixing is that the signal power is reduced in order to produce the mixing products. This can be minimized by increasing the wavelength spacing between the channels. A second effect occurs if the layout of the operating wavelengths is not carefully done, since some of these mixing products will occur at or near some of the operating wavelengths. For example, if the channel wavelengths are equally spaced, then some of the mixing products will fall directly atop the other channels. (For example, if three signals are present, then the mixing frequencies are $f_{ijk} = f_i + f_j - f_k$; if $f_k - f_j = f_j - f_i = \Delta f$, then some of the mixing frequencies will fall directly on the signal frequencies. This is called "four-wave mixing" since three waves interfere to provide a fourth wave.) The mixing products (also called the *crosstalk*) can destructively interfere with the signal, causing major signal effects for even small amounts of mixing products. Careful allocation of the channel spacings can eliminate this overlapping of the mixing products, allowing their removal by receiver filters tuned to the channel wavelengths. In addition, four-wave mixing is maximized by operating at or near the zero-dispersion wavelength; it tends to be minimized by operating away from this wavelength (contrary to the assumption made for single-channel systems.

3.5   Cables

The goal in making a cable [51, 52] incorporating an optical fiber is to provide strength and protection while minimizing cable volume and weight and to avoid adding appreciable optical losses. Additionally, it may be necessary to incorporate power carrying conductors in the cable for some applications, such as underseas cables [53–62]. Generally, a list of desirable properties in a cable would include:

- minimized stress-produced optical losses,

- high tensile strength,

- immunity to water vapor penetration,

- stability of characteristics over a specified temperature range,

- ease of handling and installation (especially compatibility with current installation equipment), and

- low acquisition, installation, and maintenance costs.

Fiber cables can be as simple or complicated as the application requires. Typically, a cable will consist of some of the following elements:

- The optical fibers, which may be single fibers or multiple fibers.

- A *buffering material*, which is a soft substance placed around the fiber to isolate it from radial compressions and other localized stresses applied to the cable. If the fiber is able to move within the buffering material while responding to strain in the cable, then the fiber is said to be *loose-buffered*. If the fiber is rigidly constrained to its position within the cable, then the fiber is said to be *tight-buffered*.

- Strength members are high tensile-strength materials that provide the longitudinal strength of the fiber. Separation of the strength member role from the information carrying role is one of the primary advantages of fiber cables, as the selection of each material can be optimized. Typically, high-strength, low-weight materials, such as Kevlar, can be used. The tensile strength of a fiber cable is the sum of the strengths of the individual parts of the cable, that is,

$$T = S \sum_i E_i A_i , \qquad (3.132)$$

where T is the tensile load, S is the maximum allowed strain or elongation (e.g., 1%), E_i is the Young's modulus of the i-th component, and A_i is the cross-sectional area of the i-th component.

If there is a requirement to carry power down the length of the wire, copper conductors or copper-coated high-strength wires could also be utilized. One potential problem is elongation of the cable. A typical fiber will break at an elongation greater than 1%, while a typical stress member can elongate on the order of 20% before breaking. This disparity can result in the requirement to helically coil the fiber within the center region of the cable to allow the cable to elongate without breaking the fiber. A cable-making machine has this capability to wind the fiber, buffer material, and strength members at the proper pitches without producing microbends in the fiber and without inducing a torque in the cable that would hinder the installation process.

- Filler yarns, added to take up space between strength members and also to provide a degree of buffering.

- A *jacket*, to provide abrasion protection; waterproofing; protection from rodents, fish, and other gnawing animals; resistance to chemical reaction; and other environmental protection. The outer jacket largely determines the installation properties, as it must have the correct friction properties to allow convenient installation.

Typical cable installations involve a wide range of environments including ducts, aerial stringing from posts, burial in trenches, underwater installation, and laying the fiber on the top of the ground.

- Figure 3.22 shows a representative cable. The outer jacket is a polyurethane plastic with the right properties for fire and smoke retardation, jacket smoothness for fiber drawing, and color for fiber identification. The next layer is a set of Kevlar fibers for load-carrying. These high-strength, low-weight fibers are frequently found in fiber cables. The fiber itself is surrounded by a protective plastic jacket that isolates the fiber from the other parts of the cable design.

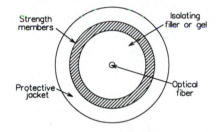

Figure 3.22 Typical fiber-optic cable structure. Representative outside diameters are 2 mm to 3 mm.

- Figure 3.23 shows a representative cable designed for aerial stringing from post to post. Multiple fibers are present in the cable along with two conducting members. The fibers are in a filling gel.

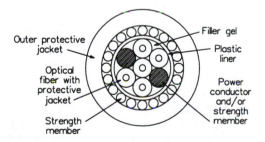

Figure 3.23 Typical aerial fiber-optic cable structure. Representative outside diameters are 6 mm to 7 mm.

- Figure 3.24 on the next page shows a representative cable designed for burial in trenches. More outer protective layers are evident.

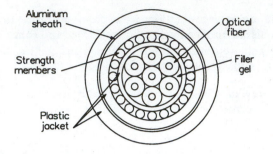

Figure 3.24 Typical fiber-optic cable structure for burial installation. Representative outside diameters are 9 mm to 10 mm.

- Figure 3.25 on the following page shows a representative cable designed for short-distance, undersea transmission. Here, copper-clad steel wires are used both to conduct electrical power through the cable and to provide cable strength for the harsh installation and repair environment. (It should be noted that the presence of electrical power in the cable can provoke defensive behavior from sharks and fish in certain installation areas. The cable may require extra protective layers.)

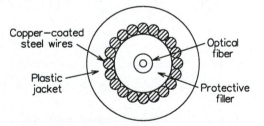

Figure 3.25 Typical short-distance undersea cable structure. Representative outside diameters are 2 mm to 3 mm.

3.6 Summary

In this chapter we introduced more of the parameters used to describe a fiber's performance. The losses of the fiber are due to various factors and predict a loss minimum occurring at 1550 nm for silica fibers—a wavelength that is currently being exploited by the industry. This minimum occurs in silica-based fibers and can be reduced only by going to different materials, with their attendant fabrication problems. Losses can also be incurred by bending losses (e.g., the added losses due to incorporating a fiber into a cable. Various attenuation measurement techniques have also been described.

The bandwidth or data rate limitations of an optical fiber are due to the various dispersion factors that can cause a pulse to spread. In multimode fibers, modal- and material-dispersion effects dominate. In single-mode fibers, modal dispersion is absent and material and waveguide dispersions are significant. Near 1300 nm, the combination of material dispersion and waveguide dispersion can produce zero dispersion. This region produces the minimum dispersion of the fiber. With more complicated refractive-index profiles this dispersion minimum can be shifted in wavelength to occur at 1550 nm in dispersion-shifted fibers or a relatively flat region of small dispersion can be achieved in dispersion-flattened fibers.

Nonlinearities have become important now that we have combinations of sources of relatively high power, small mode-field diameters, optical amplifiers that can regenerate a signal's power, and long propagation distances that allow the effects to accumulate. In single-channel systems, Brillouin scattering limits the maximum power that can be carried. Its threshold is lower than that associated with Raman scattering. Self-phase modulation can spread the spectrum of an optical pulse, causing increased material and waveguide dispersion. For multichannel systems, four-wave mixing limits the power that be carried in each channel. Cross-phase modulation can cause crosstalk between the channels, limiting the achievable signal-to-noise ratio.

Finally, we have briefly described the elements of a fiber cable and some of the properties that determine the cable's composition.

3.7   Problems

1. Find an expression that relates α (in units of dB/km) to α_p (in units of m^{-1}).

2. An optical fiber has a loss of 1 dB/km. Find the corresponding value of the loss coefficient α_p (in units of m^{-1}).

3. The optical power after propagating through a fiber that is 450 m long is reduced to 30% of its original value. Calculate the fiber loss α in dB/km.

4. Consider an OTDR that injects a laser pulse with a pulse width of τ seconds into the test fiber at time $t = 0$.

   (a) Show that, at time $t = t_1$, the light returning to the input was backscattered from a segment of the fiber that is Δz long, $\Delta z = (c/n_1)(\tau/2)$

   (b) Show that, at $t = t_1$, the edge of the fiber segment (of size Δz) furthest from the input is a distance of $(c/n_1)(t_1/2)$ away from the input.

   (c) The size of Δz is the *resolution* of the OTDR. Calculate the resolution if the pulse width is 50 ns.

   (d) Calculate the pulse width required to have an OTDR resolution of 1 m in a fiber with an index of 1.45..

   (e) Calculate the round-trip time for light that is backscattered from a fiber segment located 10 km from the input of the test fiber.

5. Consider a fiber with an attenuation coefficient of α (per unit length) that has a rectangular pulse of power P_i and width τ coupled into it. The scattered light power from a segment of the fiber that is Δz long is given by $P(z)\alpha_s \Delta z$, where $P(z)$ is the incident light power at the location of the fiber segment and α_s is the scattering loss coefficient for material ($\alpha_s \approx 0.7$ km^{-1} for glass fibers). Not all of the scattered light, however, is guided by the fiber back to the input end of the test fiber; only a fraction, S, is. The value of S depends on the fiber properties as [23]

$$S_{\mathrm{MM}} \;=\; \left(\frac{3}{16}\right)\left(\frac{\mathrm{NA}^2}{n_1^2}\right)\left(\frac{g}{g+1}\right)$$

for multimode fibers and

$$S_{\mathrm{SM}} \;=\; \left(\frac{3}{16}\right)\left(\frac{2\lambda}{\pi n_1 \,\mathrm{MFD}}\right)$$

for single-mode fibers, where NA is the fiber numerical aperture, g is the fiber profile parameter, and MFD is the mode-field diameter. Hence, the backscattered light that is collected by the fiber is $SP(z)\alpha_s\,\Delta z$.

If Δz is given by the expression given in the previous problem, show that the backscattered power arriving back at the input at time t_1 is given by

$$P(t_1) \;=\; P_iS\alpha_s\left(\frac{\tau}{2}\right)\left(\frac{c}{n_1}\right)e^{-\alpha\left(\frac{c}{n_1}\right)t_1}.$$

(Note that there is a tradeoff between the power returned to the input and the resolution of the OTDR. To minimize the resolution, we want to shorten the pulse width τ; this reduces the backscattered power returned to the OTDR detector.)

6. Consider a step-index multimode fiber with NA $= 0.2$, $n_1 = 1.5$, $\alpha_{\text{fiber}} = 0.6$ dB·km^{-1}, and $\tau = 50$ ns.

   (a) Using the results of the previous problem, calculate the ratio of the backscattered power to the input power for $t_1 = 0$ (i.e., for light backscattered from the segment of the fiber located right at the input of the test fiber).

   (b) Calculate the ratio of the backscattered power to the input power for a segment of fiber located 10 km from the input of the fiber.

   (c) Calculate the ratio of the backscattered power from the segment located 10 km from the input to the backscattered power from the segment located right in front of the input.

   (d) The *dynamic range*, DR, (in dB) of an OTDR is defined as

   $$\text{DR} \;=\; 10\log\left(\frac{P_{s\text{ max}}}{P_{s\text{ min}}}\right),$$

   where $P_{s\text{ max}}$ is the power of the strongest backscattered signal that can be detected at the OTDR receiver and $P_{s\text{ min}}$ is the power of the weakest signal that can be detected. If the dynamic range of the detector is 20 dB, calculate the maximum operating range of the OTDR, assuming that fiber loss is the only loss encountered (i.e., there are no connector or splice losses). You may assume that the coupler that connects the laser to the OTDR fiber pigtail (and the detector to the pigtail) has a 3 dB loss for each pass through it.

7. Consider a single-mode fiber operating at 1300 nm with a mode-field diameter of 9 μm, $n_1 = 1.5$, and $\alpha = 0.5$ dB·km^{-1}. Assume, again, that $\tau = 50$ ns. Repeat the calculations of the previous problem.

8. A break occurs in a fiber with a loss of 3 dB/km. The output power from an OTDR set used to locate the break is 250 mW and the detected echo power is 1 μW. Approximately 10 dB of loss are encountered in coupling the OTDR signal into the fiber and the returned signal encounters 6 dB of loss at the optical splitter (see Fig. 3.4 on page 44). The reflectivity of a perpendicular break in the fiber is approximately 4%; the average reflectivity of a nonperpendicular break is about 0.5%. Using the latter value of reflectivity, calculate the distance to the break in the fiber.

9. The Sellmeier equations give us an expression for $n(\lambda)$ based on the resonances of the absorbing atoms or molecules. The equation for silica is

$$n(\lambda) \;=\; \sqrt{1+\sum_{k=1}^{3}\frac{G_k\lambda^2}{\lambda^2-\lambda_k^2}}. \tag{3.133}$$

The values of G_k and λ_k represent the resonant wavelengths and relative strengths of the resonance and are found in the Table 3.2: Values of the parameters for other materials are found in optical handbooks.

λ_k (nm)	G_k
68.4	0.69617
116.2	0.40794
9896.2	0.89748

Table 3.2 Table of resonant wavelengths and strengths for silica.

(a) Using a computer, plot $n(\lambda)$ for values of λ between 600 nm and 1600 nm.

(b) Using a computer, plot $(\lambda^2 \frac{d^2 n}{d\lambda^2})$ for values of λ between 600 nm and 1600 nm.

10. (a) Calculate the pulse spread $\Delta\tau$ for a 1 km single-mode fused silica step-index fiber if $\lambda = 1.0\mu m$, $\Delta\lambda/\lambda = 0.12\%$, $V = 1.5$, $n_1 = 1.453$, and $n_2 = 1.450$.

(b) Calculate the pulse spread $\Delta\tau$ of a multimode step-index fiber that is 1 km long and has $n_1 = 1.453$ and $n_2 = 1.438$ with $V = 38$.

11. Consider a fused silica single-mode fiber that is 1 km long. Find the material dispersion component of the pulse spread if $\Delta\lambda = 3.0$ nm and the operating wavelengths are 800 nm, 900 nm, 1300 nm and 1500 nm.

12. Calculate the material dispersion pulse spread $\Delta\tau_{mat}$ at $\lambda = 820$ nm, $\Delta\lambda = 3$ nm, $L = 2$ km, for a fused silica single-mode fiber.

13. Consider an 0.80 km fiber made of fused silica with a step-index. The following applies: $\lambda = 1.0$ μm, $\Delta\lambda/\lambda = 0.12\%$, $V = 38$, $n_1 = 1.453$, and $n_2 = 1.438$.

(a) Calculate the pulse spread due to group delay.

(b) Calculate the pulse spread due to material dispersion.

14. Using a computer, plot $b(V)$, $d(Vb)/dV$, and $V\, d^2(Vb)/dV^2$ over a range of $V = 0$ to $V = 3$.

15. Consider a 9/125 single-mode fiber with $n_1 = 1.48$ and $\Delta = 0.22\%$ as described in the waveguide dispersion example calculation on page 51. Let $L = 1000$ and $\Delta\lambda = 1$ nm.

(a) Plot $\Delta\tau_{wg}$ for wavelengths ranging from 1200 nm to 1400 nm.

(b) Plot $\Delta\tau_{mat}$ for the same range of wavelengths.

(c) Plot the sum of the material pulse spread and the pulse spread due to waveguide dispersion.

16. A fiber has a loss of 0.22 dB/km.

(a) Find the effective length of the a long fiber using the long-length approximation.

(b) How long must the fiber actually be to have the approximate value of the effective length be within 5% of the actual value?

17. A fiber chain has a total length of 10,000 km.

(a) Find the effective length if there are 100 segments of 100 km length.

(b) Find the effective length if there are 300 segments of 33 km length.

18. Consider a chain of fibers with optical amplifiers interspersed. The total length is 10,000 km, the data rate is 2.5 Gb/s, and the desired SNR is 200. At an amplifier spacing of 1 km, 40 μW of power is required. Using a computer plot the required input power (in mW) vs. the amplifier spacing. Use a log-log plot, allowing the spacing to range from 1 km to 1000 km. You may assume that $n_{sp} = 1$.

19. Using Eq. 3.106 on page 68, prove that a set 1-ps pulses that are initially launched together will separate in about 30 cm of fiber.

20. Consider a single-mode fiber with an MFD of 10 μ and a loss of 0.2 dB/km at 1550 nm. Find the power limit determined by the Raman scatter.

21. Calculate the Brillouin scatter limit for the fiber of the previous problem for a bit rate of 1 Gb/s.

22. A silica fiber with an MFD of 11 μm has a 10 ps light pulse in it. If the operating wavelength is 1550 nm and the peak power is 10 W, find the critical length of the pulse in this fiber.

23. We want to compute the fraction of a tensile load that is carried by the coating of a fiber. The fraction of the total stress σ_{total} that is carried by the coating is found from

$$\frac{\sigma_{coating}}{\sigma_{total}} = \frac{E_{coating} A_{coating}}{E_{coating} A_{coating} + E_{fiber} A_{fiber}}$$

where $E_{coating}$ and E_{fiber} are the Young's modulus of the coating and fiber material, respectively, and $A_{coating}$ and A_{fiber} are the cross-section areas of the coating and fiber, respectively.

Consider a 62.5/125 fiber that has a polymer coating that is 0.05 mm thick surrounding it. The Young's modulus of the coating polymer is 350 MPa and of glass is 71.9 GPa.

(a) Calculate the fraction of the total applied stress that is carried by the coating.

(b) Calculate the fraction of the total applied stress that is carried by the fiber.

(c) Suppose that the fiber were double-coated with two different concentric coatings. What do you think would be the expression for the fraction of the total stress carried by the interior coating?

References

1. G. Schick, K. A. Tellefsen, A. J. Johnson, and C. J. Wieczorek, "Hydrogen sources for signal attenuation in optical fibers," *Optical Engineering*, vol. 30, no. 6, pp. 790–801, 1991.

2. P. J. Lemaire, "Reliability of optical fibers exposed to hydrogen: prediction of long–term loss increase," *Optical Engineering*, vol. 30, no. 6, pp. 780–789, 1991.

3. D. C. Tran, George H. Sigel, Jr., and B. Bendow, "Heavy metal fluoride glasses and fibers: A review," *J. Lightwave Technology*, vol. LT–2, no. 5, pp. 566–586, 1984.

4. B. Bendow, H. Rast, and O. H. El-Bayoumi, "Infrared fibers: an overview of prospective materials, fabrication methods, and applications," *Optical Engineering*, vol. 24, no. 6, pp. 1072–1080, 1985.

5. S. Sakaguchi and S. Takahashi, "Low-loss fluoride optical fibers for midinfrared optical communication," *J. Lightwave Technology*, vol. LT–5, no. 9, pp. 1219–1228, 1987.

6. S. R. Nagel, "Fiber material and fabrication methods," in *Optical Fiber Telecommunications II* (S. E. Miller and I. P. Kaminow, eds.), pp. 121–215, New York: Academic Press, 1988.

7. R. Stolen, "Nonlinear properties of optical fibers," in *Optical Fiber Telecommunications* (S. Miller and C. Chynoweth, eds.), pp. 125–150, New York: Academic Press, 1979.

8. R. Stolen, "Nonlinearity in fiber transmission," *Proc. IEEE*, pp. 1232–1236, 1980.

9. A. R. Chraplyvy, "Limitations on lightwave communications imposed by optical-fiber nonlinearities," *J. Lightwave Technology*, vol. 8, no. 10, pp. 1548–1557, 1990.

10. G. P. Agrawal, *Nonlinear Fiber Optics: Second Edition*. San Diego: Academic Press, 1995.

11. F. P. Kapron, "Fiber-optic test methods," in *Fiber Optics Handbook for Engineers and Scientists* (F. C. Allard, ed.), pp. 4.1–4.54, New York: McGraw-Hill, 1990.

12. M. Artiglia, G. Coppa, P. DiVita, M. Potenza, and A. Sharma, "Mode field diameter measurements in single-mode optical fibers," *J. Lightwave Technology*, vol. 7, no. 8, pp. 1139–1152, 1989.

13. K. Petermann and R. Kuhne, "Upper and lower limits for the microbending losses in arbitrary single-mode fibers," *J. Lightwave Technology*, vol. LT–4, no. 1, pp. 3–7, 1986.

14. M. Miyamoto, M. Sakai, R. Yamauchi, and K. Inada, "Bending loss evaluation of single-mode fibers with arbitrary core index profile by far-field pattern," *J. Lightwave Technology*, vol. 8, no. 5, pp. 673–677, 1990.

15. J. Senior, *Optical Fiber Communications: Principles and Practice*. Englewood Cliffs, NJ: Prentice Hall, 1985.

16. P. Kaiser and D. B. Keck, "Fiber types and their status," in *Optical Fiber Telecommunications II* (S. E. Miller and I. P. Kaminow, eds.), pp. 29–54, New York: Academic Press, 1988.

17. M. Artiglia, G. Coppa, P. D. Vita, H. Kalinowski, and M. Potenza, "Simple and accurate microbending loss evaluation in generic single-mode fibers," *Proc. 12th ECOC (Barcelona, Spain)*, vol. 1, pp. 341–344, 1986.

18. M. Artiglia, G. Coppa, P. D. Vita, H. Kalinowski, and M. Potenza, "Bending loss characterization in single-mode fibers," *CSELT Technical Report*, vol. XV, no. 6, pp. 411–415, 1987.

19. Charles H. Gartside III, P. D. Patel, and M. R. Santana, "Optical fiber cables," in *Optical Fiber Telecommunications II* (S. E. Miller and I. P. Kaminow, eds.), pp. 217–261, New York: Academic Press, 1988.

20. M. Barnoski and S. Personick, "Measurements in fiber optics," *Proc. IEEE*, vol. 66, no. 4, pp. 429–441, 1978.

21. C. J. Cannell, R. Worthington, and K. Byron, "Measurement techniques," in *Optical Fibre Communication Systems* (C. Sandbank, ed.), pp. 106–155, New York: Wiley, 1980.

22. L. Cohen, P. Kaiser, and C. Lin, "Experimental technique for evaluation of fiber transmission loss and dispersion," *Proc. IEEE*, vol. 68, no. 10, pp. 1203–1209, 1980.

23. B. L. Danielson, "Optical time-domain reflectometer specifications and performance testing," *Applied Optics*, vol. 24, no. 15, pp. 2313–2322, 1985.

24. M. Tateda and T. Horiguchi, "Advances in optical time-domain reflectometry," *J. Lightwave Technology*, vol. 7, no. 8, pp. 1217–1224, 1989.

25. V. C. So, J. W. Jiang, J. A. Cargil, and P. J. Vella, "Automation of an optical time domain reflectometer to measure loss and return loss," *J. Lightwave Technology*, vol. 8, no. 7, pp. 1078–1083, 1990.

26. P. Healey, "Review of long wavelength single-mode optical fiber reflectometry techniques," *J. Lightwave Technology*, vol. LT–3, no. 4, pp. 876–886, 1985.

27. J. King, D. Smith, K. Richards, P. Timson, R. Epworth, and S. Wright, "Development of a coherent OTDR instrument," *J. Lightwave Technology*, vol. LT–5, no. 4, pp. 616–624, 1987.

28. G. Keiser, *Optical Fiber Communications, Second Edition*. New York: McGraw-Hill, 1991.

29. D. Gloge, "Dispersion in weakly guiding fibers," *Applied Optics*, vol. 10, no. 11, pp. 2442–2445, 1971.

30. D. Gloge, "Weakly guiding fibers," *Applied Optics*, vol. 10, no. 10, pp. 2252–2258, 1971.

31. B. J. Ainslie and C. R. Day, "A review of single-mode fibers with modified dispersion characteristics," *J. Lightwave Technology*, vol. LT–4, no. 8, pp. 967–979, 1986.

32. W. Gambling, H. Matsumura, and C. Ragsdale, "Zero total dispersion in graded-index single-mode fibers," *Electronics Letters*, vol. 15, pp. 474–476, 1979.

33. W. Gambling, H. Matsumura, and C. Ragsdale, "Mode dispersion and profile dispersion in graded-index single-mode fibres," *Microwaves, Optics, and Acoustics*, vol. 3, pp. 239–246, 1979.

34. U. Paek, G. Peterson, and A. Carnevale, "Dispersionless single-mode lightguides with α index profiles," *Bell Sys. Technical J.*, vol. 60, pp. 583–598, 1981.

35. K. I. White, "Design parameters for dispersion-shifted triangular profile single-mode fibers," *Electronics Letters*, vol. 18, pp. 725–727, 1982.

36. K. Yamauchi, M. Miyamoto, T. Abiru, K. Nishide, T. Ohashi, O. Fukada, and K. Inada, "Design and performance of gaussian-profile dispersion-shifted fibers manufactured from the VAD process," *J. Lightwave Technology*, vol. LT–4, no. 8, pp. 997–1004, 1986.

37. V. Bhagavatula, J. Lapp, A. Morrow, and J. Ritter, "Segmented-core fiber for long-haul and local-area-network applications," *J. Lightwave Technology*, vol. 6, no. 10, pp. 1466–1469, 1988.

38. A. H. Cherin, *Introduction to Optical Fibers*. New York: McGraw-Hill, 1983.

39. J. E. Midwinter, *Optical Fibers for Transmission*. New York: Wiley, 1979.

40. R. Olshansky and D. Keck, "Pulse broadening in graded-index optical fibers," *Applied Optics*, vol. 15, no. 12, pp. 483–491, 1976.

41. R. Olshansky, "Multiple–α index profiles," *Applied Optics*, vol. 18, pp. 683–689, 1979.

42. D. Gloge, E. A. Marcatili, D. Marcuse, and S. D. Personick, "Dispersion properties of fibers," in *Optical Fiber Telecommunications* (S. E. Miller and A. G. Chynoweth, eds.), pp. 101–124, New York: Academic Press, 1979.

43. D. Marcuse and H. Presby, "Effects of profile deformations on fiber bandwidth," *Applied Optics*, vol. 18, pp. 3758–3763, 1979.

44. R. Olshansky, "Propagation in glass optical waveguides," *Review of Modern Physics*, vol. 51, pp. 341–367, 1979.

45. R. G. Waarts, A. Friesem, E. Lichtman, H. H. Yaffe, and R.-P. Braun, "Nonlinear effects in coherent multichannel transmission through optical fibers," *Proc. IEEE*, vol. 78, no. 8, pp. 1344–1368, 1990.

46. T. Li, "The impact of optical amplifiers on long-distance lightwave telecommunications," *Proc. IEEE*, vol. 81, no. 11, pp. 1568–1579, 1993.

47. J. Thiennot, F. Pirio, and Jean–Baptiste Thomine, "Optical undersea cable systems trends," *Proc. IEEE*, vol. 81, no. 11, pp. 1610–1623, 1993.

48. H. Taga, N. Edagawa, S. Yamamoto, and S. Akiba, "Recent progress in amplified undersea systems," *J. Lightwave Technology*, vol. 13, no. 5, pp. 829–840, 1995.

49. C. R. Pollock, *Fundamentals of Optoelectronics*. Chicago: Richard D. Irwin, Inc., 1995.

50. R. W. Tkach, A. R. Chraplyvy, F. Forhieri, A. H. Gnauk, and R. M. Derosier, "Four-photon mixing and high-speed WDM systems," *J. Lightwave Technology*, vol. 13, no. 5, pp. 841–849, 1995.

51. M. Schwartz, P. Gagen, and M. Santana, "Fiber cable design and characterization," *Proc. IEEE*, vol. 68, no. 10, pp. 1214–1219, 1980.

52. M. M. Ramsay, "Fiber-optic cables," in *Fiber Optics Handbook for Engineers and Scientists* (F. C. Allard, ed.), pp. 2.1–2.50, New York: McGraw-Hill, 1990.

53. A. Adl, T.-M. Chien, and T.-C. Chu, "Design and testing of the SL cable," *J. Lightwave Technology*, vol. LT–2, no. 6, pp. 824–832, 1984.

54. S. R. Nagel, "Review of the depressed cladding single-mode fiber design and performance for the SL undersea system application," *J. Lightwave Technology*, vol. LT–2, no. 6, pp. 792–801, 1984.

55. D. Paul, K. H. Greene, and G. A. Koepf, "Undersea fiber optic cable communications system of the future: Operational, reliability, and systems considerations," *J. Lightwave Technology*, vol. LT–2, no. 6, pp. 414–425, 1984.

56. P. K. Runge and P. R. Trischitta, "The SL undersea lightwave system," *J. Lightwave Technology*, vol. LT–2, no. 6, pp. 744–753, 1984.

57. R. E. Wagner, "Future 1.55–μm undersea lightwave systems," *J. Lightwave Technology*, vol. LT–2, no. 6, pp. 1007–1015, 1984.

58. P. Worthington, "Cable design for optical submarine systems," *J. Lightwave Technology*, vol. LT–2, no. 6, pp. 833–838, 1984.

59. H. Fukinuki, T. Ito, M. Aiki, and Y. Hayashi, "The FS–400M submarine system," *J. Lightwave Technology*, vol. LT–2, no. 6, pp. 754–760, 1984.

60. Y. Niiro and H. Yamamoto, "The international long-haul optical-fiber submarine cable system in Japan," *IEEE Communications Magazine*, vol. 24, no. 5, pp. 24–32, 1986.

61. P. Runge and N. S. Bergano, "Undersea cable transmission systems," in *Optical Fiber Telecommunications II* (S. E. Miller and I. P. Kaminow, eds.), pp. 879–909, New York: Academic Press, 1988.

62. P. R. Trischitta and D. T. Chen, "Repeaterless undersea lightwave systems," *IEEE Communications*, pp. 16–21, March 1989.

Chapter 4

Splices, Connectors, Couplers, and Gratings

4.1 Introduction

Since fibers are available in lengths typically ranging up to a few kilometers, the fiber-optic system user must have means of interconnecting or joining lengths of fibers in a way that offers low *insertion loss* (i.e., the additional loss introduced by the connection), high strength, and simplicity of installation. The same properties are also required for the repair or expansion of a fiber-optic link. A *splice* [1, 2] is a permanent joining of fibers; a *connector* allows a demountable connection between fibers or between a fiber and a source or detector. Fiber joints are also characterized by their reflections, since the operation of single-frequency lasers can be upset by reflected light entering the laser cavity. These reflection losses are called the *return losses* of the joint.

Another need is to split the power in a fiber into two or more fibers or to combine the power from several fibers into a single fiber. Devices that perform this operation are *couplers*, and they too have an insertion loss. In this chapter we will also describe the various couplers available.

Glass permanently changes its index of refraction slightly when exposed to intense ultraviolet light. This effect has been used to "write" periodic index structures in fiber cores with side illumination from lasers. The structure offers wavelength-dependent reflectivity and has been used to make wavelength-sensitive filters, reflectors and other devices. We describe the grating structures and some of their applications in this chapter.

4.2 Joining Losses

The causes of loss in a fiber connection or splice have been subdivided into two areas, those that depend on the properties of the fibers being joined (the *intrinsic* sources of loss) and those due to factors external to the fiber parameters (the *extrinsic* sources of loss), such as fiber misalignment. We use the *coupling efficiency*, η, to describe the fraction of the incident power that is transmitted through a joint. It is given by

$$\eta = \frac{P_{\text{out}}}{P_{\text{in}}}, \tag{4.1}$$

where P_{in} is the optical power in the fiber on the input side and P_{out} is the optical power in the fiber at the output side of the connector. In general, the coupling efficiency is *not* the same in both directions. The *joint loss*, L_j, is the dB equivalent of η, given by

$$L_j = -10 \log \left(\frac{P_\text{out}}{P_\text{in}} \right). \tag{4.2}$$

(Since the output power will be less than or equal to the input power, the joint loss will be a positive value.)

4.2.1 Multimode Fibers

The optical power coupled from one fiber to another is limited by the lesser of the number of modes in each. Optimum coupling occurs (in either direction) when the number of modes is matched. The number of modes N in a fiber is [3]

$$N = k^2 \int_0^a \text{NA}^2(r)\, r\, dr, \tag{4.3}$$

where $k = 2\pi/\lambda$, a is the radius of the fiber, and $\text{NA}(r)$ is the generalized numerical aperture for both step-index and graded-index fibers. This equation reduces to

$$N = k^2 \text{NA}^2(0) \int_0^a \left[1 - \left(\frac{r}{a} \right)^g \right] r\, dr. \tag{4.4}$$

Hence we see the factors of interest include fiber radius a, axial numerical aperture $\text{NA}(0)$, and index gradient g. The usual approach taken is to isolate the effects as if they were independent and, then, to add the dB losses to provide an estimate of the overall losses.

Additionally, the losses depend on the *power distribution* among the modes of the fiber. Computations usually assume that the power is uniformly distributed over all of the modes of the fiber; in reality, the power may be unevenly distributed due to the launch conditions or due to the modal effects in very long fibers that leave most of the power in the lower-order modes of the fiber. Such power distribution effects also play a role in the extrinsic factors of loss. For example, the losses encountered due to an axial misalignment of the fibers depend on the distribution of power in the modes. Light that is concentrated in the lower modes will be more sensitive to axial displacements than fibers that have the light uniformly distributed. Because of the computational difficulties of including these effects, the approximation is to assume uniform illumination in the fiber that excites all modes uniformly.

We now want to consider the effects of joining mismatched fibers.

- Numerical aperture effects—The coupling efficiency depends on the numerical apertures of the fibers as

$$\eta_{\text{NA}} = \begin{cases} \left(\dfrac{\text{NA}_r(0)}{\text{NA}_e(0)} \right)^2 & \text{if } \text{NA}_r(0) < \text{NA}_e(0) \\ 1 & \text{if } \text{NA}_r(0) \geq \text{NA}_e(0), \end{cases} \tag{4.5}$$

  where $\text{NA}_r(0)$ and $\text{NA}_e(0)$ are the axial numerical apertures of the receiving and emitting fibers, respectively. The formula assumes equal radii and index profiles of both fibers.

- Fiber radius effects—The coupling efficiency for fibers of differing core radii is given by

$$\eta_r = \begin{cases} \left(\dfrac{a_r}{a_e} \right)^2 & \text{for } a_r < a_e \\ 1 & \text{for } a_r \geq a_e, \end{cases} \tag{4.6}$$

where a_r and a_e are the radii of the core of the receiving and emitting fiber, respectively. The fiber numerical apertures and axial index profiles are assumed equal.

- Index profile effects—The coupling efficiency due to differences in the index of refraction profile is given by

$$\eta_g = \begin{cases} \dfrac{g_r(g_e + 2)}{g_e(g_r + 2)} & \text{for } g_r < g_e \\ 1 & \text{for } g_r \geq g_e, \end{cases} \tag{4.7}$$

where g_r and g_e are the refractive-index profile values of the receiving and emitting fibers, respectively. Figure 4.1 shows a plot of this equation.

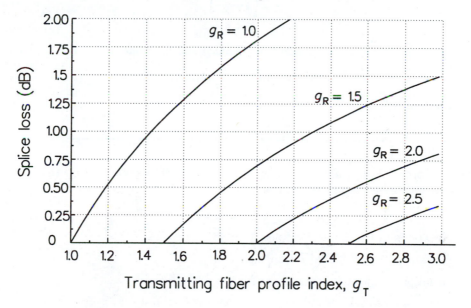

Figure 4.1 Typical calculated losses due to refractive-index profile mismatches.

- Combined effects—Having found the individual intrinsic coupling efficiencies, we estimate the total intrinsic coupling efficiency η_{total} as

$$\eta_{\text{total}} \approx \eta_{\text{NA}}\, \eta_r\, \eta_g \,. \tag{4.8}$$

To express the combined losses in dB, we find

$$\begin{aligned} L_{\text{total}}(\text{dB}) &= -10 \log \eta_{\text{total}} \approx -10 \log(\eta_{\text{NA}}) - 10 \log(\eta_r) - 10 \log(\eta_g) \\ &\approx L_{\text{NA}} + L_r + L_g \,. \end{aligned} \tag{4.9}$$

$$\underline{\hspace{5cm}\infty\hspace{5cm}}$$

Example: Calculate the coupling efficiency that can be expected in coupling a 50/125 SI (emitting) fiber with an NA of 0.15 to a 62.5/125 GI ($g = 2$) receiving fiber with an NA = 0.20.

<u>Solution:</u> Since $NA_r > NA_e$,

$$\eta_{NA} = 1. \tag{4.10}$$

Since $a_e < a_r$,

$$\eta_r = 1. \tag{4.11}$$

Since $g_e(=\infty) > g_r$,

$$\eta_g = \frac{g_r(g_e + 2)}{g_e(g_r + 2)} = \frac{g_r\left(1 + \dfrac{2}{g_e}\right)}{(g_r + 2)} \tag{4.12}$$

$$= \frac{2\left(1 + \dfrac{2}{\infty}\right)}{(2 + 2)} = 0.5.$$

The total efficiency is the product of the individual contributions,

$$\eta = \eta_{NA}\eta_r\eta_g = (1)(1)(0.5) = 0.5. \tag{4.13}$$

The total losses are $-10\log(0.5) = 3$ dB.

---------------------------------- ∞ ----------------------------------

- Effects of other fiber parameters—There are a variety of loss mechanisms that are functions of the fiber fabrication process and its quality control. These include

  - ellipticity of the core,
  - variations in the index of refraction profile,
  - concentricity of the core within the cladding,
  - variation in the core diameter, and
  - other factors that depend on fabrication tolerances.

  Since these factors are determined by the manufacturing process, the user has little control over them, except to establish specifications and inspection procedures.

Of these factors of joint loss, the differences in core radii and numerical aperture have the dominant effects, while mismatches in the core ellipticity and the refractive index profile play lesser roles (for the same values of fractional mismatch).

4.2.2 Misalignment Effects

The effects just discussed are properties of the fiber. We now turn consideration to factors that are under the control of the connector designer (and to some degree, the user of the connectors). These effects are primarily due to the misalignment of the fibers [1, 4–6] and they determine the mechanical tolerances required of the connector or splicer to meet a given loss allocation.

- Lateral displacement effects—The misalignment to which a connection is most sensitive is *lateral displacement*, shown in Fig. 4.2a. Mechanical misalignments of the fibers cause losses because of the areas of the fiber core do not overlap sufficiently. In the analysis of misalignments the usual assumptions are that the fibers have equal radii, index profiles, and

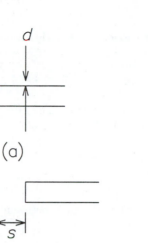

Figure 4.2 Types of misalignments: (a) lateral displacement, (b) longitudinal separation, and (c) angular misalignment.

numerical apertures to isolate the effects of the misalignment. The fiber is also assumed to have uniform power distribution across its area. For a step-index fiber the coupling efficiency depends on the lateral misalignment distance d as [3]

$$\eta_{SI} = \frac{2}{\pi} \cos^{-1}\left(\frac{d}{a}\right) - \frac{d}{\pi a}\sqrt{1 - \left(\frac{d}{2a}\right)^2}. \tag{4.14}$$

This result is derived from a calculation of the overlapping areas of two equal circles with centers separated by a distance, d. A similar calculation for a graded-index fiber is complicated by the radial dependence of the numerical aperture. The result of the calculation [3] based on a parabolic index ($g = 2$) predicts a coupling efficiency of

$$\eta_{GI} = \frac{2}{\pi}\left[\cos^{-1}\left(\frac{d}{2a}\right) - \sqrt{1 - \left(\frac{d}{2a}\right)^2}\left(\frac{d}{6a}\right)\left(5 - \left(\frac{d^2}{2a^2}\right)\right)\right]. \tag{4.15}$$

For $d/a < 0.4$ we can use the approximation [3]

$$\eta_{GI} \approx 1 - \frac{8d}{3\pi a} \tag{4.16}$$

A different form for the loss in a graded-index fiber that is valid for any value of g is found in Ref. [7] as

$$\eta_{GI} \approx 1 - \left(\frac{2d}{\pi a}\right)\left(\frac{g+2}{g+1}\right). \tag{4.17}$$

• Longitudinal displacement effects—The effects caused by a pure *longitudinal displacement* (Fig. 4.2b) are due to the fact that some of the light has spread beyond the area of the receiving fiber, as shown in Fig. 4.3. For the step fiber, we have [3]

$$\eta_{SI} = \left(\frac{1}{1+(s/a)\tan\theta_c}\right)^2, \tag{4.18}$$

where s is the separation distance between the fiber ends and θ_c is the critical angle of acceptance $(= \sin^{-1}NA)$.

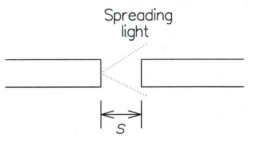

Figure 4.3 Spreading of light in longitudinally separated fibers.

———————————————— ∞ ————————————————

Example: Calculate the coupling efficiency for a 50/125 SI fiber if the longitudinal displacement is 10% (i.e., $s/a = 10\%$) and $NA = 0.2$.
Solution: The critical angle is found by

$$\theta_c = \sin^{-1}(NA) = \sin^{-1}(0.2) = 11.54°. \tag{4.19}$$

The coupling efficiency is found as

$$\eta_{SI} = \left(\frac{a}{a+s\tan\theta_c}\right)^2 \tag{4.20}$$

$$= \left[\frac{25 \times 10^{-6}}{(25 \times 10^{-6})+(25 \times 10^{-7})(\tan 11.54°)}\right]^2$$

$$= (0.980)^2 = 0.960 = 96\%.$$

———————————————— ∞ ————————————————

Reference [7] gives a formula for this coupling efficiency for a step-index fiber as

$$\eta_{SI} \approx 1 - \frac{s\sqrt{2\Delta}}{4a}. \tag{4.21}$$

No similar formula for the losses due to longitudinal displacement seems to be available for a graded-index fiber.

- Angular misalignment effects—The effects of *angular misalignment* are shown in Fig. 4.2c. Reference [7] gives the loss due to angular misalignment ϕ in a fiber with an arbitrary index profile g as

$$\eta \approx \frac{\sin \phi}{\sqrt{2\pi\Delta}} \left(\frac{\Gamma\left(\frac{2}{g}+2\right)}{\Gamma\left(\frac{2}{g}+\frac{3}{2}\right)} \right), \tag{4.22}$$

where $\Gamma(x)$ is the Gamma function (found in books of advanced math tables).

- Combined misalignment effects—Again, we combine the misalignment effects by multiplying the individual coupling efficiencies or adding the coupling losses in dB.

4.2.3 Single-Mode Fiber Joints

The mode-field diameter (MFD) of a single-mode fiber is physically useful because it relates to the performance of the fiber in an optical system. Specifically, the MFD determines the sensitivity of the joint losses at a connector or splice to misalignment. The MFD will also determine the sensitivity of the fiber to excess losses due to microbends and macrobends. In addition, the cutoff wavelength and the dispersion properties of the fiber can be inferred from a spectral measurement of the MFD. The following sections discuss some of these dependencies.

Joint losses occur at splices and connectors. Since the power transmitted through a joint depends on the fields supported on each side, the shape and diameter of the fields are expected to be key parameters in determining the losses at the joint.

Reference [8] gives an expression for the coupling loss (in dB) for two single-mode fibers with unequal mode-field diameters between the transmitting fiber and the receiving fiber (W_1 and W_2, respectively) with a lateral offset d, a longitudinal offset s, and an angular offset ϕ. This single-mode loss, based on a Gaussian-wave assumption, is

$$L = -10 \log \left[\left(\frac{16 n_1^2 n_3^2}{(n_1 + n_3)^4} \right) \left(\frac{4\sigma}{q} \right) e^{-\frac{\rho u}{q}} \right], \tag{4.23}$$

where λ is the source wavelength, n_1 is the refractive index of the fiber cores (assumed to be the same on both sides of the joint), n_3 is the refractive index of the medium between the fibers (if applicable),

$$\sigma = \left(\frac{W_2}{W_1} \right)^2, \tag{4.24}$$

$$k = \frac{2\pi n_3}{\lambda},$$

$$\rho = (kW_1)^2,$$

$$F = \frac{d}{kW_1^2} \quad \text{(a lateral offset parameter)},$$

$$G = \frac{s}{kW_1^2} \quad \text{(a longitudinal offset parameter)},$$

$$q = G^2 + (\sigma + 1)^2, \text{and}$$

$$u = (\sigma + 1)F^2 + 2\sigma FG \sin\phi + \sigma(G^2 + \sigma + 1)\sin^2\phi.$$

Equation 4.23 on the page before can be used to predict the losses expected due to each of the extrinsic alignment parameters or for any given combination of the parameters. An alternative approach is presented in Ref. [9]

4.2.4   Reflection Losses

The coupling efficiency at a perpendicular interface is given by the *Fresnel reflection loss* formula

$$\eta_{\text{reflection}} = 1 - \left(\frac{n_1 - n_2}{n_1 + n_2} \right)^2 , \tag{4.25}$$

where n_1 is the index of one medium and n_2 is the index of the other medium. The reflection losses (in dB) are found by calculating $-10 \log(\eta)$. The reflection losses are the same regardless of the direction of travel of the light. The losses at an air-glass interface ($n_{\text{glass}} = 1.5$, $n_{\text{air}} = 1.0$) in a connector can add 0.2 dB loss ($n_{\text{glass}} = 1.5$, $n_{\text{air}} = 1.0$) for each fiber face. These losses can be eliminated by the application of an index-matching gel to the c onnector or through the use of an index-matching epoxy in making a permanent joint.

The light that is not transmitted at an interface is reflected. Reflected light can upset the operating characteristics of a source (particularly, a single-frequency source). Optical amplifiers can also amplify reflected light, disturbing the operation of these amplifiers. The measure of the reflection properties of a fiber join is its *return loss*, defined by

$$L_{\text{return}} = -10 \log \left(\frac{P_{\text{reflected}}}{P_{\text{incident}}} \right) . \tag{4.26}$$

The return loss can be minimized by

- using index-matching material between the fiber ends,

- by using physical-contacting fiber ends,

- by using end-faces that are cut at non-perpendicular angles, and/or

- using optical isolators.

The *optical isolators* are single-mode fiber devices that utilize polarized light and optical rotators to manufacture a unidirectional optical device. Connectors with an air-gap can have a typical return loss of about 14 dB. A reasonable return loss of a physical-contact joint is 30 dB; higher return losses ($\sim$40–50 dB) are also possible in connectors optimized for this parameter. A superior joint utilizing angled ends in physical contact can have return losses of 60 dB or larger.

4.2.5   Fiber End Preparation

The expressions for coupling efficiency or loss all assume that the end of the fiber is a perfect transmitter. Any pits or imperfections in the end of the fiber will scatter light into higher angles, many of them lying outside of the numerical aperture of the receiving fiber. Similarly, the end-faces must be parallel to the propagation axis in order to ensure uniform illumination or reception. (Usually the ends are perpendicular to the axis.) End preparation techniques have evolved to ensure a smooth perpendicular end of the fiber for splicing or connecting.

- The *grinding-and-polishing technique* uses progressively finer abrasives to smooth and polish the ends. The perpendicularity of the fiber end is assured by the use of a mechanical fiber holder that aligns and holds the fiber perpendicular to the polishing surface. This technique is labor- and time-intensive.

- The *score-and-break technique* uses traditional glass-cutting methods to provide a uniform surface. A fiber is placed under mild tension as in Fig. 4.4 and is scratched with a scoring blade. The scratch creates a stress concentration at the tip of the scratch. The tension is then increased and the crack tip propagates across the fiber. If the curvature of the fiber and the application of the tension are carefully controlled, then the crack will propagate perpendicular to the fiber axis and create a clean smooth break. Improper control of the crack propagation can cause the formation of a *lip* or a *hackle* on the fiber end, as shown in Fig. 4.5 on the following page. Microscope inspection of the fiber end is necessary to detect such problems that would require a new scoring. A number of tools have been commercially developed to implement this technique, which has the advantage of taking very little time for an experienced user.

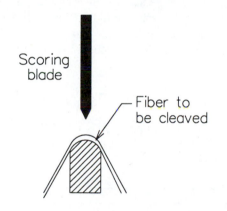

Figure 4.4 Score-and-break technique of cutting fiber with smooth ends.

4.3   Splices

Three types of splices [1, 10–12] have proved popular in fiber technology:

- Perhaps the most popular splice is the *fusion splicing technique* [12], illustrated in Fig. 4.6 on the next page. The fusion splicer uses micro-manipulators to bring the prepared ends of the fiber into close alignment. The ends are then heated, typically with an electric arc, until they grow molten and fuse together. As the joint cools, the surface tension at slightly misaligned fibers will pull the fibers into alignment. In these splicers the arc voltage is kept low, until the fiber ends are rounded, thereby avoiding bubble formation, and then increased to complete the fusing process. Losses caused by splices made with this technique are typically a few tenths of a dB, with the primary problem being reduced fiber strength in the region near the joint. This reduced strength (about 60% of the strength before making

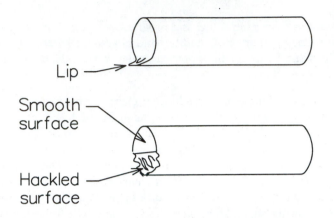

Figure 4.5 Improper fiber ends.

the joint) is caused by the development of microcracks on the fiber surface in the region that has been stripped of its buffering and by chemical changes in the glass due to the heating. This reduction of strength is countered by the use of a high-strength wrapping placed around the spliced region, which protects the fiber and provides stress relief for the splice region.

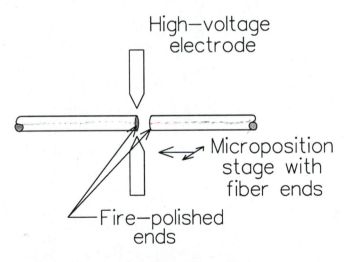

Figure 4.6 Fusion splicing technique.

- The *V-groove splice* (Fig. 4.7) uses a V-shaped groove as an alignment aid to bring the two fibers into mechanical alignment, before applying an adhesive, such as epoxy, or being fastened in place with a cover plate. The channel can be made in plastic, silicon, ceramic, or metals. A variation on this technique, called the *loose-tube splice*, uses the corner of a rectangular tube as the alignment aid, as shown in Fig. 4.8 on the facing page. Since this technique uses the outside surface of the fiber as a reference surface, it is susceptible to losses due to variations in core ellipticity, core concentricity, and core size. The fiber ends require preparation before splicing. Losses as low as several hundredths of a dB have been reported with these techniques. Note, however, that splicing fibers of unequal diameters by this technique will result in unacceptable misalignments.

- Another type of splice, shown in Fig. 4.9 on the next page, uses an *elastic material* to bring the fibers into alignment. With a central circular hole slightly smaller than the fiber diameter, the restoring forces will center the fiber (again with respect to the outside surface). The advantages of this technique are that a wide range of fiber diameters can be inserted into the device and fibers of unequal diameters can be aligned.

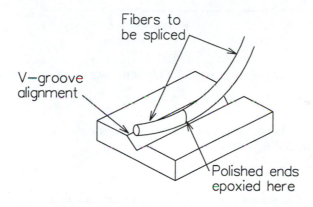

Figure 4.7 V-groove splicing technique.

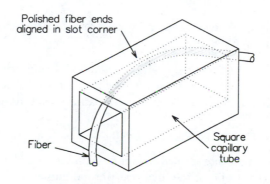

Figure 4.8 Loose-tube splice.

4.4 Connectors

Optical-fiber connectors [6, 13] are designed to allow disconnection and reconnection. The goal is to provide a low-insertion-loss connector that will provide reliable reproducible connections. Most connector designs incorporate the fiber into a precision alignment aid that then plugs

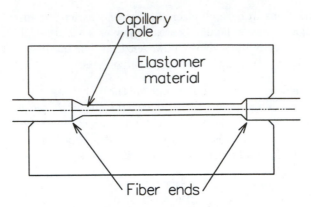

Figure 4.9 Splice using elastic material. (The fibers are shown only partially inserted.)

into a receptacle in the connecting piece. Various environmental factors can make the design increasingly difficult, including

- dust levels,

- pressure differentials across the connector (as in aircraft bulkheads or underwater connectors),

- presence of water vapor and water, both around the connector and within the fiber cable (as in underwater connectors), and

- connector mating requirements (e.g., is a male-female arrangement satisfactory or is a hermaphrodite connector required?).

The different types of connectors are generally divided according to the alignment aid that is used. In application, however, the user is concerned with the fiber size accepted by the connector, the connector type (i.e., SMA, BNC, or other standard and nonstandard fastener types), and the losses of the connector.

Different technologies are used to align the fibers.

- One technique uses a precision-drilled hole in a cylinder, called a *ferrule*. The ferrule then fits an *alignment sleeve* to bring the fiber ends into alignment, as shown in Fig. 4.10 on the facing page. The fiber is stripped of its protective jacket and buffering and inserted into the housing and out through the hole in the ferrule. After epoxying the fiber into the ferrule, the end is prepared by the grind-and-polish technique. The main problem is to carefully center the hole in the alignment sleeve and to make the hole slightly larger than the outside diameter of the fiber. Ferrules are commonly made of aluminum (for low-cost connectors), stainless steel, or ceramics.

- A similar technique developed by Bell Laboratories, the *biconic connector*, injection-molds the alignment element (Fig. 4.11). This element, with the shape of a biconical taper, is designed to mate with the housing so that the fiber is centered.

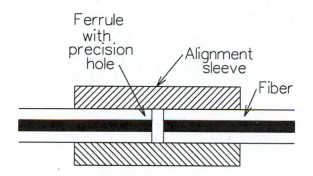

Figure 4.10 Connector made with alignment hole in a ferrule. The sleeve aligns the ferrules that contain the fibers.

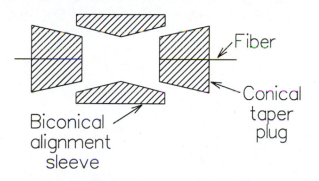

Figure 4.11 Biconical taper connectors (cross section).

- Another approach is the *expanded-beam connector*, shown in Fig. 4.12 on the next page. A microlens is inserted at the fiber's end to collimate the beam. The receiving fiber has a similar collimator to receive the beam and focus it on the core of the fiber. The collimation reduces the requirements on the lateral and longitudinal alignments required, at the penalty of increasing the angular alignment required. (This is a desirable trade-off, since the angular tolerance is more easily controlled.) The lenses are either gradient-index lenses of the proper length to form a collimator or some other structure that will receive the fiber and hold it in alignment.

4.4.1   Commercial Connectors

Several connector types have become classic connectors that are widely used in the field. For patent and proprietary rights reasons, several manufacturers have invented their own type of connectors that are used in their systems. Frequently these connector designs are "second-sourced" or cross-licensed with other connector companies, expanding their impact. Reliability issues [14] require that special attention be paid to the materials used and the method of attaching the connectors to the fiber cable. Typical insertion losses for connectors range from several tenths of a dB to a few dB in value.

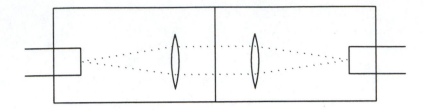

Figure 4.12 An expanded-beam connector.

- The *SMA connector* (Fig. 4.13) was borrowed from the RF field and has proved to be a popular connector for multimode fibers. A single-mode version is also available. The fiber is inserted into a ferrule that matches the outside diameter of the fiber and is epoxied in place. The fiber tip is polished (along with the end of the mounting barrel).

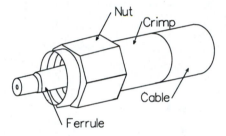

Figure 4.13 An SMA connector.

- The *biconic connector* (Fig. 4.14) was developed by AT&T (American Telephone and Telegraph) and is in wide use in their single-mode systems. It uses a molded, ground plastic or ceramic plug to achieve the close tolerances required for a single-mode connector. It is available in both a single-mode version and a multimode version.

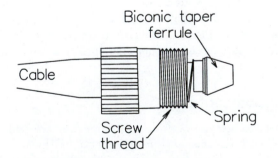

Figure 4.14 A biconic connector.

- The *ST connector*, developed by AT&T, is also widely used in single-mode systems. (ST is a registered trademark.) It is also available for multimode systems as well. The connector, shown in Fig. 4.15, features a spring-loaded bayonet clip. It requires no polishing (i.e., the fiber is cleaved before insertion) and is fairly easy to terminate.

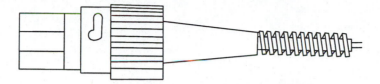

Figure 4.15 An ST connector.

- The *FC (or "D3") connector* (Fig. 4.16) is an NTT (Nippon Telephone and Telegraph)

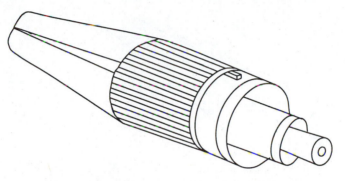

Figure 4.16 An FC connector.

design for single-mode fibers. The D3 connector is an NEC (Nippon Electronics Corporation) clone of the FC connector. It too is a spring-loaded connector with a screw-on nut. It uses a metal ferrule to align the fiber.

- The *FC/PC connector* is an offshoot of the FC connector. It uses a pure ceramic ferrule for increased alignment accuracy over the metal/ceramic one used in the FC connector. It has a spring-loaded physical contact mechanism to allow the fiber ends to make contact. It is primarily used for long-haul telecommunications applications, where the increased performance and cost are justified.

- The *D4 connector* was designed by NEC. It is similar to the D3 connector, but smaller. Figure 4.17 on the following page shows a cross-section of this type of connector.

- The *SC connector* (Fig. 4.18 on the next page) is a newer plastic-case connector that uses a push-pull configuration. It has a ceramic ferrule in the center for fiber alignment.

- The *FDDI connector* (Fig. 4.19 on page 101) is a dual-fiber connector, designed for use in FDDI data links. (The FDDI network is discussed in detail in Chapter 8.)

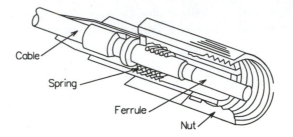

Figure 4.17 Cross section of a D4 connector.

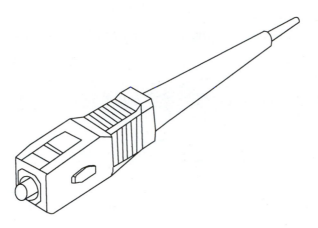

Figure 4.18 An SC connector.

4.5   Splice/Connector Loss Measurement

We have mentioned several times that the losses associated with a splice or connector depend on many variables—especially the optical-power launch conditions, the type of source used, and the characteristics of the fiber on either side of the joint. The experimental setup shown in Fig. 4.20 on the next page can be used to measure the losses in a fiber splice or connector [15]. We measure the power at the input and output of the connector (P_1 and P_2) and define the *insertion loss L_s* as

$$L_s = -10 \log \left(\frac{P_2}{P_1} \right) . \tag{4.27}$$

With a short length of fiber, we find that the loss is strongly dependent on the launching condition of the source as reflected by the numerical aperture of the source. With a sufficient length of the fiber, however, we find that the connector has been isolated from the effects of the source launching conditions due to the mode conversion effects of the fiber. Hence we require that an *equilibrium mode distribution* be established in the illuminating fiber. Two techniques have evolved for this.

- One technique is to include a long length of fiber as a "pigtail" to allow the equilibrium pattern to be established before the light reaches the connector.

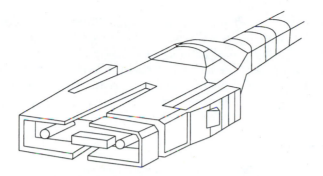

Figure 4.19 An FDDI connector.

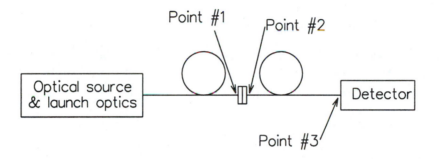

Figure 4.20 Experimental setup for measuring splice losses.

- The other technique is to include an *equilibrium mode simulator*, which usually is a short length of fiber undergoing a serpentine path that allows ample opportunity for mode mixing. A frequently used equilibrium mode simulator is a 1 meter (or so) length of of fiber wrapped around a 1.6 cm diameter spool (often called a "mandrel").

Splices, in turn, can have an effect on the measured losses of a length of multimode fiber that follows a splice due to the splice's effect on the mode distribution (i.e., the splice will redistribute the power in the modes of the fiber). Observation of this effect involves measurements of power (P_1 and P_3) in the setup of Fig. 4.20. The measured loss $-10 \log(P_3/P_1)$ is frequently greater than the sum of the insertion loss and the fiber loss. Since this additional loss is due to the splice rather than the fiber, the loss is properly added to the splice loss rather than the fiber loss. The total loss of the splice or connector is

$$L_{\text{s total}} = -10 \log \left(\frac{P_3}{P_1} \right) - \alpha L \,, \tag{4.28}$$

where αL is the product of the fiber loss coefficient and the fiber length. The excess loss $\delta L_{\text{s excess}}$ of the splice is the difference between $L_{\text{s total}}$ and the insertion loss L_s of the splice,

$$\delta L_{\text{s excess}} = L_{\text{s total}} - L_s \,. \tag{4.29}$$

For a splice with reasonable extrinsic losses (e.g., good alignment) between identical fibers, $L_{s\ total}$ is usually greater than L_s by a factor of two or so. The conclusion is that, while the insertion loss L_s is a well defined local loss quantity, the true total splice loss $L_{s\ total}$ is highly variable and depends on not only the splice but also the length, mode mixing properties, and modal attenuation of the receiving fiber. Usually L_s is the parameter quoted for a splice as measured with a long transmitting fiber and a short receiving fiber. This value, however, can be only partially representative of the splice's effect in a system.

∞

Example: (a) We want to measure a connector's loss. The power in the fiber at the connector input is 100 μW; the output power immediately after the connector is 83.2 μW. Calculate the insertion loss of the connector.

Solution: The insertion loss is

$$L_s = -10\log(P_{out}/P) = -10\log(83.2/100) = 0.8\ \text{dB}. \tag{4.30}$$

(b) When 1.8 km of 1.5 dB/km fiber is added after the connector, the measured output power at the end of the added fiber is 35.5 μW. Calculate the excess loss of the connector.

$$\begin{aligned}
\delta L_{s\ excess} &= -10\log(P_{out}/P) - L_s - \alpha L \tag{4.31} \\
&= -10\log\left(\frac{35.5}{100}\right) - (0.8) - (1.5)(1.8) \\
&= +4.50 - 0.8 - 2.7 = 1.0\ \text{dB}.
\end{aligned}$$

∞

4.6 Couplers

In fiber-optic systems there can be applications where it is desirable to combine separate optical signals or to divide the optical signal. Such *multiplexing* and *demultiplexing* tasks are handled by *optical couplers* [11]. Usually it is desired that each output share the signal equally, although it is possible, in some designs, to weight the *coupling fraction* between a source line and an output line. If one input equally feeds N output lines, then the *power splitting factor* $L_{pwr\ split}$ (in dB) is

$$L_{pwr\ split} = -10\log\left(\frac{1}{N}\right) = 10\log(N). \tag{4.32}$$

This loss is expected and is unavoidable. Any *extra* losses, however, are included in the *insertion loss*, L_{insert}, of the device,

$$L_{insert} = -10\log\left(\frac{\sum_{j=1}^{N} P_j}{P_i}\right), \tag{4.33}$$

where P_i is the power into the i-th terminal and the summation is of all of the power out of the other ports of the device. Of more utility is a splitting matrix that provides the measured or specified loss L_{ij} for an output j when the splitter is excited at input i.

4.6.1 Coupler Descriptions

Since couplers can be used to do more than just split and combine light, specific terms are used to describe their functions. It is instructive to consider the terminology applied to the various coupling functions as they illustrate potential applications of the devices.

- A *splitter* (Fig. 4.21a) splits or divides an input signal into two or more output channels. The fraction of the input optical power that is delivered to any output arm is under the designer's control. For a two-port splitter standard designs are 50:50 (i.e., 50% of the input light is found in each output port) and 90:10. Other ratios are available by custom order.

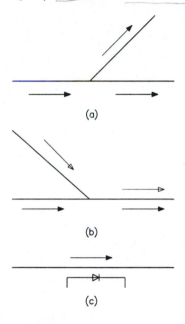

Figure 4.21 Examples of optical couplers: (a) splitter or tee coupler, (b) a combiner, and (c) a monitor.

- A *polarizing splitter* also splits the signals into two output channels, but the polarizations in each output are orthogonal. These devices work only for single-mode fibers.

- A *combiner* (Fig. 4.21b) combines two or more input channels into one output channel. Many (but not all) passive devices are reciprocal, so a splitter can sometimes be used as a combiner as well.

- A *monitor* (Fig. 4.21c) is a splitter that couples very little light (e.g., 1%) into the monitor port. A detector is placed at the output of the monitor port to measure the light output. Knowing the coupling ratio, we can establish the light level in the fiber.

- A *directional coupler* (Fig. 4.22a) is a nonreciprocal device that isolates one input from the other.

- A *multiplexer* (actually a *wavelength multiplexer*) is a combiner that joins two or more source signals of differing wavelengths. A *demultiplexer* (actually a *wavelength demultiplexer*) splits the signals at the receiving end according to wavelength. Figure 4.22b

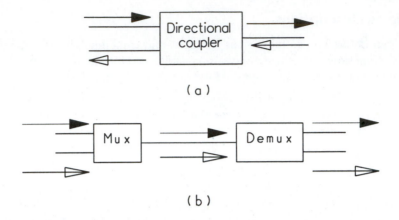

Figure 4.22 More examples of optical couplers: (a) directional coupler and (b) multiplexer and demultiplexer.

illustrates their application in a unidirectional wavelength-division multiplexing (WDM) system. (Wavelength-division multiplexing is considered in more detail in Chapter 9.)

- A *circulator* is a three-terminal device that allows light entering at one terminal (terminal #1) to be coupled out of another terminal (terminal #2) with (ideally) 100% efficiency. Light entering terminal #2 is coupled out of the third terminal (#3) with (ideally) 100% efficiency. These devices would allow a bidirectional link to work without incurring the 25% coupling efficiency of using a 2×2 splitter at each end.

4.6.2   Coupler Fabrication

Three different technologies have developed in the fabrication of these coupling elements.

1. The most popular technology is the *biconical taper coupler* (or *fused coupler*). If the claddings of two (or more) fibers are partially removed and the fibers are placed in close proximity over some length, then some light will couple from one fiber into the other(s). The fraction that couples can be controlled by the thickness of the remaining cladding and the length of the region where the fibers are in proximity. As shown in Fig. 4.23, this type of coupler can be made by taking a group of fibers with the claddings exposed (i.e., with no protective jacketing), applying tension, and heating the junction. The coupling fraction is controlled by the amount of tension and the time of heating. Surprisingly, equal coupling can be achieved for all fibers with low insertion loss. In excess of 100 fibers have been formed into a star coupler by this technique.

2. The second technique uses mode-mixing rods (Fig. 4.24 on the next page) as the coupling mechanism. A *mode-mixing rod* is glass rod of a few millimeters diameter with sufficient length to allow the light from all input locations to fully expand to (ideally) uniformly illuminate the end of the rod. All output fibers are uniformly excited. In addition to this transmissive configuration, the rod can be cut in half, a mirrored reflecting surface can be applied, and the output fibers can be moved to the input side of the device to make a reflective system.

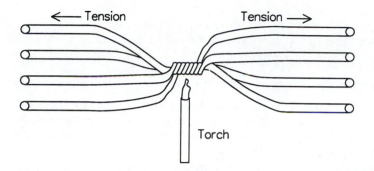

Figure 4.23 Fabrication of a biconical taper coupler.

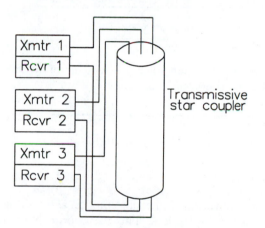

Figure 4.24 A star coupler made with a mode-mixing rod.

3. The third technology uses mirrors with limited apertures placed in strategic locations to intercept a fraction of the illumination beam emitted from the source fiber. By properly shaping the curvature of the mirror and locating the output fibers properly, the output fibers will capture the focussed light, which is a fraction of the incident light, thereby performing the coupling function.

4.6.3 Typical Coupler Specifications

The losses of a coupler consist of the desired *splitting loss* and the undesired *excess loss* that exists beyond the splitting loss. Typical excess losses are on the order of 1 dB. Most coupler losses are specified in a matrix that relates each input fiber to each output.

4.7 Fiber Grating Devices

It is possible to change the index of refraction in core of fiber by exposing it to ultraviolet light (e.g., a pulsed or CW UV laser). The effect can be strengthened (i.e., the exposure time can be reduced) "hydrogen-loading" the fiber, i.e., exposing the fiber to a high-pressure hydrogen environment for several hours. Increases in the refractive index of a few tenths of a percent are possible.

───────────── ∞ ─────────────

Example: Suppose that a certain ultraviolet exposure can increase the core index by 0.1%. Assuming plane-wave propagation, find the length of the fiber that will induce an extra phase shift of π radians. Assume that the initial value of the core index is 1.46 and that the wavelength is 1550 nm.

<u>Solution:</u> The phase shift incurred by a plane wave is $\phi = kz$ where $k = 2\pi n z/\lambda$. For our case we have

$$\Delta\phi \;=\; k'z - kz = \frac{2\pi[(n + \Delta n) - n]z}{\lambda} = \frac{2\pi \Delta n\, z}{\lambda} \tag{4.34}$$

$$z \;=\; \frac{\lambda\,\Delta\phi}{2\pi\,\Delta n} = \frac{(1550 \times 10^{-9})(\pi)}{2\pi(0.001)(1.46)} = 5.31 \times 10^{-4}\ \text{m} = 531\ \mu\text{m}\,.$$

Hence, we see that even this small change in refractive index of the core can cause a significant phase shift over a relatively short length of fiber.

───────────── ∞ ─────────────

This effect can be used as a basis for a new class of fiber devices, a fiber grating.

4.7.1 Grating Structures

A grating structure is a spatially varying index of refraction along the core of a fiber. The simplest grating is a periodic $\cos^2(2\pi z/\Lambda)$ variation. This variation can be made with a shadow mask of the proper transmittance or, more commonly, by interfering two plane waves with an angle θ between their propagation directions. These plane waves can be formed with a phase mask that has a relief profile necessary to generate two strong plane-wave components. Extended patterns (up to tens of centimeters in length) can be written with a step-and-repeat process, moving either the mask and laser or the fiber. The phase masks are now available commercially (as are custom-made fibers with the gratings written into fiber). The bare fiber is exposed to the ultraviolet light from the side through the phase mask by an ultraviolet laser. Various lasers, both pulsed and continuous have been used, including frequency-doubled dye lasers, pulsed excimer lasers, and cw frequency-doubled argon lasers. Long gratings are made with a step-and-repeat operation.

These periodic grating structures have a narrow reflection band. Relatively high sidelobes are seen with these periodic gratings (see Fig. 4.25a). (Sidelobes are the regions of localized maxima of transmission appearing away from the main peak of the response.) More complicated patterns are also written. An "apodized" pattern tapers the grating strength (i.e., the difference between the maximum refractive index and the minimum) at the ends of the grating. Apodized gratings offer the advantage of having lower-strength sidelobes outside of the passband region of the filter (see Fig. 4.25b). Apodized gratings are made by reducing the laser strength (or exposure time) in the regions near the ends of the grating.

Chirped gratings have an increasing spatial frequency along the length of the grating. These gratings have the advantage of offering a wider passband than can be achieved with a periodic grating (see Fig. 4.25c). Chirped gratings can also be apodized to flatten the ripple response in the passband (and lowering the sidelobe peaks) (see Fig. 4.25d). Chirped gratings also have the

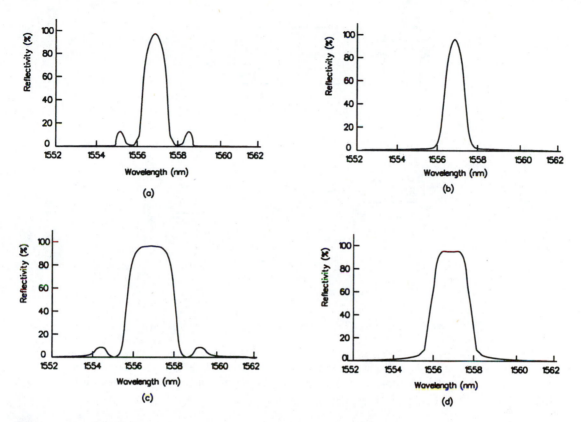

Figure 4.25 Representative spectral responses of (a) a uniformly spaced grating, (b) an apodized, uniformly spaced grating, (c) a "chirped" grating, and (d) an apodized "chirped" grating.

property of introducing a wavelength-dependent delay in the reflected light. As seen in Fig. 4.26a, short-wavelength light is reflected from the regions of close grating spacing; long-wavelength light is reflected from the grating region with the larger periodicity. Chirped grating are written by moving the phase mask and/or laser beam while writing the grating.

4.7.2   Grating Applications

Fiber grating are developing many applications. These include

- Wavelength sensitive reflectors and/or transmitters

  - Add/drop devices for WDM systems: A WDM (wavelength-division multiplexed) system (see Chapter 9) carries several signals, each at a different wavelength. Filters that reflect only a certain wavelength region (one channel) while passing all of the other wavelengths (all of the other channels) are in great demand for this application since it allows an individual channel to be dropped from the fiber. The major goal of these devices is to be able to separate out narrow wavelength regions (<1 nm wide) from the multichannel signal.

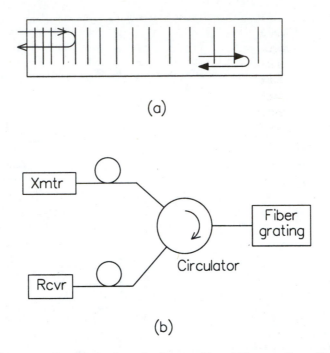

(a)

(b)

Figure 4.26 (a) Reflection of different wavelengths from a chirped fiber grating. (Long wavelengths are reflected from regions of larger spacing.) (b) Implementation of dispersion compensation with a fiber grating.

- Reflectors for fiber lasers and fiber amplifiers: Lasers often need a high reflectivity reflector to operate. With the possibility of making fiber amplifiers (as described in Chapter 7 on optical links), it is possible to surround regions of optical gain with in-fiber reflectors, especially fiber gratings, that are integrated right into the fiber. The narrow band of high reflectivity allows the operating wavelength of the fiber laser to be precisely controlled, with a minimum $\Delta\lambda$.

- Reflectors for external-cavity lasers: As we will see in Chapter 5 on sources, lasers that have a small distance between their reflectors will operate as multimode sources. One way of obtaining single-frequency operation is to increase the distance between the reflectors, until only a single mode can be supported. As shown in Fig. 4.27 on the facing page, a Fabry-Perot semiconductor laser has an antireflection coating applied on one side, as does the end of the fiber. The fiber grating then acts as one of the resonator mirrors. (The other mirror is still the back side of the semiconductor.) These external-cavity lasers exhibit improved single-frequency operation and improved frequency stability when pulsed, compared to the conventional semiconductor laser.

• Dispersion compensation devices: A chirped grating, with its wavelength-dependent delay, can be used to cancel the material (or chromatic) dispersion in a fiber. We have seen in our discussion of material dispersion that the long wavelengths in the source spectral distribution travel faster than the slow wavelengths. Depending on the design of the fiber, there

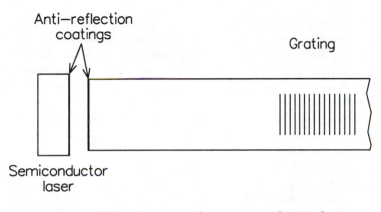

Figure 4.27 Fiber grating used in an external cavity laser.

is a zero-dispersion wavelength. Operation at wavelengths shorter than this wavelength leads to negative-valued dispersion values (in a region of operation called *normal dispersion*. (See Fig. 3.11 on page 52.) In this region the long wavelengths are, indeed, faster than the shorter wavelengths. At wavelengths longer than the zero-dispersion wavelength, the material dispersion gives positive values of dispersion. In this region the fiber design causes the short wavelengths to travel faster than the long wavelengths; this results in *anomalous dispersion*. (The signs of the dispersion are not intuitive because of the presence of a minus sign in the material dispersion formula. See Eq. 3.32 on page 47.)

For the normal dispersion where the long wavelengths are faster, we can insert a fiber grating into a link as shown in Fig. 4.26b. A circulator is used since the device is operating in the reflection mode. (Recently, use of gratings in transmission has been demonstrated.) The chirped grating is arranged so that the high spatial frequency end is closest to the circulator. The grating reflects short wavelengths first at the high spatial frequency end and long wavelengths further down the grating as was shown in Fig. 4.26a. The time delays are in the reverse order as occurs in fiber chromatic dispersion, and the grating can be designed to cancel (or partially cancel) the chromatic dispersion. Gratings as long as 10 cm can be used to cancel the dispersion of a few hundred kilometers of fiber. Preliminary studies have shown that the optimum grating location is in the center of the link.

The design and application of fiber gratings is currently under intense research investigation. Offering optical properties in a small package with little insertion loss (because the fiber grating properties are so similar to the link fiber), they should prove to be useful additions to the designer's set of tools to build fiber systems.

4.8 Summary

We have seen that the losses incurred at a fiber joint—whether a splice or a connector—depend on many factors. These include the fiber geometry (core ellipticity, core-cladding concentricity, area mismatches, etc.), the waveguide characteristics of the fiber (NA, index profile), the mechanical alignment of the fiber (lateral and longitudinal displacement, angular misalignment), the power distribution within the fiber (due to excitation conditions or mode conversion within long fibers), and the fiber end-face quality (i.e., scratches or the presence of lips or hackles). Despite the large number of parameters, commercial manufacturers have proved quite successful in producing connectors and splicing techniques with acceptable low losses for most applications.

The return loss of a connector or splice is a measurement of the fraction of light that is reflected. In today's links with precision sources and optical amplifiers, this parameter plays an increasingly important role.

With fiber-optic couplers, we are able to combine and separate the light in a fiber. The primary parameters are the excess insertion loss and the splitting loss of the coupler. Couplers can also be made to be wavelength sensitive.

Fiber gratings are currently interesting devices under intense research investigation. Their primary advantage is that they can be built into a fiber system with negligible insertion loss. Wavelength-sensitive transmission and reflection characteristics have led to their being incorporated in wavelength filters, fiber lasers and amplifiers, and a wide variety of other applications.

4.9   Problems

1. Consider two fibers with the properties given in the table below. Assuming perfect alignment, calculate the splicing losses for...

   (a) ...light going from fiber #1 into fiber #2, and

   (b) ...light going from fiber #2 into fiber #1.

Parameter	Fiber 1	Fiber 2
n_1	1.45	1.48
Δ	1.5%	1.2%
a	50 μm	30 μm
g	1.80	2.00
NA	0.25	0.22

2. Consider two 50/125 μm SI fibers. Calculate the coupling coefficient if the fibers are laterally misaligned by 5% (i.e., $d/a = 0.05$).

3. Consider a connector that joins two single-mode fibers with the same mode-field *radius* of 4 μm at 1300 nm. The core index of each fiber is 1.47; the index of the medium between the fiber ends is $n = 1$.

   (a) Using a computer, plot the connector loss (in dB) as a function of lateral displacement d for $0 \leq d \leq 8$ μm. Assume that the longitudinal offset and the angular misalignment are zero.

   (b) Using a computer, plot the connector loss (in dB) as a function of longitudinal displacement s for $0 \leq s \leq 8$ μm. Assume that the lateral offset and the angular misalignment are zero.

   (c) Using a computer, plot the connector loss (in dB) as a function of angular misalignment ϕ for $0 \leq \phi \leq 1.0°$. Assume that the lateral and longitudinal offsets are zero.

4. (a) Sketch a bidirectional link that uses an ideal 2×2 coupler at each end. Show that only 25% of the transmitter power from one end will reach the receiver at the other end.

   (b) Sketch a bidirectional link that uses an ideal circulator at each end. Show that ideally 100% of the transmitter power will reach the receiver.

References

1. S. C. Mettler and C. M. Miller, "Optical fiber splicing," in *Optical Fiber Telecommunications II* (S. E. Miller and I. P. Kaminow, eds.), pp. 263–300, New York: Academic Press, 1988.

2. P. Morra and E. Vezzoni, "Fiber-optic splices, connectors, and couplers," in *Fiber Optics Handbook for Engineers and Scientists* (F. C. Allard, ed.), pp. 3.1–3.86, New York: McGraw-Hill, 1990.

3. G. Keiser, *Optical Fiber Communications, Second Edition.* New York: McGraw-Hill, 1991.

4. J. Cook and P. K. Runge, "Optical fiber connectors," in *Optical Fiber Telecommunications* (S. E. Miller and A. G. Chynoweth, eds.), pp. 483–497, New York: Academic Press, 1979.

5. J. Dalgleish, "Splices, connectors, and power couplers for field and office use," *Proc. IEEE*, vol. 68, no. 10, pp. 1226–1232, 1980.

6. J. Dalgleish, "A review of optical fiber connection technology," in *Optical Fiber Technology II* (C. Kao, ed.), pp. 206–212, New York: IEEE Press, 1981.

7. D. Marcuse, D. Gloge, and E. A. Marcatili, "Guiding properties of fibers," in *Optical Fiber Telecommunications* (S. E. Miller and A. G. Chynoweth, eds.), pp. 37–100, New York: Academic Press, 1979.

8. S. Nemota and T. Makimoto, "Analysis of splice loss in single-mode fibers using a gaussian field approximation," *Optical Quantum Electronics*, vol. 11, no. 5, pp. 447–457, 1979.

9. J. Meunier and S. I. Hosain, "An accurate splice loss analysis for single-mode graded-index fibers with mismatched parameters," *J. Lightwave Technology*, vol. 10, no. 11, pp. 1521–1526, 1992.

10. D. Gloge, A. H. Cherin, C. M. Miller, and P. W. Smith, "Fiber splicing," in *Optical Fiber Telecommunications* (S. E. Miller and A. G. Chynoweth, eds.), pp. 455–482, New York: Academic Press, 1979.

11. M. K. Barnoski, ed., *Fundamentals of Optical Fiber Communications, Second Edition.* New York: Academic Press, 1981.

12. Y. Kato, S. Sekai, N. Shibata, S. Tachigama, and Y. Toda, "Arc-fusion splicing of single-mode fibers, 2: A practical splicing machine," *Applied Optics*, vol. 21, no. 11, pp. 1916–1921, 1982.

13. W. Young and D. Frey, "Fiber connectors," in *Optical Fiber Telecommunications II* (S. E. Miller and I. P. Kaminow, eds.), pp. 301–325, New York: Academic Press, 1988.

14. W. C. Young, "Introduction to reliability-related problems in optical fiber connectors," *Optical Engineering*, vol. 30, no. 6, pp. 821–823, 1991.

15. F. P. Kapron, "Fiber-optic test methods," in *Fiber Optics Handbook for Engineers and Scientists* (F. C. Allard, ed.), pp. 4.1–4.54, New York: McGraw-Hill, 1990.

Chapter 5

Optical Transmitters

5.1 Introduction

In this chapter we will consider the optical sources currently used in fiber-optic systems, describe their operational parameters, and consider coupling between the source and the fiber.

The sources currently used with fiber optics are semiconductor light sources [1–11], either *light emitting diodes* (LED) or *semiconductor lasers*. These sources enjoy a combination of usable properties in size, wavelength availability, power, linearity, simplicity of modulation, low cost, and reliability that make them suitable for this application. Sources are currently classified according to wavelength. Short-wavelength sources produce light from 500 to 1000 nm and are typically a ternary blend of semiconductors such as GaAlAs. Long-wavelength sources operate in the region of 1200 to 1600 nm in an effort to minimize fiber losses and dispersion effects and, for reasons dealing with the physical properties of semiconductor materials, are made up of quaternary semiconductor blends such as InGaAsP.

The devices are forward-biased semiconductor *pn* junctions. The materials for making the diodes are selected from semiconductors allowing direct transitions and are typically doped much more heavily than an electronic diode. The mechanism for producing light requires current densities of relatively high value compared to most other electronic devices. Charges in the current flow tend to spread themselves widely apart as they cross the junction, lowering the current density. To overcome this spreading effect, current flow is confined to a small area. To help confine the light to a low-loss portion of the device, other doped layers are added on either side of the *pn* junction. The resulting junctions, called *homojunctions* or *heterojunctions* depending on their material composition, have served to increase the overall operating efficiency of the devices. Temperature effects are important in semiconductors, this overall increase in the efficiency of the sources has led to increased stability and reliability as well.

5.2 Light Generation by Semiconductors

Light can be generated by a radiative recombination of an electron and a hole within the semiconductor. Electrons and holes can also recombine nonradiatively, producing heat within the semiconductor, but no light. The fraction of the total recombinations that occur radiatively is expressed by the *internal quantum efficiency* η_i of the device. It is given by

$$\eta_i = \frac{R_r}{R_r + R_{nr}}, \tag{5.1}$$

where R_r is the number of radiative recombinations per second and R_{nr} is the number of non-radiative recombinations per second. An efficient semiconductor light generator (i.e., one with a high internal quantum efficiency) will have many more radiative recombinations than nonradiative recombinations. This is accomplished by choosing the semiconductor material properly and by flooding the light-generating region with charge carriers. This flooding with charge carriers is done by using energy barriers to confine the carriers in the vicinity of a forward-biased junction.

When a *pn* junction is forward biased, holes will be injected from the *p* material into the *n* material and electrons will be injected from the *n* material into the *p* material. After crossing the *pn* junction, the injected *minority carriers* will find themselves in the presence of majority carriers of the opposite charge and will recombine. The energy possessed by the charge-carrier pair before the recombination process can be converted into either electromagnetic radiation (a *radiative transition*) or into heat (a *nonradiative transition*). The energy E produced by the recombination is approximately equal to the band-gap energy E_g of the material. If the energy E of a radiative transition corresponds to an optical frequency ($E = h\nu$), then light is generated. (Some materials produce rf and microwave radiation.) For an optical source we want the radiative transitions to totally dominate the nonradiative emissions.

In general, there are two types of radiative recombinations. In a *spontaneous emission*, the light produced is incoherent, randomly polarized, and randomly directed. These recombinations are used in light-emitting diodes. The second radiative recombination requires the presence of a stimulating light wave to trigger the recombination. The light produced by this *stimulated emission* is in phase with the stimulating light (i.e., coherent light is produced), of the same polarization, and in the same direction. Lasers make use of these stimulated emissions to produce their light.

Recombination of a charge-carrier pair requires that the vector momentum of the carriers be conserved. (The momentum of a particle is $m\vec{v}$; the momentum of a photon is $h\vec{k}$, where $|\vec{k}|$ is $2\pi/\lambda$ and the direction of $\vec{k}$ is the direction of emission.) In silicon and germanium, it is difficult to meet the momentum conservation requirement without a third momentum vector from an intermediary phonon (i.e, some of the heat that is within the semiconductor). This is a so-called *indirect transition* and is so inefficient that we do not make optical sources from either silicon or germanium.

Some semiconductors that are made from alloys of elements in columns III and V of the periodic chart and, columns II and VI, do not require this third momentum term and produce a *direct transition* The direct transition in these materials is much more probable than the indirect transition in silicon and germanium. Some examples of the alloys used to make sources for fiber-optic communications are mixtures of gallium (Ga), aluminum (Al), and arsenic (As) and mixtures of indium (In), gallium (Ga), arsenic (As) and phosphorus (P). Other materials can be used to make sources for non-fiber applications.

The diode devices are made by growing layers of semiconductor material on top of one another, using the process of liquid-phase epitaxy. In this process, a substrate crystal is exposed to a liquid solution containing the proper mixture of components. Under precise temperature control, the liquid precipitates on the substrate crystal. If done properly, the crystalline structure of the layer mimics that of the substrate or layer below it. This requires that the spacing of the atoms of the grown layer (the so-called *lattice spacing*) be close to the lattice spacing of the lower layer. (Too much difference in the lattice spacing leads to mechanical strain between the layers and a subsequent large increase in the nonradiative transitions.) Layers are grown on top of each other, with the thickness of each layer being precisely controlled by the temperature of the media and the time duration of the growth process. Other techniques, including vapor-phase epitaxy, molecular-beam epitaxy, and metal-organic chemical vapor beam deposition are also being studied and used to construct optical-diode sources [9, 12].

5.2.1   Wavelength and Material Composition

The wavelength of the emitted light is determined by the *bandgap energy* E_g of the material. (The bandgap energy is the energy required to create a hole-electron pair in the material. This energy is recovered when the hole and the electron recombine.) The relationship between bandgap energy and nominal wavelength is

$$\lambda = \frac{hc}{E_g}, \tag{5.2}$$

where λ is the nominal emitted light wavelength, h is Planck's constant ($= 6.63 \times 10^{-34}$ joules·sec), and E_g is the bandgap energy (usually specified in eV, where 1 eV $= 1.6 \times 10^{-19}$ joules). If the wavelength is specified in units of μm and the bandgap energy is in units of eV, the relationship reduces to

$$\lambda(\text{in } \mu\text{m}) = \frac{1.240}{E_g(\text{in eV})}. \tag{5.3}$$

Since the bandgap energy depends on the material composition, the wavelength of the emitter is determined by the material composition of the emitter. (The wavelength is also a function of operating temperature of the device. The wavelength is shifted toward a longer wavelength at a rate of 0.6 nm/C by temperature increases [9].)

Some short-distance, inexpensive systems use LEDs emitting in the visible region of the spectrum, typically at the red color of 665 nm. These sources are made of GaP. Other sources work in the infrared region between 800 and 930 nm. These sources are made out of alloy semiconductors with the composition $Ga_{1-x}Al_xAs$ where the subscript indicates a variable concentration in the alloy. Long-wavelength sources operating at the minimum-dispersion wavelength of 1300 nm and at the minimum-loss wavelength (for silica fibers) of 1550 nm are made of InGaAsP.

Long-Wavelength Sources

An alloy of $In_xGa_{1-x}As_yP_{1-y}$ is used for operation between 1000 and 1700 nm. The designer must decide the values of x and y. The choice is constrained by two factors, the desired wavelength and the spacing of atoms (the *lattice spacing*) in each of the materials. This latter constraint is important as semiconductor fabrication demands that the lattice spacing be equal (within a tolerance of 0.1% [10]) on either side of a semiconductor junction. This is so that junctions can be grown that will have the necessary mechanical and thermal properties to allow crystal growth without introducing numerous defect sites at the interface. Mismatched lattice spacings will usually cause unacceptable strains and dislocations across the junction, resulting in low optical generation efficiency or low lifetime. (Recently, a new family of lasers, the *strained quantum-well laser* [13–15], has been developed that purposely slightly mismatches the atomic lattice spacing across a junction to achieve an operating advantage in output power, operating efficiency, or spectral purity.)

The goal of the designer in choosing the device's semiconductor alloys is to achieve operation at the desired wavelength while matching the lattice spacings across the junctions. Figure 5.1 shows the regime of possible alloys of gallium (Ga), aluminum (Al), arsenic (As), and phosphorus (P) that can be used in fabricating some optical sources. The vertical axis is the lattice spacing of the alloy molecules; the horizontal axis is the bandgap energy of the alloy or, equivalently, the emitted wavelength. The boundaries of the shaded area comprise ternary semiconductor alloys (i.e., three-element alloys). For example, the line between the corners labeled InAs and GaAs is the alloy $In_x Ga_{1-x} As$ for values of x ranging from 0 to 1. The interior region shows the quaternary alloys of $In_x Ga_{1-x} As_y P_{1-y}$.

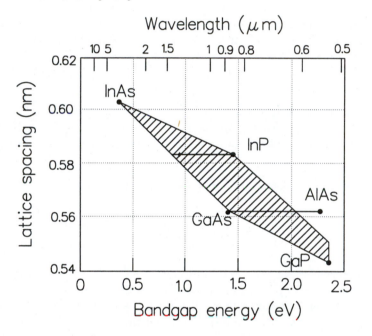

Figure 5.1 Composition relations for InGaAsP and GaAlAs emitters.

Most long-wavelength quaternary devices are built on a substrate of InP. We find the InP corner of the shaded figure on the right side and observe that the lattice spacing is slightly larger than 0.58 nm. The lattice spacing constraint means that we are required to build our junctions on a line of constant lattice spacing (i.e., a horizontal line through the InP point as drawn in the figure). Material science studies [9] show that the bandgap energy in an $In_x Ga_{1-x} As_y P_{1-y}$ alloy along this line is determined by the alloy fraction y as

$$E_g(\text{in eV}) = 1.35 - 0.72y + 0.12y^2 \tag{5.4}$$

and that, for this relationship, the alloy fraction x is related to y by

$$x = \frac{0.4526}{1 - 0.031y}. \tag{5.5}$$

For an operating wavelength of 1300 nm, we find (see the problems at the end of this chapter) that $y = 0.589$ and $x = 0.461$ and that the emitting alloy should be constructed as $In_{0.461} Ga_{0.539} As_{0.589} P_{0.411}$.

Short-Wavelength Sources

Also shown on Fig. 5.1 on the facing page is the line for GaAlAs, extending from the GaAs corner of the shaded area to the AlAs point noted in the figure. Fortuitously, this line is almost horizontal, indicating little change in the lattice spacing. For $Ga_xAl_{1-x}As$, the bandgap energy E_g (in eV) depends on the alloy fraction x as [16]

$$E_g(\text{in eV}) = 1.424 + 1.266x + 0.266x^2 . \tag{5.6}$$

The value of x is constrained to the region $0 < x < 0.37$ because the transition becomes indirect for higher values of x.

5.2.2   Typical Device Structure

The typical light emitting semiconductor is fabricated with four primary layers, grown on top of a substrate. (Other layers may be added for other purposes.) As shown in Fig. 5.2 for a long-wavelength source, the substrate is usually InP and the layers consist of (1) an n-type InP buffer layer, (2) thin active region of p-type InGaAsP, (3) a p-type cladding layer of InP, and (4) a heavily doped p^+-type layer of InGaAs. (Typical dimensions of these layers are given later, in discussions of particular sources.) The layers are added to help confine the current carriers and the generated light in the vertical direction. (Techniques to confine the current carriers and light in the horizontal direction are discussed later.)

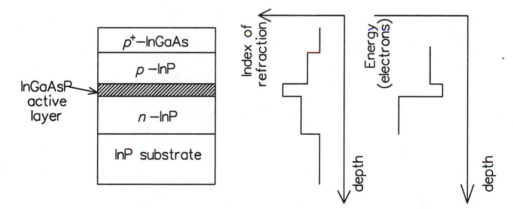

Figure 5.2 Typical diode structure consisting of four layers on a substrate, index of refraction variation through the layers, and energy well for electrons.

Current-Carrier Vertical Confinement

Current-carrier confinement is desirable because the radiative recombination efficiency increases with current density. To consider the confinement of the carriers, we want to focus our attention on the active layer and its adjacent layers. As seen in the right side of Fig. 5.2, electrons will be injected from the n-type InP material into the active regions (seen as a drop in energy on the diagram). The electrons face an energy barrier at the interface between the active region and the p-type material below it. Few electrons will have the energy to climb this barrier and, so, the electrons will be confined to the active region. A similar energy barrier between the active layer and the n-type material exists for the holes that are injected from the p-type material into the active layer; the holes are again trapped in the active region. This boundary structure on either side of the active layer is called a *double heterostructure*.

Light-Wave Vertical Confinement

Light confinement in the vertical direction is desired because the efficiency of the stimulated emission process depends on the strength of the optical field. Without any confinement mechanism, the optical field will quickly diverge out into the inactive material and lose its power to the absorption in the material. The light wave can be confined by shaping the indices of refraction of the layers on either side of the active region. As seen in the center of Fig. 5.2 on the page before, the index of refraction of the active region is greater than the index of the two adjacent layers. The two outer layers (the p^+ layer and the InP substrate) have lower indices of refraction than their adjacent layers. This step-wise variation in the refractive indices forms a light guide that serves to confine the light field to the active region and its adjacent layers. Figure 5.3 shows a side view of the fundamental spatial mode in the presence of the confining structure. While an appreciable portion of the field extends beyond the active region, the field would be considerably wider in the absence of any confinement. The fraction of the field that is within the active region is a function of the thickness of the active layer d and the height of the index mismatches at each interface [1].

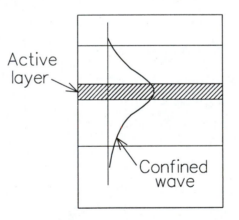

Figure 5.3 Side view of an optical field confined in the vertical direction. The portion of the field lying outside of the active region is less than it would be in the absence of any confinement.

5.3   Light Emitting Diodes

5.3.1   LED Configurations

Two configurations [7, 9, 17, 18] of light emitting diodes have become popular, surface emitters and edge emitters. The surface emitters are widely used in multimode fiber systems, since the wide-angle beams that they produce are more efficiently coupled into multimode fibers; the edge emitters are used in both single-mode and multimode systems since they provide a tighter emission pattern that is more efficiently coupled into fibers with low numerical apertures.

5.3.2 Surface-Emitting LEDs

The *surface-emitting LED* (also known as a *SLED* or *Burrus emitter*) is illustrated in Fig. 5.4 and Fig. 5.5.

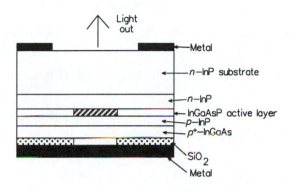
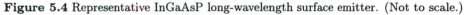

Figure 5.4 Representative InGaAsP long-wavelength surface emitter. (Not to scale.)

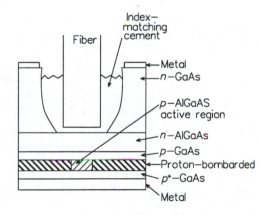
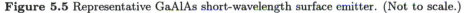

Figure 5.5 Representative GaAlAs short-wavelength surface emitter. (Not to scale.)

In the InGaAsP device, the four layers grown on the InP substrate consist of [9]

1. an n-type InP buffer layer that is 2 to 5 μm thick,

2. a p-type InGaAsP active layer that is 0.4 to 1.5 μm thick,

3. a p-type InP layer that is 1 to 2 μm thick, and

4. a p^+-type InGaAs "cap" layer that is about 0.2 μm thick. (This latter layer is to help reduce the metal-to-semiconductor contact resistance.)

The device consists of the double heterojunction structure around the active region. The light is emitted from a circular planar region of the active layer, usually 20 to 50 μm in diameter [9]. As mentioned earlier, it is desirable to keep the current density as high as possible in the active region. This is done vertically through the use of heterojunctions, as explained earlier. Confinement of the current carriers in horizontal dimension can be done by a variety of techniques, which include:

- adding a layer of dielectric insulation, such as SiO_2, with a hole etched through to allow current flow in a limited area (Fig. 5.4 on the preceding page),

- using proton bombardment to create a high-resistivity region outside of the boundaries of the active region to minimize current flow through this region (Fig. 5.5),

- etching away the surrounding material to form a central *mesa structure* that isolates the active region, or

- diffusing some zinc into the central region of the material to form a low-resistivity region that provides a channel for the current flow.

Light from the active light-emitting region can be collected from either side of the device (i.e., through the substrate or through the other side). In a GaAlAs device, the GaAs substrate will absorb appreciable light, so a well is etched out of the substrate to allow the fiber to approach the active region more closely (Fig. 5.5). In InGaAsP devices, the InP substrate does not appreciably absorb the light, so the well can be omitted.

Because of the high refractive index of the semiconductor materials ($n = 3.4$ for InP, $n = 3.6$ for GaAs), there is a high reflection loss if the light is coupled into air. Additionally, the critical angle for a semiconductor-air interface is only 15°, thereby causing significant power to be reflected back into the semiconductor. One potential solution to both problems is the use of an index-matching epoxy (as in Fig. 5.5) to join a fiber pigtail to the source. A solution to the refraction problem in surface emitting devices is the formation of a domed output surface on the device which has less refractive losses due to the geometry of the interface. For a semiconductor-air interface, the coupling efficiency is approximately $1/[n(n + 1)^2]$; while for an LED with a hemispherical output dome the coupling efficiency is approximately $[2n/(n + 1)^2]$—an improvement of $2n^2$. For a fiber, however, this improvement can be transitory since the effective area of the source is also magnified by n^2. Since any increase in effective emitter area beyond the core area of the fiber is ineffective, improvement in the output light level occurs only when the original source area is smaller than the fiber core. (The improved efficiency is real, however, for LEDs used in non-fiber applications.)

Output Beam Pattern: Surface Emitting LEDs

Since the emitting region of the surface emitting LED is circularly symmetric, the emitted *beam pattern* will also be symmetric with a 60° (typical) half-cone angle beam divergence. Here, *beam divergence* is defined as the angular spread of the emitted beam as measured in the far-field of the beam. The angular spread is measured at the points where the power is decreased to one-half the maximum on-axis power (i.e., the -3 dB optical power point). Both the full-angle and the half-angle can be used to describe the beam divergence; the user must take note of the specified test conditions.

———————————— ∞ ————————————

Example: An optical beam is found to be -10 dB from its on-axis value of power at a measured angle of $75°$.

a) If the angular power dependence is $P(\theta) = P_0 \cos^n \theta$ where θ is the angle measured from the on-axis position and P_0 is the on-axis power, find the value of n.

Solution:

$$\frac{P(\theta)}{P_0} = \cos^n \theta \tag{5.7}$$
$$10^{10/10} = 0.1 = \cos^n 75°$$
$$\log(0.1) = n \log(\cos 75°)$$
$$n = \frac{-1}{-0.587} = 1.703\,.$$

b) Calculate the full-angle beam divergence of this source.

Solution: We need the angle where P/P_0 is reduced to a value of $1/2$.

$$\frac{P}{P_0} = \cos^{1.703} \theta \tag{5.8}$$
$$\frac{1}{2} = (\cos \theta)^{1.703}$$
$$\cos \theta = (0.5)^{1/1.703} = 0.665$$
$$\theta = 48.3° \text{ (half-angle divergence)}\,.$$

———————————— ∞ ————————————

5.3.3 Edge-Emitting LEDs

Edge-emitting LEDs remove the light along an axis transverse to the current flow, as shown in Fig. 5.6 on the next page.

A representative edge-emitting LED structures is shown in Fig. 5.7 on the following page. The four layers grown on the substrate are similar to the surface-emitting devices except that the active layer is much thinner, being 0.05 to 0.25 μm thick (compared to 0.4 to 1.5 μm for the surface-emitting device).

The insulating SiO_2 layer has a stripe hole in it that guides the current down into the active region, laterally confining the current. The typical width of the active region of an edge emitter (chosen to match the core size of typical fibers) is 50 to 70 μm [16]. The length of the region is typically 100 to 150 μm.

The heterojunctions on either side of the active region play an additional role in the edge emitter; they act as waveguides to help confine the light. The active region is designed to have a high index of refraction with the materials on either side of the active layer having a lower index of refraction. The substrate and the p^+-doped layer, lying further away from the active layer, have an even lower index value. The combined structure of the five layers makes an optical waveguide. The light that is generated in the active region is confined to the region made up by the active layer and its two adjacent layers. This optical confinement, along with the charge carrier confinement of the heterojunction, increases the efficiency of the optical generation process.

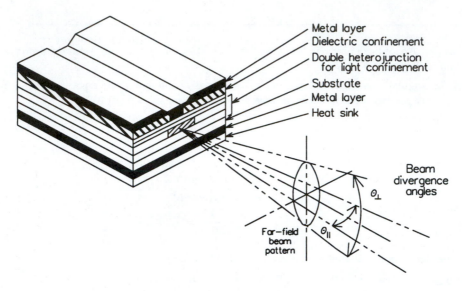

Figure 5.6 Representative edge emitter and far-field beam pattern.

Beam Pattern: Edge-Emitting LEDs

Because of the asymmetry of the rectangular-shaped active region at the emitting end of the device, the far-field pattern will be elliptical (as shown in Fig. 5.6) and two angles will be required to describe the beam divergence. Because of diffraction effects, the beam perpendicular to the junction will have the larger divergence (i.e., the larger divergence is aligned along the shortest dimension). A typical value for the half-angle in this direction is 60°. The angle parallel to the junction will have the smaller value, typically a 30° half-angle. Besides their asymmetric output beam, edge emitters will typically produce less power (approximately 1/2 to 1/6 the power)

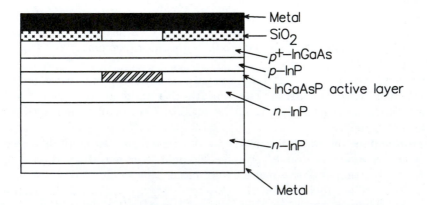

Figure 5.7 Representative InGaAsP edge-emitting LED structures.

than surface emitters. We will find, however, that the amount of light coupled into a fiber is comparable for both devices, due to offsetting differences in coupling efficiency.

5.3.4   LED Output Power Characteristics

Typically, LEDs used in fiber communications produce power levels of several microwatts or tens of microwatts in the fiber. For low drive currents, the output power is a linear function of current, as shown in Fig. 5.8. As the drive current becomes larger, eventually the output power saturates (i.e, levels off). The *linearity* of the diode refers to the linearity of the curve that relates the output power to the drive current. Linearity is important to allow faithful analog modulation of the output power by direct current modulation.

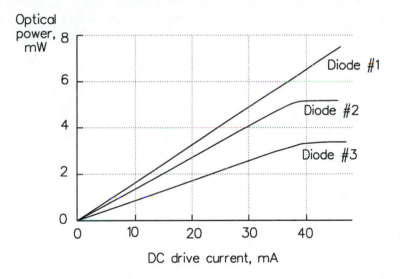

Figure 5.8 Output power vs. drive current for a typical LED source.

The amount of power that can be produced for any given drive current is a function of the *operating efficiency* of the LED. The operating efficiency of a diode is the measure of its ability to convert electrical input power into optical output power. This efficiency is determined by three factors.

- The first factor is the ratio of radiative transitions to nonradiative transitions that occur in the active layer of the material. Nonradiative transitions rob the optical generation process of carriers. Nonradiative recombination tends to occur at crystal defects, especially at the boundaries between the active layer and its two surrounding layers. The ratio of radiative recombination to nonradiative recombination is maximized by

  1. having a high current density in the active region, as was previously discussed, and

  2. by careful layer growth to remove boundary defects.

- The second factor is the amount of light absorption that occurs in the active region of the emitter. This is *self-absorption*. It can be important when the nonradiative transitions described in the previous factor are minimum and when the doping levels of the active region are high [16]. This absorption is minimized by keeping the active layer of the device thin, as is done in most modern double-heterojunction devices. Studies [16] indicate that an optimum thickness can be calculated for both surface emitters and edge emitters as a function of doping levels and nonradiative recombination rates.

- The third factor is the reflection losses (i.e., Fresnel losses) and refraction losses that occur at the semiconductor output face, as described previously.

5.3.5 LED Spectral Width

The *spectral width* of the source is important because it determines the contribution to material dispersion, as seen in Chapter 3. Lower spectral width will allow increased data rate if material dispersion effects are the limiting factor. The spectral width $\Delta\lambda$ of LEDs can be approximated by [19]

$$\Delta\lambda'[\mu\text{m}] \approx 1.45\lambda'^2[\mu\text{m}]\,(kT)'\,[\text{eV}] \tag{5.9}$$

where λ' is the nominal LED operating wavelength *in units of* μm and kT' is the product of Boltzmann's constant and the junction temperature *in units of* eV. We note that the linewidth generally increases as the square of the wavelength; long-wavelength LEDs will have appreciably more linewidth than short-wavelength LEDs. Typical LED spectral widths for a GaAlAs source are on the order of a few to several 10s of nm. The spectral width of an InGaAsP surface-emitting LED can be about 100 nm, while that of an equivalent edge emitter is typically 60 to 80 nm [9]. Heating the device or operating at increased drive currents can increase the spectral width of a source.

5.3.6 LED Modulation Bandwidth

The speed of response of an LED (i.e., how fast an LED can respond to changes in its driving current) depends on the time duration of the charge carriers in the active region. The length of time that the minority charge carriers remain in the active region is determined by the *lifetime* of the carriers τ_{lifetime}. If the drive current is modulated by a sinusoidal signal (added to a dc bias), the power output P_{out} of the laser follows a modulation frequency response of

$$P_{\text{out}} = \frac{P_0}{1 + \omega^2\tau_{\text{lifetime}}^2}, \tag{5.10}$$

where P_0 is the unmodulated power and ω is the modulation frequency.

We now want to measure the modulation response of the laser diode. An optical detector converts the incident optical power into a current that is proportional to the optical power. The *electrical* power P_{elec} from the detector is proportional to the square of the electrical current out of the detector and, so, it is proportional to the square of the optical power (i.e., $P_{\text{elec}} \propto i^2 \propto P_{\text{optical}}^2$). So, to measure the modulation response of the laser diode, we measure the frequency response of the electrical power out of the detector (assuming that the detector bandwidth exceeds the laser bandwidth). If we record the 3-dB frequency $\omega_{3-\text{dB electrical}}$ of the detected electrical power, we will be measuring the frequency where $P_{\text{out}}^2/P_0^2 = 1/2$. Hence, we find

$$f_{(3-\text{db electrical})} = \frac{1}{2\pi\tau_{\text{lifetime}}}. \tag{5.11}$$

(Another definition of the modulation bandwidth has been that frequency where $P(\omega)/P_0 = 1/2$; the user must determine which definition has been used to describe the bandwidth.)

The lifetime τ_{lifetime} depends on several parameters. For a lightly doped active layer (doping levels on the order of 2×10^{17} atoms·cm^{-2}) and a thin active-region thickness d, the nonradiative recombination processes are dominated by the nonradiative recombinations that occur at the layer interfaces. The radiative recombination component of the lifetime depends on the carrier density in the active region.

- When the drive current is low, the radiative lifetime is independent of the drive current level and linearly dependent on the doping level of the active region.

- When the drive current is high, the radiative carrier lifetime is proportional to $\sqrt{d}$ and inversely proportional to $\sqrt{J}$, where d is the active layer thickness and J is the current density in the active region [16].

Increasing the modulation bandwidth requires reducing the carrier lifetime. For low-drive-current LEDs the modulation bandwidth can be improved by increasing the doping level. At high current levels the bandwidth can be improved, assuming that nonradiative recombinations are not dominating the process, by reducing the value of d (i.e., decreasing the thickness of the active region), increasing the current density in the device (i.e., increasing the drive current, or decreasing the active area of the device, or both), or all of these.

LED Rise-Time

An alternative approach to the frequency-domain approach to device response, is to work in the time domain. The *rise time!device* of a device can determine the maximum modulation rate of the source. If the fall-time of the LED is negligible compared to the rise-time, then the maximum modulation rate is given approximately by the inverse of the rise-time. Several factors combine to determine the rise-time capability of an LED.

The rise-time can be limited by capacitance C_s associated with the active region. Since charge is present in this region, it represents a capacitive charge-storage medium. (We will neglect other parasitic capacitances that might be present.) The capacitance associated with this charge storage in an LED is in the range of 350 to 1000 pF [16]. Keiser [16] gives an expression for the rise-time t_r (defined as the time required for the device to transition from 10% of its final light output value to 90% of its final value) as

$$t_r = 2.20 \left(\frac{2kTC_s}{qI_p} + \tau_{\text{lifetime}} \right), \tag{5.12}$$

where k is Boltzmann's constant, T is the device temperature, C_s is the junction space-charge capacitance, and I_p is the size of the current step function used to drive the device. This relationship illustrates that the rise-time can be reduced by reducing the capacitance and by maximizing the drive current. (The high-current limit is seen to be $2.20\tau_{\text{lifetime}}$.) In fact, the rise-time of the device can be minimized by overdriving the diode with a current wave form that momentarily exceeds its pulsed value. The fall-time can also be minimized by providing a momentary negative bias to the device when it is first turned off. (The negative bias clears the space charge from the active region.)

LED Power-Bandwidth Trade Off

Keiser [16] shows that the power-bandwidth product for an LED can be expressed as

$$P \, \Delta f = \frac{hc}{2\pi q \lambda} \frac{J}{\tau_{\text{lifetime}}} \tag{5.13}$$

where τ_{lifetime} is the radiative lifetime of the minority carriers in the active layer and J is the current density. At a given drive current, therefore, the user must trade off speed and power (i.e., a fast LED is a low-power LED).

5.3.7   LED Summary

Due to the combination of their operating properties (especially their wide beam divergence, relatively low power coupled into a fiber and relatively large spectral width), LEDs are typically suitable for systems using multimode fibers requiring less than 50 Mb·s^{-1} of information rate.

5.4   Laser Diode Sources

Laser diode sources [4, 6–10, 20–25] produce more power than an LED, have a narrower spectrum, and can couple more power into a fiber. The structure of a laser diode is much like that of an edge-emitting LED. (While some surface-emitting lasers have been constructed [16], the majority of diode lasers are edge emitters.) The principal difference between an edge-emitting LED and the edge-emitting laser is that, in the laser, the active region is thinner vertically and narrower horizontally. In addition, multilayer reflectors are added to the ends of the structure to provide optical feedback. (This feedback raises the optical-field strength to ensure that the stimulated emissions dominate the spontaneous emissions in the laser. The mirror structure also serves to reduce the beam divergence of the emitted pattern and to narrow the spectrum of the light output.) Double heterojunctions are used to confine both the charge carriers and the optical fields in the vertical direction. Additional structures are incorporated to confine the current and the light laterally.

5.4.1   Gain-Guided Lasers

Several techniques are incorporated into the structure of the devices to horizontally confine the electrical carriers to a narrow region. Figure 5.9 on the next page shows a typical *stripe-geometry* laser diode. Here a layer of SiO$_2$ has been fabricated and a narrow stripe has been etched through the SiO$_2$ layer, followed by the deposition of the metal. The stripe can range from 5 to 20 μm in width and from 150 to 500 μm in length. The emitting region in the active layer is formed with a width that is slightly larger than the stripe (allowing for some slight current-spreading mechanisms).

In the emitting region under the stripe, the index of refraction is slightly higher than the laterally adjacent regions, because of the presence of the current carriers in that region. This slight rise in the refractive index forms a lateral waveguiding structure. (The vertical waveguide structure is formed by the heterojunction materials.) In this way, the generated light is guided both vertically and laterally and is more confined as it propagates down the length of the diode.

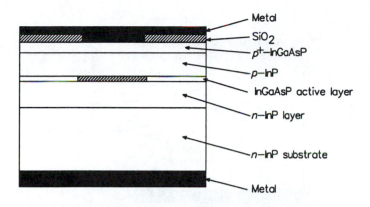

Figure 5.9 Representative stripe-geometry laser.

The confinement minimizes absorption in the non-active region of the active layer. The mechanism of providing lateral waveguiding with the change in the refractive index caused by the current carriers is called *gain-guiding*.

Other techniques can be used to provide a narrow current stripe. Proton bombardment into the semiconductor material can be used to create a high-resistivity region outside of the desired current flow region. (The protons break bonds in the semiconductor material and raise its resistivity. The lateral extent and depth of the bombardment is externally controllable.) Alternatively, a V-shaped groove that can be etched in the laser material to control the current flow. (The thinner cap layer on top of the active region has lower resistance and, so, the current flow is most intense at the bottom of the groove.)

5.4.2 Index-Guided Lasers

An alternative laser-diode structure incorporates a deliberate change in material in the lateral direction across the active layer to form a waveguide structure. The change in the index of refraction of the materials forms a lateral waveguide that confines the light to a narrower region. Several structures have been devised to perform this function:

- Figure 5.10 on the following page shows a *buried heterostructure laser* where the *n*-type InGaAsP active emitting region is surrounded to the left and right by *p*-type InP. The change in material is accompanied by a step-change in the refractive index providing the lateral waveguide. The vertical waveguide structure is again done by the heterojunctions.

- A *buried-channel substrate laser* has an etched V-groove in the substrate material . The active region "slumps" into an etched groove (i.e., is formed within the groove), physically isolating the light-emitting portion of the active layer in the lateral direction by the edges of the groove.

- A *double-channel planar buried-heterostructure laser* [9] has a channel built into either side of the light-emitting region of the active layer that isolates the region and provides the light-guiding structure.

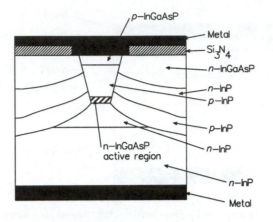

Figure 5.10 Index guiding of light in a buried heterostructure laser.

All of these lasers have a step change in the refractive index. It is also possible to guide light with a more gradual, tapered change in the index of refraction. (These structures lose part of their carrier confinement, however.) Figure 5.11 illustrates an *inverted-rib laser*. In this laser, the first layer boundary below the active layer has a channel etched into it. Because of the narrow layer-spacing the effects of this change in geometry are coupled into the active region as a gradual change in the index [9]. This index change provides a lateral light guiding structure. An alternative structure is the *ridge-guide laser* [9]. In this geometry, the shaped region is above the active region and, again, induces a waveguiding change in the index of refraction due to the coupling of the ridge effects into the active region.

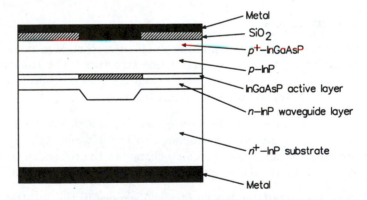

Figure 5.11 Index-guided inverted-rib laser.

Generally, index-guided lasers are superior to gain-guided lasers. They typically have a lower threshold current (i.e., the current drive where lasing begins), have better mode stability under pulsed operation, and have a narrower frequency spectrum than the gain-guided lasers.

5.4.3   Beam Patterns

The typical diode laser emitting region at the output face is 150 to 500 μm long by 5 to 20 μm across by 0.1 to 0.2 μm high. These dimensions give the beam pattern an asymmetric far-field pattern with a perpendicular beam divergence of 30 to 50° (measured perpendicular to width of the emitting region) and a parallel beam divergence of 5 to 10°. This latter value is about one-fifth the comparable value for an edge-emitting LED and implies that the laser-beam pattern is more directional than the LED. This directionality is beneficial in trying to couple the light into optical fibers.

5.4.4   Laser Power Characteristics

The drive mechanism that operates the laser is the current through the forward-biased device. A typical plot of the output power vs. the driving current amplitude is shown in Fig. 5.12. Observe that there are two linear regions of operation. When the drive current is below the threshold current, the diode is not lasing. It is operating as an LED, emitting a small amount of incoherent light. When the drive current is above threshold, the device is lasing and is producing coherent light. Increases in output power in this region are linearly related to increases in driving current. The *threshold current* is obtained by linearly extrapolating the output power curve down to the zero level of power. The threshold-current value is a key diode-laser parameter. Obviously, it is desirable to have as low a value of threshold current as possible, because the drive power required to reach threshold will only reduce the overall efficiency of the laser. Much effort has been put into reducing the threshold current. Typical values for index-guided lasers are in the range of 10 to 30 mA; for gain-guided lasers, the values are in the range of 60 to 150 mA [9].

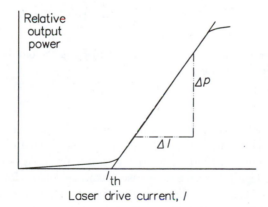

Figure 5.12 Output characteristics of a diode laser.

Conversion Efficiency

Two types of conversion efficiency can be defined for the diode laser or LED. The first is the *overall electrical conversion efficiency* (or *efficiency*), given by

$$\eta = \frac{P_{\text{out optical}}}{P_{\text{in electrical}}} = \frac{P_{\text{out}}}{V_f I_f}, \tag{5.14}$$

where η is the efficiency, P_{out} is the optical output power of the source, and P_{in} is the electrical input power. Here, we have recognized that the input power is the product of the forward voltage across the device V_f and the forward current through the device I_f.

$$\infty$$

Example: A diode laser produces 1 mW of light when driven at a forward current of 100 mA. The nominal forward voltage across the diode is 2 volts. Calculate the overall electrical conversion efficiency.

Solution: The electrical efficiency is found as

$$\eta = \frac{P_{\text{out}}}{V_f I_f} = \frac{1 \times 10^{-3}}{(2.0)(100 \times 10^{-3})} = 5 \times 10^{-3} = 0.5\% . \tag{5.15}$$

We note that the efficiency is quite low.

$$\infty$$

The second conversion efficiency is the *incremental efficiency*, given by

$$\eta_i = \frac{dP}{dI} , \tag{5.16}$$

where dP/dI is the slope of the output characteristics measured *above threshold*. The slope of the output characteristics is a measure of the efficiency of converting charge carriers into photons of light. A related efficiency is called the *external quantum efficiency* η_{ext} and is given by

$$\eta_{\text{ext}} = \frac{q}{E_g} \frac{dP}{dI} , \tag{5.17}$$

where q is the electron charge (1.6×10^{-19} coulombs) and E_g is the bandgap energy of the material (determined by the alloy composition).

Linearity

Another property of the output-power curve that affects the performance of an optical device in the operation of an analog communication system is the *linearity*. Nonlinearities associated with saturation of the output power at high drive currents or changes in the output-power level associated with jumps in lasing wavelength cause harmonic distortion and intermodulation of multiple signals carried on the link. Generally speaking, LEDs are quite linear in operation while laser diodes can suffer some linearity deficiencies in comparison. If the nonlinearities are intolerable in the desired application, several compensation techniques have been devised to ameliorate their effect. These techniques are based on feedback and other circuit procedures that sense the nonlinearity and attempt to cancel it.

5.4.5   Fabry-Perot Laser Resonator

The *optical resonator* is required to allow the light to make the equivalent of several passes through the low-gain active region. In a semiconductor laser, this optical resonator is frequently a Fabry-Perot resonator, made up of parallel front and rear surfaces of the diode as shown in Fig. 5.13 on the next page. (We will later discuss alternative resonator structures.) The parallel alignment is easily achieved by cleaving the crystal along its crystalline planes. In GaAs, the power reflectivity at the crystal interface is 32% ($n = 3.63$); in InGaAsP, it is 33% ($n = 3.71$). This reflectivity is normally high enough to eliminate the need for mirrors; however, the back surface of the laser is frequently coated with a multilayer, 100%-reflecting surface, resulting in emission from the front surface only. For higher-power lasers the front surface is also coated with a partially transmitting coating to protect the surface from forming defects due to ambient moisture.

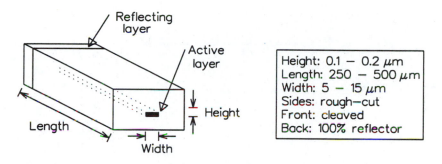

Figure 5.13 Physical structure of a laser diode.

5.4.6 Laser Modes

Just like the fiber, the introduction of the reflecting surface in the optical resonator introduces *electromagnetic modes* within the resonator.

The modes are geometrical descriptions of the waves that fit the boundary conditions imposed by the resonator boundaries. Each mode, in general, oscillates at a different frequency as (simplistically) shown in Fig. 5.14 on the following page. Each frequency is determined by the spacing of the resonator mirrors (the *longitudinal modes*) and the width and height of the waveguiding region of the diode (the *lateral modes*). The spacing of the longitudinal mode frequencies $\Delta\nu$ is determined by the separation of the mirrors and is given by

$$\Delta\nu = \frac{c}{2L} \tag{5.18}$$

where c is the velocity of light and L is the separation of the mirrors. The spacing of the frequencies of the transverse modes is a more difficult expression and will not be displayed here.

It is noted that, as the diode current is increased, the central mode grows faster than the side modes, thereby increasing the so-called *side mode suppression ratio (SMSR)* of the diode laser. This results in a spectral-power output that is more single-frequency at high drive currents. This effect is illustrated in Fig. 5.15 on the next page.

There are several disadvantages to multimode lasers.

- The linewidth of the source is broader than the linewidth of any single line. This has adverse implications for the material and waveguide dispersions of a fiber link with a subsequent reduction in the attainable bandwidth.

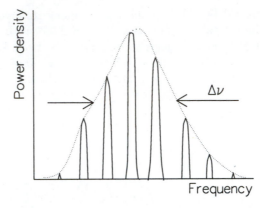

Figure 5.14 Spectral output of a multimode diode laser.

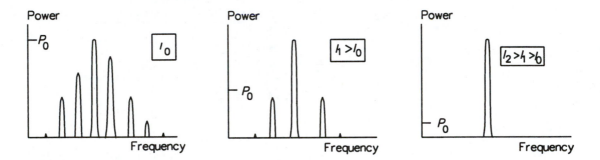

Figure 5.15 Reduction of the number of modes as drive current is increased (from left to right).

- A multimode diode laser will randomly jump from mode to mode, causing the power output to vary with time. This leads to a noise, called *mode partition noise*, in the output. (This source noise is described in more detail later in this chapter.)

- Coherent detection techniques, where the signal is combined with a frequency-locked local source, have inherent sensitivity and multichannel advantages. These techniques require a single-frequency source, rather than a multimode laser.

Single-Transverse-Mode Diode Lasers

The first step in reducing the number of modes is to reduce the number of transverse modes to one in both the vertical and horizontal directions. This is a *single-lateral-mode laser*. The higher-order modes can be eliminated by making the difference between the propagation losses of the lowest-order mode and the higher-order mode as large as possible. In this way, preference is given to the lowest-order mode in establishing oscillations.

- Higher-order transverse modes in the vertical dimension are eliminated by reducing the height of the active light-emitting region and the waveguiding layer structure.

- Similarly, to reduce the number of lateral modes, the width of the current stripe and/or the waveguiding horizontal region can be minimized. (For this reason, index-guided lasers, with their narrowly defined guiding region, tend to have fewer modes than gain-guide lasers, with their less-constrained guiding region.) A single transverse mode will be supported when the width of the emitting region is smaller than about 2 μm [10].

  Another technique to eliminate the higher-order modes in the lateral direction is to incorporate regions with high optical loss into the edges of the emitting region. Since the higher-order modes extend more into these lossy regions, they will have a higher loss and will be unable to sustain oscillation.

The number of longitudinal modes can be reduced by shortening the resonator length. This has the adverse effect of decreasing the laser power, since less gain in incurred in the shorter distance. Other techniques have become more useful in reducing the number of longitudinal modes.

5.4.7   Single-Frequency Diode Lasers

Once a laser has been made to operate in a single lateral mode, it is desirable to reduce the number of longitudinal modes. Such a laser is a *single-frequency laser* and would have a spectrum as shown in Fig. 5.16.

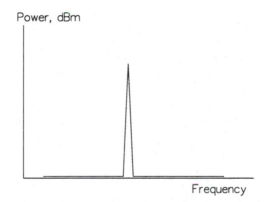

Figure 5.16 Single-frequency operation of a laser.

Once a laser is made to operate at a single frequency under steady-state conditions, it becomes important to maintain that state under pulse conditions. The charge carriers in the current pulse and the photons in each of the modes are coupled together. Under a current pulse, the light in the modes typically undergoes oscillatory behavior (called *relaxation oscillations*) causing the frequency content of the light to vary in time. In addition, the index of refraction of the lasing medium is a function of the gain of the medium. The time-varying gain leads to a time-varying index of refraction which, in turn, leads to a change of the center operating frequency of the laser. For a step-increase in current, the frequency is observed to shift in a *frequency chirp*. Techniques to reduce the observed chirp include biasing the laser in its "OFF" state just above threshold, reducing the active region volume, and running the laser continuously and using an

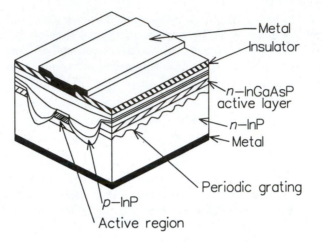

Figure 5.17 Physical structure of a distributed-feedback (DFB) laser. (Note: for simplicity, not all layers are labelled.)

external modulator (usually an electro-optic modulator) to modulate the laser beam after it has left the laser.

Once achieved, single-frequency laser operation can be easily upset by light reflected from any interface in the system back into the laser resonator. While it is possible to eliminate reflections by carefully matching indices of refraction at all interfaces or using polarization-rotating devices as reflection isolators, both techniques require user awareness of the reflection problem and careful design to remove the effect.

Techniques have been developed to produce single-frequency lasers with superior mode characteristics, including the distributed-feedback laser and the distributed Bragg reflector laser.

Distributed-Feedback Laser

Instead of concentrating the reflectivity at the ends of the laser as in a Fabry-Perot resonator, the reflection properties can be continuously distributed throughout the lasing medium, resulting in a *distributed-feedback (DFB) laser* [10, 26]. Such a laser, shown in Figure 5.17, uses a corrugated surface as the interface between two layers in the heterostructure. The periodicity of the structure, Λ, determines the wavelength of maximum interaction λ_B by

$$\lambda_B = \frac{2n\Lambda}{k}, \tag{5.19}$$

where n is the index of the mode in the laser and k is an integer that indicates the grating order. (Usually $k = 1$, but sometimes $k = 2$ is used.) Although physically removed from the active region, the periodic grating affects the propagation properties of the active region and results in an output at [10]

$$\lambda = \lambda_B \pm \frac{\lambda_B^2}{2nL_g}\,(m+1), \tag{5.20}$$

where m is an integer and L_g is the grating length. Usually the $m = 0$ modes are the ones to oscillate. While theory predicts that two modes will oscillate (i.e., one for each value of the sign), random imperfections in the cleaving at the end-facets of the device introduce asymmetric phase differences and result in only one mode oscillating.

The stability of the single-frequency DFB laser under pulsed operation can be improved by modifying the grating. Figure 5.18 shows a periodic grating structure and a grating structure

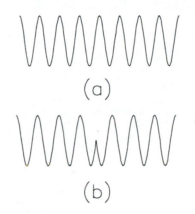

Figure 5.18 Two versions of grating structures for distributed feedback lasers: (a) periodic grating structure and (b) grating structure with a $\pi/2$ phase shift in its center.

that has a phase shift of $\pi/2$ in the middle of it. (This latter grating design is called a $\lambda/4$-shifted grating.) It has been shown [10] that this is the optimum phase shift to improve the performance of the laser. These lasers will operate at a single frequency corresponding to the wavelength, λ_B. Phase-shifted grating lasers have superior frequency stability and a narrower linewidth than non-shifted DFB laser.

DFB lasers exhibit good single-frequency operation and have little sensitivity to drive-current pulses and temperature changes.

Distributed Bragg Reflector Laser

The distributed Bragg reflector (DBR) laser also uses a corrugated interface to provide reflection at a wavelength λ_B that is determined by the grating period Λ as

$$\lambda_B = \frac{2n_e\Lambda}{l},$$

(5.21)

where n_e is the mode's propagation constant and l is an integer that indicates the order of the grating (usually $l = 1$). The corrugated sections providing the reflections are located outside of the current-pumped active region. Figure 5.19 on the next page compares the basic structures of the DFB laser (where the corrugation region overlaps the current-pumped region) and the DBR laser. The DBR laser will oscillate at the single frequency corresponding to λ_B. These lasers typically have a higher threshold current than the DFB lasers and are more susceptible to temperature variations [9]. They are also more susceptible to frequency chirp when pulsed.

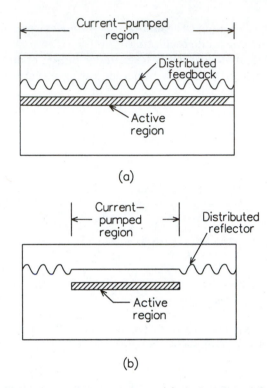

Figure 5.19 Side views of slices through two lasers. (a) A distributed feedback (DFB) laser. (b) A distributed Bragg reflector (DBR) laser. (For simplicity, the layer structure is not shown.)

Quantum-Well Lasers

Improvements in fabrication technology now allow the thickness in the active layer of the diode structure to be as small as 5 to 10 nm with smooth, defect-free interfaces. In such thin layers the electronic behavior of the electron charge carriers can no longer be modelled by their behavior in bulk material, but their quantum mechanical states must be used. The behavior of these electrons gives rise to a family of devices known as *quantum-well lasers* [10, 26]. Early developmental models of these lasers have shown nice combinations of desirable properties, including low threshold current, higher output power (and gain), narrow linewidth, frequency stability under pulsed operation, and low noise. Quantum-well lasers have been demonstrated in GaAlAs for some time [10]; work on InGaAsP devices is proceeding. Usually the long-wavelength devices are *multiple-quantum-well (MQW) lasers* where thin layers of active GaInAs are alternated with thin barrier layers of GaInAsP (forming the walls of the energy "well").

Quantum-well structures that are built with a purposeful mismatch of lattice constant (a "strained" lattice) offer the promise of improved performance over unstrained devices [10]. These *strained-layer multiple quantum-well laser* show promising results in the laboratory and continue to be explored.

Material	Wavelength	T_0
InGaAsP	1300 nm	60–70K
InGaAsP	1500 nm	50–70K
GaAlAs	850 nm	110–140K

Table 5.1 Representative range of values of empirical parameter T_0 for temperature dependence of threshold current.

5.5   Laser Temperature Dependence

One major problem with laser diodes is the temperature dependence of the threshold current. Long-wavelength sources made of InGaASP exhibit a greater temperature sensitivity than the short-wavelength sources made of GaAlAs. The temperature dependence of the threshold current I_{th} can be expressed as

$$I_{th} = I_0 e^{T/T_0} \tag{5.22}$$

where I_0 is a constant (established at a reference temperature) and T_0 is an empirical constant fit to the measured data. Typical ranges of values of T_0 are found in Table 5.1.

Laser diodes are usually elaborately heat-sinked to provide a constant operating temperature, or they incorporate a thermoelectric cooler to remove heat from the diode. The cooler capacity needs to be large enough to maintain a constant operating temperature while the diode is operating.

As an alternative, laser diode drive circuits can include temperature compensation circuitry to minimize the effects of temperature changes. A photodetector samples a small portion of the output light (either at the back facet of the laser or by splitting off a small fraction at the output) and provides a voltage proportional to the laser power. Through a feedback mechanism, the drive current can be adjusted to maintain a constant output power level. (The same circuit provides compensation for power changes due to aging of the laser diode.) Further details on temperature-compensating circuitry are contained in the discussion on electronic drive circuits, later in this chapter.

5.6   Source Reliability

Reliability [20, 23, 27–31] of source devices plays an important role in determining overall system reliability. This is particularly true since early diode lasers had low lifetimes of a few hundreds of hours.

The failure rate for any device is modelled by the *bathtub curve* (Fig. 5.20 on the next page), named because its shape resembles a transverse profile of a bathtub. After a short period of operation, certain devices will fail (called "infant failures") due to fabrication problems and other factors. Those that survive will typically be long-lived with few failures. At the end of the device lifetime, the failure rate will increase due to accumulated effects. The problems for the manufacturer and user are to minimize the infant failures (or at least be certain that they are screened out and not used), to maximize the useful lifetime of the device under the required operating conditions, and to allow for a graceful degradation in performance at the end of useful life. The optical sources are typically characterized by their useful predicted lifetime.

First we will discuss laser reliability and then LED reliability.

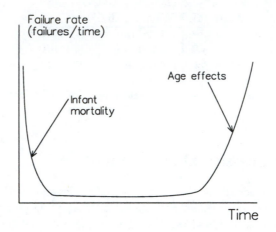

Figure 5.20 "Bathtub curve" for device failure rates.

5.6.1   Laser Reliability

A typical diode laser will have a power output that decreases exponentially in time, following a relation of the form

$$P_{\text{out}}(t) = P_i \, e^{-t/\tau_m} , \qquad (5.23)$$

where $P_{\text{out}}(t)$ is the time-dependent output power, P_i is the initial power of the laser, and τ_m is the exponential lifetime. This lifetime can be found by measuring the average time that it takes the power of a sample batch of lasers to reach a predetermined fraction of its initial value. Knowledge of τ_m allows us to extrapolate the power behavior to any amount of time.

──────────────── ∞ ────────────────

Example: It is predicted that a certain diode laser will have its power decrease to 90% of its initial value in 3 years. How many years will be required for the power to decrease to 10% of its initial power?

Solution: We begin by finding τ_m from the initial information,

$$\frac{P_{\text{out}}}{P_i} = e^{-t/\tau_m} \qquad (5.24)$$

$$0.90 = e^{-3/\tau_m}$$

$$\tau_m = -\frac{3}{\ln(0.90)} = 28.5 \text{ years}.$$

To find the time required to reduce the power to 10% of the original power, we use

$$\frac{P_{\text{out}}}{P_i} = 0.10 = e^{-t/28.5} \qquad (5.25)$$

$$\ln(0.10) = -\frac{t}{28.5}$$

$$t = -28.5 \ln(0.10) = 65.6 \text{ years}.$$

──────────────── ∞ ────────────────

———————————————— ∞ ————————————————

Example: Suppose that we want the laser in the previous example to reach 10% of its original power at 20 years of life. What fraction of the output power should be present at the end of 3 years to ensure this performance?

Solution: We need to find the required τ_m first.

$$\frac{P_{\text{out}}(20 \text{ years})}{P_i} = 0.10 = e^{-\frac{20}{\tau_m}} \tag{5.26}$$

$$-\frac{20}{\tau_m} = \ln(0.10)$$

$$\tau_m = -\frac{20}{\ln(0.10)} = 8.69 \text{ years}.$$

Finding the power after 3 years,

$$\frac{P_{\text{out}}(3 \text{ years})}{P_i} = e^{-3/8.69} = 0.708 = 70.8\%. \tag{5.27}$$

———————————————— ∞ ————————————————

Reliability in diode lasers is an issue because of the harsh electrical and optical environment encountered inside the laser's active region. A typical diode laser is subject to severe operating conditions that include current densities of 2000 to 5000 A/cm^2 within the active region and optical power densities of 10^5 to 10^6 W/cm^2. For the purpose of testing lasers it has been observed that continuous (CW) operation is the most strenuous environment and is used for most testing. The definition of when a laser has reached the end of a useful life is open to some interpretation. One method of determining this is to hold the driving current fixed and to define the end of life as that time when the output power falls below a usable amount (e.g., 1.25 mW). Another method monitors the output power and, as the power starts to fall, adjusts the drive current to a (larger) value required to maintain the original power level. For this mode of operation, the end of life occurs when the drive current reaches the rated maximum of the device. Other definitions are also used, making it difficult to compare test data. Sometimes device performance is measured below threshold to characterize the laser. This is particularly true when tests are run at high temperatures, where the amount of current required to achieve lasing is extreme enough to introduce new causes of degradation.

Laser Degradation Mechanisms

Three sources of degradation have been identified in diode lasers [27–31]:

- Facet damage—This is physical damage to the reflecting surfaces of the laser due to operation at high optical power densities. Improved cleaving techniques and the addition of protective passivating layers on the laser facets have removed this problem.

- Ohmic contact degradation—All semiconductor devices operated at high current densities and elevated temperatures exhibit deterioration in the ohmic contacts at the metal-semiconductor interface. Improved solders and heat-sinking have remedied this problem.

- Internal damage formation—This degradation mechanism is due to the formation of internal lattice defects in the semiconductor crystal. It is the least understood and hardest-to-control mechanism. These nonradiating defects initially form along the lines of crystal dislocations and have been named *dark lines* due to their appearance under a microscope. These defects were found in GaAlAs short-wavelength where they led to severe lifetime problems in early devices, resulting in a major effort to improve the fabrication of these lasers. Improved fabrication techniques have reduced the effect of these defects, and any devices with these defects are usually found by preliminary testing of the lasers. The sources that have dislocation defects are identified by running *burn-in testing* of approximately 100 hours for each device. Sources suffering from degraded performance in this test are either sufficiently defective to justify rejection or will typically be susceptible to degradation by the propagation of dark lines in the future and are also rejected. InGaAsP long-wavelength devices do not show these defects and have superior reliability performance over short-wavelength lasers.

5.6.2 Laser Testing

The testing of lasers [27–31] is not a standardized process. Manufacturers have different testing conditions, different sampling techniques, and different definitions of the end of useful life. Results are quoted in statistical form, which can also be misleading, or, worse yet, by anecdotes about particular lasers under test, as if such lasers were the norm rather than statistical oddities on the outer fringe of the distribution. As the room-temperature lifetime of laser sources becomes longer, techniques are required to accelerate the degradation processes. Two possibilities occur, to increase the drive current level or to increase the operating temperature.

High-Current Testing

Increasing the operating current level has been observed to decrease the operating life τ by the proportionality

$$\tau \propto J^{-n}, \tag{5.28}$$

where J is the current density and n is an empirically fit parameter with values typically ranging from 1.5 to 2.0. This mechanism accelerates the facet degradation more than high-temperature testing and so it is not used very often.

———————————————— ∞ ————————————————

Example: Consider a laser with a predicted lifetime of 20 years at an operating current of 100 mA.

(a) Assuming that $n = 1.75$, what would be its lifetime if the current were doubled?

<u>Solution:</u> We can use proportional relationships:

$$\tau \quad \propto \quad J^{-n} \tag{5.29}$$

$$\frac{\tau_1}{\tau_2} = \frac{J_1^{-n}}{J_2^{-n}} = \left(\frac{J_1}{J_2}\right)^{-n} = \left(\frac{I_1}{I_2}\right)^{-n} = \left(\frac{100}{200}\right)^{-1.75} = 3.36.$$

Hence,

$$\tau_2 = \frac{\tau_1}{3.36} = \frac{20}{3.36} = 5.95 \text{ years}.$$

(b) ...if the current were halved?

Solution:

$$\frac{\tau_1}{\tau_2} = \left(\frac{I_1}{I_2}\right)^{-n} = \left(\frac{100}{50}\right)^{-1.75} = 0.297. \tag{5.30}$$

$$\tau_2 = \frac{\tau_1}{0.297} = \frac{20}{0.297} = 67.3 \text{ years}.$$

Note that the nonlinear behavior of this dependence favors operation of lasers requiring long lifetimes at the lowest current level consistent with required link performance. Modest reductions in drive current can produce large increases in lifetime.

—————————————————— ∞ ——————————————————

High-Temperature Testing

Just as the lifetime is sensitive to changes in drive current, it is also sensitive to changes in temperature. High temperatures can cut the device lifetime significantly. The general form of the operating life dependence on temperature is

$$\tau \propto e^{E/kT}, \tag{5.31}$$

where k is Boltzmann's constant, T is the operating temperature, and E is an empirically determined *activation energy* parameter. While reported values of E have ranged widely (from 0.3 to 0.95 eV), a generally accepted value of 0.7 eV is used for laser sources that have survived the burn-in testing procedure described earlier [16].

Increased temperature operation is currently the preferred method of acceleration rather than increased current operation. Since high-current operation is not desirable, testing is done at moderate lasing current levels or, in some cases, at current levels below threshold with intermittent testing at lasing current drives.

—————————————————— ∞ ——————————————————

Example: Consider a laser diode with a predicted lifetime of 10 years at room temperature (300K).

(a) Calculate the predicted lifetime if the operating temperature is reduced by 10° C.

Solution:

$$\frac{\tau_1}{\tau_2} = \frac{e^{\frac{E}{kT_1}}}{e^{\frac{E}{kT_2}}} = e^{\frac{E}{k}\left(\frac{1}{T_1} - \frac{1}{T_2}\right)}. \tag{5.32}$$

A typical value of the activation energy is $E = 0.7$ eV $= 0.7(1.6 \times 10^{-19}) = 1.12 \times 10^{-19}$ joules.

$$\frac{\tau_1}{\tau_2} = e^{\frac{1.12 \times 10^{-19}}{1.38 \times 10^{-23}}\left(+\frac{1}{300} - \frac{1}{290}\right)} \tag{5.33}$$

$$= e^{\frac{1.12 \times 10^{-19}}{1.38 \times 10^{-23}}\left(-1.149 \times 10^{-4}\right)} = 0.393.$$

$$\tau_2 = \frac{\tau_1}{0.393} = \frac{10}{0.393} = 25.4 \text{ years}.$$

We observe that small decrease in operating temperature causes a large increase in the operating lifetime. Hence, we want to keep the operating temperature as low as possible.

(b) Calculate the expected lifetime if the temperature is *raised* by 10°C.

Solution:

$$\frac{\tau_1}{\tau_2} = e^{\frac{1.12 \times 10^{-19}}{1.38 \times 10^{-23}}\left(+\frac{1}{300} - \frac{1}{310}\right)} = 2.39. \tag{5.34}$$

$$\tau_2 = \frac{\tau_1}{2.39} = \frac{10}{2.39} = 4.18 \text{ years}.$$

Hence, as expected from the result of part (a), small increases in operating temperature cause a large decrease in the operating lifetime.

─────────────────────── ∞ ───────────────────────

Figure 5.21 shows some typical results of laser lifetime tests. The power output of the test lasers is plotted as a function of time, showing a history of the power variation of the laser. Typical median lifetimes for AlGaAs short-wavelength lasers under continuous wave (i.e., dc operation) are about 10^5 hrs; InGaAsP lasers have much longer median lifetimes [9].

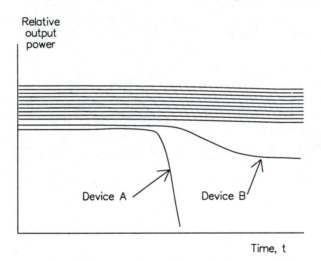

Figure 5.21 Typical laser diode lifetime test data. (The curves have been displaced for clarity.) Device A has failed; device B has lost significant output power but has not failed yet.

5.6.3   LED Reliability

Failure mechanisms in LEDs are similar to those in lasers (except that there is no facet damage in LEDs) but, because of more benign current density levels and optical power levels, degradation of LED output power is much more gradual. However, because LEDs produce less power than lasers, there is a tendency to run them at high drive-current levels. The same processing and passivating improvements that apply to lasers have also brought about improvements in LED performance.

Burn-in testing can be used to remove the devices that exhibit infant failure. Generally, LED lifetime due to normal aging degradation is not an issue in system design. (There can still be, however, freak failures of atypical devices or failure due to user errors.) Typical predicted lifetimes of LEDs, based on high-temperature testing, are in the range of 10^5 to 10^8 hours, depending on their design and fabrication properties.

5.6.4 Laser Modulation Response

Laser diodes can be intensity modulated (IM) by modulating the drive current of the device; the light can be frequency modulated (FM) by pulsing the drive current, and can be phase modulated (PM) with the use of an external modulator. We will consider only intensity modulation here.

We have seen from the output power characteristics of the laser that, once past threshold, the output power is linear with further increases in current. To pulse modulate the laser intensity, we pulse the drive current from the threshold level (or just above that value) to some larger value; the output power of the laser jumps from a small value to a larger value. For continuous bipolar signals, we need to establish a bias point partway up the power curve and allow the drive current to deviate around that bias value.

5.6.5 Modulation Bandwidth

The speed of response of the source plays a role in determining the maximum data rate that can be transferred if the fiber does not present a limit.

Laser Intensity Modulation Bandwidth

One of the important characteristics of the source for high-speed data links is the maximum modulation rate of the source. This maximum modulation rate is characterized by the modulation frequency response (i.e., a sine-wave modulating signal frequency is increased until the source response decreases). This upper limit of the frequency response is determined by (1) the RC time constant of the source (where R is the source and circuit resistance and C is determined by the parasitics of the device) and (2) the interaction of the light and charge carriers in the active region of the device. Figure 5.22 on the next page shows a representative response of an intensity-modulated diode laser. This response is characterized by a low-frequency flat portion and a high-frequency resonant peak. The exact frequency location and magnitude of this peak is dependent on the operating point and device parameters of the diode [26, 32]. The frequency width is described by the 3-dB frequency (i.e., the frequency where the response is 1/2 of the low-frequency value). While it is possible to use compensation techniques to utilize the full frequency response of the device, including the resonant peak, most applications confine themselves to the flat portion of the response to the left of the peak. Current wide-bandwidth lasers have a modulation bandwidth of more than 10 GHz.

Another effect is observed when diode lasers are driven with a pulse of current. Due to the influence of the charge carriers on the index of refraction of the active region, the optical length of the resonator changes with current density. This change in optical width appears as a linear shift in operating frequency of the source [26, 32]. This *frequency chirp* can have the detrimental effect of increasing the linewidth of the source. On the other hand, the same frequency shift can be used as a method of frequency modulating the laser. As discussed earlier, distributed feedback (DFB) lasers and distributed Bragg reflector (DBR) lasers are more resistant to this chirp effect.

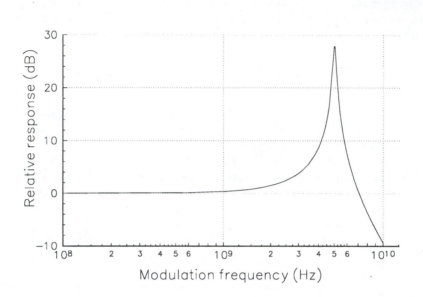

Figure 5.22 Representative frequency response of an intensity-modulated diode laser.

5.7   Source Noise

5.7.1   LED Source Noise

The LED is considered to be essentially a noise-free source, i.e., the source does not contaminate the signal with additional noise.

5.7.2   Laser Source Noise

The laser has three potential noise sources.

1. The first noise is that noise that is quantization noise due to the light being quantized into energy packets (i.e., photons). This can be observed in a weak signal and also includes the *spontaneous emission* added to the coherent light of the laser output. Such a noise is called *relative intensity noise,* commonly abbreviated as RIN. For a relatively powerful laser, this small amount of spontaneous light is overpowered by the large amount of coherent light and the noise is negligible. For weaker sources, such as can occur in portions of an analog signal transmission, the noise can be appreciable and can affect the analog link operation. For coherent communication links where the information is encoded in the frequency or phase of the optical signal, this noise can limit the link performance.

2. A second noise is *partition noise.* A pulsed multimode laser (or even some supposedly single-frequency lasers operated at high current levels) will operate at several frequencies

simultaneously, as seen in Fig. 5.14 on page 132. The power is "partitioned" among the various longitudinal spectral modes. While the *total* power can be kept constant, the power in each mode is not a steady function of time; the power distribution among modes changes with time. Each time the distribution changes, the power output undergoes fluctuation leading to a noise term on the nominally stable output. One technique for removing this noise is to use a true single-frequency laser, such as the distributed feedback (DFB) laser.

3. The third noise is associated with a multimode source when combined with a multimode fiber in a system that modifies the fiber-mode power distribution. This can occur at splices and connectors. When viewed at the end of a fiber, a coherent source will form an interference pattern consisting of the constructive and destructive interference of the modes in the fiber. For an ideal fiber with all modes excited by the source, there is no fluctuation in power as time progresses. One of the potential bad effects of splices or connectors (or an imperfect fiber) in a fiber link, however, is to redistribute the power in the fiber modes as light passes through the splice or connector fiber. Usually some of the higher modes are removed by passage through the connector or imperfection. The multimode source changes its spectral distribution of power as a function of time. The laser, in turn, excites the various fiber modes with more or less power, depending on the instantaneous power distribution. Since the power in the higher modes is removed by the splice or connector, the total source power minus that removed is time-varying. This time-varying amplitude represents noise. This noise (actually due to the combined fiber, splice or connector, and source) is attributed to the source and is called *modal noise*. Since the noise mechanism requires coherent light to form the interference pattern, one solution is to use a high power source of low coherence (or *short coherence length*, in the jargon). The superradiant LED is such a source.

5.8 Electronic Driving Circuits

As noted in the specifications, optical diode sources are operated at 1–2 volts of forward bias with variable current drive, depending on the device and the output power required. Since light is only unipolar, bipolar electrical signals must be dc-shifted to a unipolar representation before driving the optical source. The dc level can be electronically removed after detection. Both analog and digital modulation can be accomplished.

For direct-modulated digital modulation, the current is pulsed through the device. The device is biased at a low value of optical output (usually, but not always, at zero power) for a logical **0** and pulsed to higher output power for a logical **1**. The ratio of the lower power to the higher power is the *extinction ratio* of the modulation scheme. Nonzero extinction ratios incorporate a link performance penalty, usually in the maximum data rate of the link.

For an LED, zero current represents the OFF state; and a positive current value (at or below the maximum rated current of the device) is driven through the device, as shown in Fig. 5.23 on the next page. As described earlier in the section on LED rise-time, a momentary reverse bias and a momentary overdrive can speed up the device performance. For a laser, the OFF state can be biased just below threshold (Fig. 5.23b) or at some value above threshold. We do *not* want to bias the OFF state at zero current, since the diode is slower to switch ON when biased at that location.

In analog modulation, the dc bias point is placed on the linear portion of the output transfer characteristics, as shown in Fig. 5.24 on page 147. The time-varying signal is then applied about

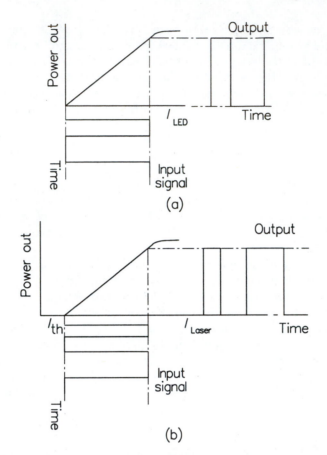

Figure 5.23 Pulse modulation for (a) an LED and (b) a laser.

the bias point. Obviously, the amplitude of the signal must be kept small enough to avoid the nonlinear portion of the transfer characteristics to avoid distorting the signal.

For low data rates and compatibility with digital circuits, LED sources and lasers with low threshold currents can frequently be driven directly from the output of a digital gate. The TTL family, with its ability to source and sink current, is particularly appropriate for direct drive. Other logic families and certain applications requiring larger drive currents use driver circuits designed to provide appreciable drive current. Figure 5.25 on page 148 illustrates one of these driving circuits.

Another frequently used technique is to place the source in the collector or emitter arm of a bipolar transistor circuit to increase the drive current as shown in Fig. 5.26 on page 148. Current-limiting resistors of the proper magnitude keep the current below the maximum current rating of the diode. Switching speeds are limited by the switching speed of the transistor. The same technique is also useful in logic gates with an open-collector output stage allowing the user to tailor the design of the diode driving circuit.

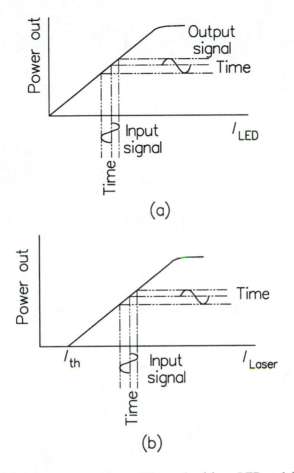

Figure 5.24 DC bias with AC modulation for (a) an LED and (b) a laser.

5.9 Source-to-Fiber Coupling

The source-to-fiber coupling problem [16, 33] is important because substantial power is lost at this interface. We wish to find the dependence of the coupling efficiency η on the fiber and source parameters that have been discussed. We define the *coupling efficiency* η as the fraction of source power that is coupled into the fiber. Mathematically, this is

$$\eta = \frac{P_f}{P_s}, \tag{5.35}$$

where P_f is the power in the fiber and P_s is the total power emitted by the source.

It should be noted that most sources used in fiber optic applications have a short length of fiber-optic cable attached (called a *pigtail*). The manufacturer specifies the optical power from the source *in the pigtail* and has optimized the coupling into the fiber. Typical optical powers coupled into the pigtail range from microwatts to a few milliwatts.

The spatial emission pattern from the source plays a key role in determining the value of

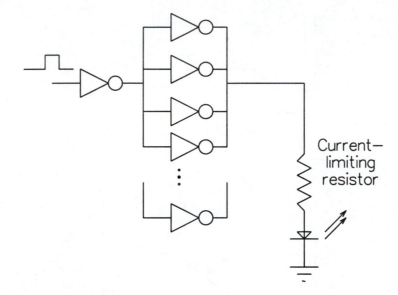

Figure 5.25 LED and laser drive using logic circuits.

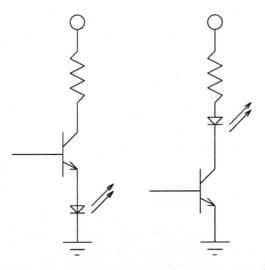

Figure 5.26 Light source drive using bipolar transistors.

efficiency. Generally, this pattern is two-dimensional and asymmetric (as we have seen) for the edge-emitting LED and laser diode. The surface-emitting LED is symmetric in output.

5.9.1   Coupling Model

A hypothetical emitter with a well-behaved symmetric emission pattern is the *Lambertian source*. Such a device has a spherical emission pattern that is symmetric and is given mathematically by

$$B(\theta, \phi) = B_0 \cos\theta, \tag{5.36}$$

where $B(\theta, \phi)$ is the radiance of the source, θ and ϕ are the polar angles as shown in Fig. 5.27 on the next page, and B_0 is the radiance measured along the normal to the emitting surface. The *radiance* is a measure of optical-power emission into a solid angle oriented around the axis defined by θ and ϕ from an infinitesimal area on the surface of the emitter. Its units are $\text{W·sr}^{-1}\text{·m}^{-2}$ where a steradian (abbreviated sr) is a measure of solid angle. As seen in Eq. 5.36, the measured radiance from a Lambertian source is a function of the angle from the source. For a more directional source, we frequently use a modified Lambertian approximation to the emission pattern measured. A dependence of

$$B = B_0 \cos^m\theta \tag{5.37}$$

is fitted to the data where m has a value larger than 1. Figure 5.28 on the next page illustrates such a distribution.

──────────── ∞ ────────────

Example: A diode laser has a measured full-angle beam divergence of 30° at the −3 dB power points. Calculate the value of m that will allow a $(\cos\theta)^m$ spatial power distribution.

Solution: The 3-dB power points are the half-power points and occur at ±15° (since the *full*-angle beam divergence is specified). We want to fit

$$\frac{P}{P_0} = \cos^m\theta \Rightarrow 0.5 = (\cos 15°)^m \tag{5.38}$$

$$\log 0.5 = m \log(\cos 15°)$$

$$m = \frac{\log(0.5)}{\log(\cos 15°)} = 19.9.$$

──────────── ∞ ────────────

An asymmetric emission distribution can also be modeled as Lambertian sources or sources with a higher dependence. Such a distribution is given by

$$B(\theta, \phi) = \frac{1}{\left(\dfrac{\sin^2\phi}{B_0 \cos^T\theta} + \dfrac{\cos^2\phi}{B_0 \cos^L\theta}\right)}, \tag{5.39}$$

where T and L are the exponents that best fit the transverse and lateral emission patterns. Typically, for edge emitters, $T = 1$ (i.e., Lambertian), and L is a higher value. For laser diodes, L can take on values of 100 or larger because of the increased directionality of the pattern. Although many practical sources have such asymmetric emission patterns, we usually estimate the losses by using formulas based on symmetric emission assumptions. Such assumptions simplify the evaluation of integrals and provide conservative results, but a more accurate solution can be obtained with the more complicated models.

In coupling the light from the source to the fiber, we can have a fiber butted up close to optical source. (This technique is called *butt-coupling*.) The source emits with a specific angular

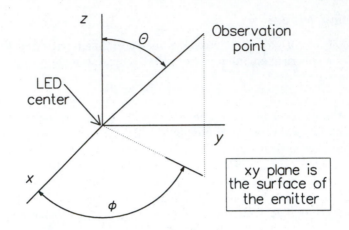

Figure 5.27 Emitter coordinate system.

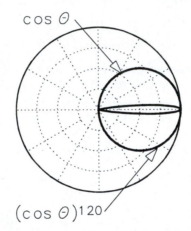

Figure 5.28 Radiance patterns for a Lambertian ($\cos\theta$) and a more directive emitter ($\cos^{120}\theta$).

radiation pattern; only some of the rays of light from the source will be accepted by the fiber—those within the *fiber acceptance cone* with a half-angle given by $\theta_{\max}$. (Here $\theta_{\max}$ is found as $\sin^{-1}$ NA, as we have seen earlier.) The power coupled into the fiber is found by the evaluation of the integral

$$P_f = \int_0^{r_u} \int_0^{2\pi} \left(\int_0^{2\pi} \int_0^{\theta_{\max}} B(\theta, \phi) \sin\theta \, d\theta \, d\phi \right) d\phi_s \, r \, dr . \tag{5.40}$$

Here the double integral within the parentheses is the amount of light accepted by the fiber from an infinitesimal area located at position r, ϕ_s. The outer integral sums up the power contributions from all of the infinitesimal emitting areas on the source. For a source smaller than the fiber core, the upper limit on the first integral would be $r_u = r_s$, the radius of the source. If the source happens to be larger than the fiber core, then the upper limit of the integral is $r_u = a$, the fiber core radius. For a step-index fiber the value of θ_{max} is a constant. For a graded-index fiber, the value of θ_{max} will vary with the radial position r of the source, complicating the evaluation of the integral. The most general case is for a graded-index fiber with a variable exponent g of the radial dependence of index of refraction variation.

Lambertian Source

Using our assumption of having a Lambertian emitter, the radiance is $B(\theta, \phi) = B_0 \cos\theta$ (one of the problems at the end of the chapter considers the case of a $\cos^m\theta$ source), and the integral becomes

$$P_f = \pi B_0 \int_0^{r_u} \int_0^{2\pi} \sin^2\theta_{max}\, d\phi_s\, r\, dr\,. \tag{5.41}$$

We recognize $\sin^2\theta_{max} = (\text{NA})^2$, giving

$$P_f = \pi B_0 \int_0^{r_u} \int_0^{2\pi} (\text{NA})^2\, d\phi_s\, r\, dr\,. \tag{5.42}$$

Step-Index Fiber Coupling

For a step-index fiber the NA is independent of position r, so it comes outside the integrals and we evaluate the double integral as

$$P_f = \pi^2 B_0 (\text{NA})^2 r_s^2 \qquad (r_s < a)\,. \tag{5.43}$$

The power from the source is

$$P_s = \pi^2 B_0 r_s^2\,, \tag{5.44}$$

giving the coupling efficiency

$$\eta = (\text{NA})^2 \qquad (r_s < a,\ \text{Step index})\,. \tag{5.45}$$

Revaluation of Eq. 5.42 for the case of $r_s > a$ gives

$$P_f = \pi^2 B_0 (\text{NA})^2 a^2 \qquad (r_s > a)\,, \tag{5.46}$$

and, since the source power is still the same, the coupling efficiency for this case will be

$$\eta = (\text{NA})^2 \left(\frac{a}{r_s}\right)^2 \qquad (r_s > a,\ \text{Step index})\,. \tag{5.47}$$

Graded-Index Fiber Coupling

For the graded-index fiber, the NA is a function of position r and is given by

$$\text{NA}(r) = \text{NA}(0)\sqrt{1 - \left(\frac{r}{a}\right)^g}\,. \tag{5.48}$$

The integral for power coupled into the fiber (Eq. 5.42) becomes

$$P_f = \pi B_0 \int_0^{r_u} \int_0^{2\pi} (\text{NA}(0))^2 \left(\sqrt{1 - \left(\frac{r}{a}\right)^g}\right)^2 d\phi_s\, r\, dr\,. \tag{5.49}$$

Fiber	$r_s \leq a$	$r_s > a$
Step index	NA^2	$NA^2 \left(\dfrac{a}{r_s}\right)^2$
Graded index	$NA^2 \left[1 - \left(\dfrac{2}{g+2}\right)\left(\dfrac{r_s}{a}\right)^g\right]$	$NA^2 \left(\dfrac{a}{r_s}\right)^2 \left(\dfrac{g}{g+2}\right)$

Table 5.2 Summary of coupling efficiencies (Lambertian emitter).

Evaluating this integral (for $r_s < a$),

$$P_f = \underbrace{\pi^2 B_0 r_s^2}_{P_s} (NA(0))^2 \left(1 - \frac{2}{g+2}\left(\frac{r_s}{a}\right)^g\right) \qquad (r_s < a). \tag{5.50}$$

The resulting coupling efficiency is

$$\eta = (NA(0))^2 \left(1 - \frac{2}{g+2}\left(\frac{r_s}{a}\right)^g\right) \qquad (r_s < a, \text{ Graded index}). \tag{5.51}$$

For the case of $r_s > a$, the power into the fiber is

$$P_f = \frac{\pi^2 B_0 (NA(0))^2 a^2 g}{g+2} \qquad (r_s > a), \tag{5.52}$$

and the coupling efficiency for this case becomes

$$\eta = (NA(0))^2 \left(\frac{a}{r_s}\right)^2 \left(\frac{g}{g+2}\right) \qquad (r_s > a, \text{ Graded index}). \tag{5.53}$$

Equations 5.45 and 5.47 and Eqs. 5.51 and 5.53 are the coupling-efficiency equations applicable to the step-index fiber and the graded-index fiber, respectively, and are summarized in Table 5.2.

───────────────── ∞ ─────────────────

Example: Consider a circular LED source with a 62.5 μm diameter. Calculate the coupling efficiencies . . .

(a) . . . into 50/125, 62.5/125, and 100/140 step-index fibers with NA = 0.20.

Solution: For a 50/125 SI fiber, $r_s > a$, so

$$\eta = NA^2 \left(\frac{a}{r_s}\right)^2 = (0.2)^2 \left(\frac{25}{31.25}\right)^2 = 0.0256 = 2.56\% \Rightarrow -15.92 \text{ dB}. \tag{5.54}$$

For a 62.5/125 fiber, $r_s = a$, so

$$\eta = (NA)^2 = (0.2)^2 = 4\% \Rightarrow -13.9 \text{ dB}. \tag{5.55}$$

For 100/140 fiber, $r_s < a$, so

$$\eta = (NA)^2 = (0.2)^2 = 4.0\% \Rightarrow -13.9 \text{ dB}. \tag{5.56}$$

(b) ...into three graded-index fibers of the same dimensions with $NA(0) = 0.2$ and $g = 1.8$.

Solution: For a 50/125 GI fiber, $r_s > a$, so

$$\eta = NA^2(0) \left(\frac{a}{r_s}\right)^2 \left(\frac{g}{g+2}\right) = (0.2)^2 \left(\frac{25}{31.25}\right)^2 \left(\frac{1.8}{1.8+2}\right) \tag{5.57}$$

$$= 0.01213 = 1.213\% \Rightarrow -19.16 \text{ dB}.$$

For 62.5/125 fiber, $r_s = a$, so

$$\eta = (NA)^2(0) \left(1 - \frac{2}{g+2}\right) = (0.2)^2 \left(1 - \frac{2.0}{1.8+2}\right) \tag{5.58}$$

$$= 0.01895 = 1.895\% \Rightarrow -17.22 \text{ dB}.$$

For 100/140 fiber, $r_s < a$, so

$$\eta = (NA)^2(0) \left[1 - \left(\frac{2}{g+2}\right) \left(\frac{r_s}{a}\right)^2\right] \tag{5.59}$$

$$= (0.2)^2 \left[1 - \left(\frac{2}{1.8+2}\right) \left(\frac{31.25}{50}\right)^2\right]$$

$$= 0.0318 = 3.18\% \Rightarrow -14.98 \text{ dB}.$$

$$\infty$$

To compare equivalent step-index and graded-index fibers, we can use the approximation for the NA or NA(0),

$$NA \approx n_1 \sqrt{2\Delta}. \tag{5.60}$$

For a step-index fiber ($g = \infty$) and $r_s = a$, we have

$$\eta(SI) = 2n_1^2 \Delta. \tag{5.61}$$

For a parabolic index fiber ($g = 2$) and $r_s = a$, the efficiency is

$$\eta(GI) = n_1^2 \Delta. \tag{5.62}$$

This leads to the conclusion that, all else being equal, a step-index fiber is twice as efficient in coupling light as a graded-index fiber (an improvement of 3 dB). This improved coupling is one of the advantages of a step-index fiber over a graded-index fiber.

Non-Lambertian Source

For a *non-Lambertian source* that is still symmetric and approximated by $(\cos\theta)^m$, the coupling efficiency for a step-index fiber with $r_s < a$ turns out to be

$$\eta_{\text{non-Lambertian}} = \left(\frac{m+1}{2}\right) NA^2, \tag{5.63}$$

showing $(m+1)/2$ greater efficiency due to the shaping of the source. This accounts for the laser diode's superiority in delivering optical power into a fiber. Note, however, that, because of the small value of Δ or NA, all fibers typically will have significant losses at the source-fiber interface.

5.9.2 Reflection Effects

The coupling efficiencies calculated above assume a perfect match of refractive index at the core-light source interface. The lack of such a match leads to additional losses due to the *Fresnel reflection losses* at the interface. Assuming perpendicular incidence, the power transmittance T at the interface between the medium and the core of a step-index fiber is

$$T = 1 - \left(\frac{n_1 - n}{n_1 + n}\right)^2 , \tag{5.64}$$

where n_1 is the index of refraction of the core and n is index of refraction of medium outside of the fiber core. Frequently a drop of index matching liquid is placed at this interface to minimize these losses. Another technique to reduce the reflection is to cut the fiber ends at a non-perpendicular angle to avoid retroreflections.

5.9.3 Lens-Coupled Fiber

As indicated in the coupling efficiency results for the butt-coupled fiber, mismatching of the source area and the area of the fiber core wastes power. The obvious solution is to physically match the size of the source and the size of the fiber. If the size of the source is *smaller* than the size of the fiber (a case that typically exists with a laser), a lens can be used to optically match the apparent size while increasing the directivity of the source by the magnification factor. (Unfortunately, no remedy exists if the size of the source exceeds the size of the fiber.) Since the source and fiber sizes are so small and one does not want to have to use a lens much bigger than this size, there arises a need for small *microlenses* to perform this coupling task. Several lens geometries have been attempted in the past; some are illustrated in Fig. 5.29. The goal of each of these schemes is to magnify the effective emitter area to match the area of the fiber core. The "bulb-end" fiber is easily made by heating the end of the fiber and letting it cool. The integral LED/lens incorporates a lens into the structure of the LED source at the time of fabrication.

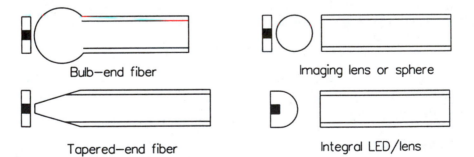

Figure 5.29 Source-fiber coupling using lenses.

The incorporation of the microlens has the potential disadvantages of adding additional fabrication steps and requiring fine resolution alignment to implement. If properly aligned, the improvement in coupling efficiency is M, the magnification of the source where $M_{\max}$ is $d_{\text{fiber}}/d_{\text{source}}$. (Improper alignment adds small, but tolerable, decreases in the coupling efficiency.)

—————————————————— ∞ ——————————————————

Example: Consider an LED source with a 50 μm diameter operating with a 100/140 SI fiber with NA = 0.15. Calculate the coupling efficiency both with and without a coupling lens assuming a Lambertian beam pattern. Calculate the optimum magnification of the lens.

Solution: Without a lens, $r_s < a$, so

$$\eta \quad = \quad \text{NA}^2 = (0.15)^2 = 2.25\% \Rightarrow -16.48 \text{ dB}. \tag{5.65}$$

With a lens, $r_s = a$ and

$$\eta \quad = \quad M(\text{NA})^2, \tag{5.66}$$

$$M_{\text{max}} \quad = \quad M_{\text{opt}} = \frac{d_f}{d_s} = \frac{100}{50} = 2,$$

$$\eta \quad = \quad 2(0.15)^2 = 4\% \Rightarrow -13.47 \text{ dB}.$$

Hence we find a 3 dB improvement in coupling that corresponds to the magnification of 2.

—————————————————— ∞ ——————————————————

5.9.4 LED Coupling

The coupling efficiency of an LED depends on whether it is a surface-emitting LED or an edge-emitting LED.

The surface-emitting LED produces the closest approximation to our ideal Lambertian emitter. Such an LED matches the theory quite well when coupled into a large-core large-NA fiber. The match is worse when the fiber has a small core and a small NA. The coupling for such fibers can be greatly improved by reducing the size of the source area during manufacture (which, in turn, requires improved heat sinking for a given drive current) and using a lens to couple the light. Increases in coupling efficiency of 3 to 5 times have been achieved with bulb-end fibers and from 18 to 20 times with tapered-end fibers [9].

The edge-emitting LED is more difficult to model due to its asymmetric emission pattern. The narrower emission pattern (in one dimension) allows greater coupling efficiency for butt-coupled fibers. Optical techniques using cylindrical lenses to improve the coupling on each axis have been attempted [9], but the improvement in coupling has been marginal in light of the more complicated lens arrangement. Bulb-end fibers and tapered fibers can improve the coupling efficiency with these edge-emitting LEDs.

5.9.5 Laser Coupling

Laser sources, like edge-emitting diodes, have asymmetric beam patterns (although the patterns are much narrower than the LED patterns). An additional difficulty is incurred when trying to couple the light into a small-core single-mode fiber. (Core diameters can range from 5 to 9 μm.) The small size of the core increases the sensitivity of the coupling efficiency to misalignment. Lens elements are frequently used to match the small active source area to the size of the fiber core. Both microlenses and *graded-index lenses* (also called "GRIN lenses") have been used [9], as well as cylindrical lenses. While the use of lenses helps to match the source and fiber core areas, it also has the effect of increasing the sensitivity of the coupling coefficient to lateral misalignment errors. In addition, the presence of reflections from the lens elements and the fiber back into the laser can upset the frequency stability of single-frequency single-mode fiber.

5.10   Summary

We have found that the laser holds significant operational advantages over the LED except for cost and temperature sensitivity. The spectral width of the laser source is appreciably lower for the laser diode; hence, this source is more suitable for use with a single-mode fiber where dispersion is to be minimized. The optical power delivered into a pigtail is higher for the laser, due to the higher coupling efficiency enjoyed by the laser because of its narrower emission pattern. The laser also enjoys an order-of-magnitude more speed, thereby increasing the potential data rate of an optical link. The cost ratio is about a factor of two in favor of the LED source, due to the increased fabrication quality requirements needed to produce today's low-threshold, high-reliability devices. In the short-wavelength region, then, LEDs offer substantial cost advantages for links operating at modest data rates (<50 Mb·s^{-1}) using multimode fibers (both step-index and graded index fibers). At higher data rates or with single-mode fibers, laser diodes are the source of choice.

In the long-wavelength region of the spectrum, InGaAsP sources are available as both LEDs and lasers. Since most sources at the long wavelengths are to operate with single-mode fibers, early work concentrated on laser diode emitters. Long-wavelength LEDs are now commercially available and are finding applications for short-distance moderate-data-rate links, such as the proposed Fiber Distributed Data Interface (FDDI) (described in Chapter 8). While InGaAsP sources have shown slower degradation than GaAlAs devices, their temperature sensitivity is greater. Since the spectral width of a source increases as the square of the central wavelength, the spectral width of these long-wavelength sources is about 2.5 times that of an 850 nm source. Single-frequency operation is important to minimize dispersion. Such techniques as distributed feedback and multiple cavity combinations have been investigated to produce single-frequency operation. Mode hopping, or the jumping of the emitted light from one mode to another, is also an undesirable effect, caused by multimode operation and reflections of the light from interfaces in front of the source. Single-mode operation by control of the width of the emitter has been effective in reducing multimode effects. (Another technique to reduce these effects is to reduce the coherence of the source, using a low-coherence wide-frequency source called a *superluminescent LED*). Polarization rotating and blocking devices called *optical isolators* are frequently used to avoid the reflection effects.

The current challenges in laser sources for fiber-optic applications are in achieving narrow spectral width, stabilizing the operating wavelength against the frequency chirp when the laser is pulsed, lowering the threshold current to values of a few to several milliamperes, and rapidly tuning the wavelength.

5.11   Problems

1. Find ...

    (a) ...the minimum wavelength of a GaAlAs source.

    (b) ...the maximum wavelength of a GaAlAs source.

2. Using the alloy fraction formulas (Eqs. 5.4 and 5.5 on page 116), find the material composition for ...

    (a) ...a 1.3 μm source.

    (b) ...a 1.55 μm source.

3. Consider two $Ga_{1-x}Al_xAs$ laser sources with $x = 0.02$ and $x = 0.09$. Find the bandgap energy and peak wavelength for these devices.

4. We want to estimate the transmission at air-material interfaces for high refractive-index materials like InP and GaAs and air $(n = 1)$. Find T in percent and in dB for ...

   (a) ... InP $(n = 3.4)$.

   (b) ... GaAs $(n = 3.6)$.

5. Show that the LED bandwidth is given by Eq. 5.11 on page 124.

6. An optical source is selected from a group of devices specified as requiring a mean time of 5×10^4 hours for the output power to degrade by -3 dB. If the device emits 5 mW at room temperature at the start of a test, what will be its emission power after (a) 1 month, (b) 1 year, (c) 5 years, and (d) 10 years?

7. A group of laser devices have operating lifetimes of 3.5×10^4 hours at 60 C and 6700 hours at 90 C. Find the activation energy for these devices and calculate the expected lifetime at 20 C.

8. A laser diode has a lateral beam divergence of 30° (full angle) and a perpendicular (to the emitting junction) beam divergence of 60°. What are the values of L and T associated with this beam pattern?

9. An LED has a circular emitting region with a radius of 20 μm. The pattern is assumed Lambertian with an axial radiance of 80 W·cm^{-2}·sr^{-1} at a 100 mA drive current.

   (a) Calculate the coupling efficiency and the optical power coupled into a step-index fiber with a 140 μm core diameter and NA $= 0.20$.

   (b) Repeat the calculation for a 50 μm diameter graded-index fiber with a parabolic-index profile, a center index of 1.48, and $\Delta = 1\%$.

10. On the same graph, plot the coupling efficiency as a function of r_s (from 0 to 50 μm) for the following step-index fibers:

    (a) ... core diameter of 50 μm and NA $= 0.15$

    (b) ... core diameter of 100 μm and NA $= 0.20$

    In what regions could a lens improve the coupling efficiency?

11. Using a computer, plot the emission patterns (on a polar plot) of a Lambertian emitter and a laser emitter with $m = 10$.

12. (a) Show that the power coupled into a step-index optical fiber from a source (with $r_s < a$) with a radiance of $B(\theta) = B_0 \cos^m \theta$ is given by

$$P_f = \frac{2\pi A_s B_0 \left[1 - (\cos^{m+1} \theta_{max})\right]}{m+1}.$$

(5.67)

   (b) Show that the total power from the source is given by

$$P_s = \frac{2\pi A_s B_0}{m+1}.$$

(5.68)

   (c) Calculate the coupling efficiency and show that for small NA the answer becomes

$$\eta = \frac{(m+1)\,\mathrm{NA}^2}{2}.$$

(5.69)

   Note: For a graded-index fiber with $r_s = a$, the coupling efficiency is

$$\eta = \left(\frac{(m+1)\,\mathrm{NA}^2}{2}\right)\left(\frac{g}{g+2}\right).$$

(5.70)

13. A step-index fiber is excited by a laser with an emission pattern approximated by $(\cos\theta)^7$. Calculate the improvement factor relative to a Lambertian source.

14. Consider a fiber ($a = 30$ μm, $\Delta = 1.5\%$, $g = 1.95$, and $n_1 = 1.45$) excited by a surface emitting LED ($B_0 = 2.0 \times 10^2$ W·cm^{-2}·sr^{-1} and radius of 50 μm). Calculate ...

    (a) ...the power emitted by the source,

    (b) ...the power coupled into the fiber, and

    (c) ...the coupling losses (in dB).

15. Consider an LED with a radius of 20 μm and a circularly symmetric Lambertian emission pattern with an on-axis radiance of 100 W·cm^{-2}·sr^{-1} at a drive current of 100 mA.

    (a) Calculate the power coupled into a step-index fiber with a core diameter of 100 μm and a numerical aperture of 0.22.

    (b) Calculate the change in coupling efficiency (in dB) if the source size is tripled, the drive current is reduced to 50 mA, and the NA of the fiber is reduced to 0.11 while keeping the fiber diameter constant.

16. (a) Calculate the loss due to reflections in an interface that has the light go from GaAs ($n = 3.5$), through an air gap ($n = 1.0$), and into a fiber ($n = 1.5$).

    (b) Calculate the transmission if the air is replaced with a medium (such as epoxy) that matches the index of the glass.

References

1. H. Kressel and J. K. Butler, *Semiconductor Lasers and Heterojunction LEDs*. New York: Academic Press, 1977.

2. C. A. Burrus, H. Craig Casey, Jr., and T. Li, "Optical sources," in *Optical Fiber Telecommunications* (S. E. Miller and A. G. Chynoweth, eds.), pp. 499–556, New York: Academic Press, 1979.

3. A. Bergh and J. Copeland, "Optical sources for fiber transmission systems," *Proc. IEEE*, vol. 68, no. 10, pp. 1240–1246, 1980.

4. P. Selway, A. Goodwin, and P. Kirby, "Semiconductor laser light sources for optical communications," in *Optical Fibre Communication Systems* (C. Sandbank, ed.), pp. 156–183, New York: Wiley, 1980.

5. H. Kressel, "Electroluminescent sources for fiber systems," in *Fundamentals of Optical Fiber Communications* (M. F. Barnoski, ed.), pp. 187–255, New York: Academic Press, 1981.

6. J. Bowers and M. Pollock, "Semiconductor lasers for telecommunications," in *Optical Fiber Telecommunications II* (S. E. Miller and I. P. Kaminow, eds.), pp. 509–568, New York: Academic Press, 1988.

7. T. P. Lee, C. A. Burrus, Jr., and R. H. Saul, "Light-emitting diodes for telecommunications," in *Optical Fiber Telecommunications II* (S. E. Miller and I. P. Kaminow, eds.), pp. 467–507, New York: Academic Press, 1988.

8. P. Shumate, "Lightwave transmitters," in *Optical Fiber Telecommunications II* (S. E. Miller and I. P. Kaminow, eds.), pp. 723–757, New York: Academic Press, 1988.

9. P. K. L. Yu and K. Li, "Optical sources for fibers," in *Fiber Optics Handbook for Engineers and Scientists* (F. C. Allard, ed.), pp. 5.1–5.61, New York: McGraw-Hill, 1990.

10. T.-P. Lee, "Recent advances in long-wavelength semiconductor lasers for optical communication," *Proc. IEEE*, vol. 79, no. 3, pp. 253–276, 1991.

11. G. P. Agrawal, ed., *Semiconductor Lasers: Past, Present, and Future*. New York: American Institute of Physics, 1995.

12. J. Long, R. Logan, and R. Kerlicek, Jr., "Epitaxial growth methods for lightwave devices," in *Optical Fiber Telecommunications II* (S. E. Miller and I. P. Kaminow, eds.), pp. 631–670, New York: Academic Press, 1988.

13. T. Otoshi and N. Chinone, "Linewidth enhancement factor in strained quantum well lasers," *IEEE Photonics Technology Letters*, vol. 1, p. 117, 1989.

14. U. Koren, M. Oron, M. Young, B. Miller, J. de Miguel, G. Raybon, and M. Chien, "Low threshold highly efficient strained quantum well lasers at 1.5 μm wavelength," *Electronics Letters*, vol. 26, p. 465, 1990.

15. T. Tanbun-Ek, R. Logan, N. Olsson, H. Temkin, A. Sargent, and K. Wecht, "High output power 1.48–1.51 μm continuously graded index separate confinement strained quantum well lasers," *Applied Physics Letters*, vol. 57, p. 224, 1990.

16. G. Keiser, *Optical Fiber Communications, Second Edition*. New York: McGraw-Hill, 1991.

17. D. Botez and M. Ettenberg, "Comparison of surface and edge emitting LEDs for use in fiber optic communications," *IEEE Trans. Electron Devices*, vol. ED–26, pp. 1230–1238, 1979.

18. D. Marcuse, "LED fundamentals: Comparison of front- and edge-emitting diodes," *IEEE J. Quantum Electronics*, vol. QE–13, no. 10, pp. 819–827, 1977.

19. B. E. Saleh and M. C. Teich, *Fundamentals of Photonics*. New York: John Wiley and Sons, 1991.

20. H. Casey and M. Panish, "Heterostructure lasers: Part A–Fundamental principles and Part B–Materials and operating characteristics," in *Heterostructure Lasers*, New York: Academic Press, 1978.

21. M. Panish, "Heterojunction injection lasers," *Proc. IEEE*, vol. 64, no. 10, pp. 1512–1540, 1976.

22. G. H. B. Thompson, *Physics of Semiconductor Laser*. New York: Wiley, 1980.

23. N. Dutta and C. Zipfel, "Reliability of lasers and LEDs," in *Optical Fiber Telecommunications II* (S. E. Miller and I. P. Kaminow, eds.), pp. 671–687, New York: Academic Press, 1988.

24. J. C. Cartledge, "Theoretical performance of multigigabit-per-second lightwave systems using injection-locked semiconductor lasers," *J. Lightwave Technology*, vol. 8, no. 7, pp. 1017–1022, 1990.

25. S. Oshiba and Y. Tamura, "Recent progress in high-power GaInAsP lasers," *J. Lightwave Technology*, vol. 8, no. 9, pp. 1350–1356, 1990.

26. A. Yariv, *Optical Electronics, Fourth Edition*. New York: Holt, Rinehart and Winston, 1991.

27. H. Ettenberg and H. Kressel, "The reliability of (AlGa)As cw laser diodes," *IEEE J. Quantum Electronics*, vol. QE–16, no. 2, pp. 186–196, 1980.

28. M. Hirao, K. Mizuishi, and M. Nakamura, "High-reliability semiconductor lasers for optical communications," *IEEE J. on Selected Areas in Communications*, vol. SAC-4, no. 9, pp. 1494–1514, 1986.

29. A. Goodwin, I. Davies, R. Gibb, and R. Murphy, "The design and realization of a high reliability semiconductor laser for single–mode fiber–optical communications links," *J. Lightwave Technology*, vol. 6, no. 9, pp. 1424–1434, 1988.

30. Y. Nakajima, H. Higuchi, Y. Sakakibara, S. Kakimoto, and H. Namizaki, "High-power high-reliability operation of 1.3-μm p-substrate buried crescent laser diodes," *J. Lightwave Technology*, vol. LT–5, no. 9, pp. 1263–1268, 1987.

31. M. Fukuda, "Laser and LED reliability update," *J. Lightwave Technology*, vol. 6, no. 10, pp. 1488–1495, 1988.

32. R. S. Tucker, "High-speed modulation of semiconductor lasers," *J. Lightwave Technology*, vol. LT–3, no. 6, pp. 1180–1192, 1985.

33. K. Kawano, H. Miyazawa, and O. Mitomi, "New calculations for coupling laser diode to multimode fiber," *J. Lightwave Technology*, vol. LT–4, no. 3, pp. 368–374, 1986.

Chapter 6

Optical Receivers

6.1   Introduction

The *optical receiver* is a combination of the optical detector, electronic preamplifier, and the electronic processing elements that recover information sent on the optical signal. The design and implementation of the receiver portion of the system is most difficult, because the receiver could be working with the weakest optical signal and we do not want to contaminate the signal with noise.

Figure 6.1 on the following page shows a block diagram of a representative optical digital data receiver. The *optical detector* converts the modulated optical input into an electronic signal for further processing. Because the optical signal is typically weak, the next step is to amplify the signal with the *preamp*. It is crucial to minimize the noise added by this amplifier. We will find that the lowest-noise preamplifiers lack the bandwidth to handle the high-data-rate signals used in fiber communications; therefore, the *equalizer* works in combination with the preamp to restore the required bandwidth. The equalizer can also be used to help alleviate the problems caused by data spilling out into adjacent bit periods because of pulse spreading. Following the equalizer, the signal is boosted further with the *postamplifier*, frequently with some sort of automatic gain control that adjusts the gain subject to the strength of the signal. The *filter* following the postamp removes unwanted frequency components that might have been generated by the signal processing to this point. In some low-data-rate optical receivers, the detection is done asynchronously, in which a comparator is used to decide whether a pulse is present or not. This type of data recovery assumes that the pulses have sharp rise times and fall times. For optimum performance in high-data-rate links, the data clock is encoded into the transmitted signal and is recovered at the receiver by the *clock recovery* circuit. The recovered clock is fed into the data *decision circuit*, where the decision is made whether the voltage represents a logical **1** or a logical **0** (at the optimum sampling time provided by the recovered clock signal). Based on the results of the decision, the output is the recovered data stream, perhaps containing some errors.

In signal recovery the issue of noise is of paramount importance, since the presence of noise leads to errors in the recovered data. In communications systems the noise can be introduced by the transmitter (as already discussed in the previous chapter), the channel, the detector, and the electronic signal processing elements. In fiber transmission, the channel noise is assumed to be zero due to the fiber's imperviousness to electromagnetic interference. The noise from the

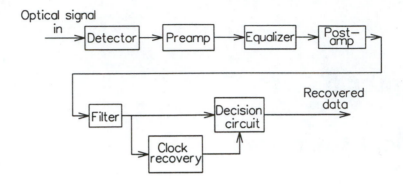

Figure 6.1 Block diagram of a representative optical receiver for a digital data link.

optical detector is different from that of radio and other electronic detectors in that it is signal-dependent. The noise of the electronic signal processing elements is the same as other electronic applications and the results of prior studies have been adapted to the fiber-receiver application.

In this chapter we will first consider the optical detectors used in fiber-optic links, the pin photodiode and the avalanche photodiode. For each detector, we discuss the physical operating principles of the devices, define the parameters used to characterize them, explore their noise performance, and consider limitations on operating speed.

We then consider digital receiver design in terms of the amplifier noise, optimization of the output signal-to-noise ratio, and the requirements for an equalization amplifier.

The sensitivity of an optical receiver is a function of the noise properties of the detector *and* the preamplifier. Therefore, we will express the receiver performance in terms of the noise performance of the detector and the noise figure of the preamplifier.

6.2   Optical Detectors

An optical detector or *photodetector* converts the optical input power into a current output. The ideal photodetector would be highly efficient, add no noise to the signal, respond uniformly to all wavelengths, would not limit the signal speed, and would be perfectly linear. Additionally, it would be small, electronically compatible with integrated circuits, reliable, and inexpensive. From the range of available photodetectors, including photoconductors, phototransistors, vacuum photoemissive devices, and pyroelectric devices, only the semiconductor photodiode [1–8] meets the latter set of properties sufficiently well to be considered for use in fiber-optic links.

6.2.1   Photodiodes

Physical Principles

The photodiode is a *pn* junction operated in a reverse-biased circuit, as shown in Fig. 6.2. The incident light must penetrate into the *depletion region* of the *p* and *n* material where the free carriers have been removed. The light is absorbed in this depletion region and delivers its energy to the material. If the absorbed energy is sufficient, a hole-electron pair can be created. The charges will be separated by the electric field existing across the depletion region, and swept to

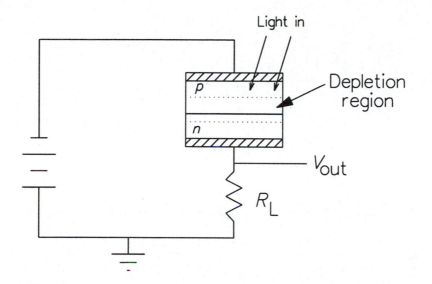

Figure 6.2 Schematic of a *pn* photodiode.

opposite sides of the depletion region. The motion of this pair of charge carriers is sensed by the outside circuitry and the net effect is the motion of one charge quantum (1.6×10^{-19} coulombs) through the external load. The number of hole-electron pairs per second so freed is linearly dependent on the power of the optical field and, hence, the electric current is proportional to the optical power.

Spectral Response

The amount of energy required to free a hole-electron pair is the *bandgap energy* of the material. Table 6.1 on the next page gives the bandgap energy for a variety of detector materials. Since the energy of a *photon* of light is $h\nu$ (where h is Planck's constant [6.63×10^{-34} joule·sec] and ν is the optical frequency), the energy condition requires

$$h\nu = \frac{hc}{\lambda} \geq E_g. \tag{6.1}$$

From this condition we observe that any given material will exhibit a *long-wavelength cutoff*, since the photons will not have sufficient energy to free a hole-electron pair whenever the wavelength exceeds λ_{max}, where

$$\lambda_{\mathrm{max}} = \frac{hc}{E_g}. \tag{6.2}$$

From this relation we expect all materials to show an abrupt cutoff in performance for wavelengths exceeding λ_{max}. Since silicon has λ_{max} of 1.13 micrometers ($E_g = 1.1$ eV), it is a suitable detector material for the short-wavelength sources previously discussed, but not for the long-wavelength sources. Although other materials are also suitable (e.g., Ge, InGaAs, GaSb, GaAlSb, and others), silicon is used primarily in the short-wavelength region because of its superior properties. (Detectors made of GaAs are also used in this region since the higher carrier velocity of this

Material	Bandgap energy (eV) at 300K
GaAs	1.43
GaSb	0.73
GaAs$_{0.88}$Sb$_{0.12}$	1.15
Ge	0.67
InAs	0.35
InP	1.35
In$_{0.53}$Ga$_{0.47}$As	0.75
In$_{0.14}$Ga$_{0.86}$As	1.15
Si	1.14

Table 6.1 Bandgap energies (in eV) at 300K for representative photodiode materials.

material can produce higher speed devices and the detector can be integrated with high-speed GaAs electronic devices to make an integrated electronic receiver.) In the long-wavelength portion of the spectrum, silicon is ineligible, so detectors are primarily made of InGaAs.

The penetration of the light down into the depletion region of the device is governed by the absorption properties of the material. The *absorption coefficient,* $\alpha(\lambda)$, for a few materials is shown in Fig. 6.3 on the facing page. Note the effect of the cutoff wavelength, causing a negligible absorption coefficient at long wavelengths. At short wavelengths, the value of the absorption coefficient increases dramatically. The amount of incident power that is absorbed in the depletion region of depth w, assuming that the layer begins at a depth d below the device surface, is given by

$$P(w) = P_i e^{-\alpha d} \left(1 - e^{-\alpha w}\right) \left(1 - R_f\right) , \tag{6.3}$$

where $P(w)$ is the power absorbed in the depletion region, P_i is the incident power, α is the absorption coefficient at the wavelength of interest, w is the depletion layer depth, and R_f is the power reflectivity of the detector's surface. The high value of α at short wavelengths causes little power to penetrate into the depletion region. At short wavelengths, therefore, we find a diminished response from the diode due to this increased surface absorption.

———————————————— ∞ ————————————————

Example: Consider a layer 1 μm thick located 3 μm below a planar air-silicon surface. Calculate the fraction of the incident power absorbed in the layer ...

(a) ...if $n_{\text{silicon}} = 3.5$ and $\alpha = 10^3$ cm^{-1} ($= 10^5$ m^{-1}) at the wavelength of interest.

Solution: We begin by finding the reflection coefficient

$$R_f = \left(\frac{n-1}{n+1}\right)^2 = \left(\frac{3.5-1}{3.5+1}\right)^2 = 0.309 . \tag{6.4}$$

We then find the fraction of the incident power that is absorbed as

$$\begin{aligned}
\frac{P(w)}{P_i} &= e^{-\alpha d} \left(1 - e^{-\alpha w}\right) \left(1 - R_f\right) \tag{6.5} \\
&= \left(e^{-10^5(3\times10^{-6})}\right) \left(1 - e^{-10^5(10^{-6})}\right) (1 - 0.309) = 4.87\% .
\end{aligned}$$

(b) ...if $\alpha = 10^4$ cm^{-1} (10^6 m^{-1}) ?

<u>Solution:</u> For the second case we have

$$\frac{P(w)}{P_i} = e^{-10^6(1\times10^{-6})}\left(1 - e^{-10^6(10^{-6})}\right)(1 - 0.309) = 2.18\%.\tag{6.6}$$

——————————— ∞ ———————————

In an effort to increase the effective depth, w, of the depletion region, a layer of lightly doped material (i.e., *intrinsic material*) can be formed between the p and n layers of the diode. This layer adds its depth to the depletion layer to increase the *active region* of the diode considerably. This increased depth also has the secondary benefit of reducing the capacitance of the diode. With the addition of this layer, the diode becomes a *pin diode*, as shown in Fig. 6.4 on the next page. The increased efficiency of these devices has made them more popular as optical receivers than are *pn* photodiodes.

Sensitivity

Optical detectors convert the input optical power into an output current from the detector. The sensitivity of a diode detector is measured by the *responsivity*, $\mathcal{R}$, of the device given by

$$\mathcal{R} = \frac{I_{\text{out}}}{P_{\text{in}}},\tag{6.7}$$

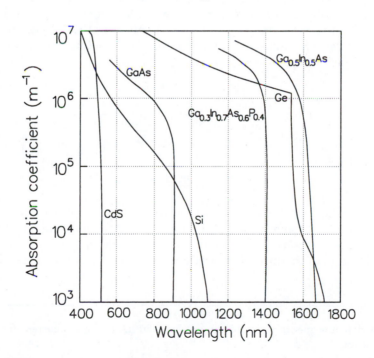

Figure 6.3 Absorption coefficient vs. wavelength for representative materials. (After T.P. Lee and T. Li, "Photodetectors," in *Optical Fiber Telecommunications*, S.E. Miller and A.G. Chynoweth, eds., pp. 593–626, (Academic Press, New York, 1979).)

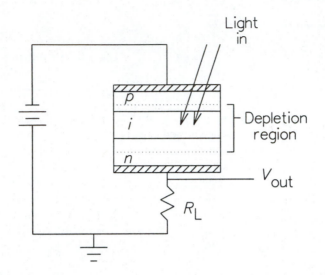

Figure 6.4 Structure of a pin photodiode.

where I_{out} is the current out of the detector and P_{in} is the incident optical power. Another parameter used to describe the detector's sensitivity is the *quantum efficiency* η which is the number of hole-electron pairs generated per photon. A value less than 1 indicates that not every photon is generating carriers. Since the number of incident photons per second is $P_i\lambda/hc$ and the number of carriers flowing per second is given by I/q, we have

$$\eta = \frac{\text{carrier pairs} \cdot \text{s}^{-1}}{\text{photons} \cdot \text{s}^{-1}} = \frac{\dfrac{I_{out}}{q}}{\dfrac{P_{in}}{h\nu}} = \frac{I_{out}hc}{qP_{in}\lambda}. \tag{6.8}$$

The responsivity is then

$$\mathcal{R} = \frac{\eta q\lambda}{hc}. \tag{6.9}$$

Figure 6.5 on the facing page shows the responsivity of three typical devices plotted against wavelength. Equation 6.8 is shown as dotted lines for constant quantum efficiency. The long-wavelength falloff in the responsivity is due to the energy deficiency of the photons; the short-wavelength falloff is due to the increased surface absorption effects that were discussed earlier.

——————————— ∞ ———————————

Example: A detector operating at 850 nm produces 80 μA of output current for a 500 μW input light beam.

(a) Calculate the responsivity of the detector.

Solution:

$$\mathcal{R} = \frac{I}{P_i} = \frac{80 \times 10^{-6}}{500 \times 10^{-6}} = 0.16 \text{ A/W} = 160 \text{ mA/W}. \tag{6.10}$$

(b) Calculate the quantum efficiency.

Solution:

$$\eta = \frac{Ihc}{qP_i\lambda} = \frac{hc\mathcal{R}}{q\lambda} = \frac{(6.63 \times 10^{-34})(3.0 \times 10^8)(0.16)}{(1.6 \times 10^{-19})(850 \times 10^{-9})} = 0.234 = 23.4\%. \qquad (6.11)$$

∞

Sensitivity—An Alternative Approach

Since the noise of an optical device plays a major role in the design of an optical receiver, it is instructive to understand in more detail the mechanisms and description of the generation of this noise. We begin with a description of a fundamental noise source, the quantization of light and charge. In this description we assume that all other noise sources are zero (i.e., we eliminate all other sources from consideration by assumption). We will also assume that we have a perfect decision process. If we free even a single charge-carrier pair with our light pulse (representing a logical **1**), we assume that we will detect the charge and decode that a **1** was transmitted. Conversely, a logical **0**, represented by the absence of any light (or charge-carrier pairs), will be perfectly identified as a **0**.

The possible noise is due to the quantized nature of the light or, equivalently, the discrete generation of charge carriers. This discrete generation of carriers means that, with a weak incident signal, the output current will not be an exact replica of the ideally predicted current, but will deviate from it by a (hopefully) small amount. This deviation is one type of noise in the output current. The interesting feature of this noise (found in optical devices) is that it is dependent on

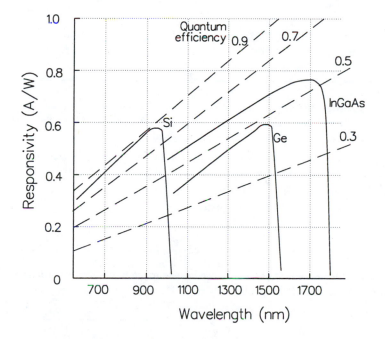

Figure 6.5 Responsivity of typical devices.

both the signal level (i.e., *signal dependent*) and the environment (e.g., temperature and device dependent).

The random generation of the charge-carrier pairs by the photons has been modeled [4] by the classical *Poisson random process* with a time-varying rate. In this model the probability $\rho(t)\,dt$ that a charge-carrier pair is generated in a time interval dt, between t and $t + dt$, is given by

$$\rho(t)\,dt = \frac{\mathcal{R}p(t)}{q}\,dt = \frac{\eta\lambda}{hc}p(t)\,dt\,, \qquad (6.12)$$

where η is the quantum efficiency. (Implicit in the use of this model are the assumptions that the generation of each pair of carriers is independent of all other carrier generations and that only one pair of carriers is generated in the interval dt.)

From these Poisson process assumptions, some conclusions follow. The total number of carriers generated in a finite interval of time from t to $t + T$ is a random variable. The average number of carriers $\overline{N}$ generated in this time interval is given by

$$\overline{N} = \frac{\eta\lambda}{hc}\int_t^{t+T} p(t)\,dt = \frac{\eta\lambda}{hc}E\,, \qquad (6.13)$$

where E is the total energy of the pulse in the interval between t and $t+T$ (given by the integral in Eq. 6.13). The probability that the number of charges created, N, will equal a specific number, n, in this interval is predicted by [9]

$$P(N = n) = \frac{\overline{N}^n e^{-\overline{N}}}{n!} = \frac{\overline{N}^n e^{-\frac{\eta\lambda E}{hc}}}{n!}\,, \qquad (6.14)$$

where $P(N = n)$ is the probability that a total of n charge carrier pairs will be generated in an interval T seconds long.

──────────────── ∞ ────────────────

Example: To illustrate the application of this probability, suppose that we wish to know the amount of energy required in the pulse to have a probability of 10^{-9} or smaller that a logical **0** will be detected when we have transmitted a logical **1**.

Solution: Assuming that a logical **0** is the creation of zero charge carrier pairs, we want to calculate the energy required to make $P(n = 0) < 10^{-9}$. From Eq. 6.14, we want to solve the inequality,

$$P(n = 0) = \exp\left(-\frac{\eta\lambda E}{hc}\right) < 10^{-9}\,. \qquad (6.15)$$

Solving for E gives

$$E > 21\frac{hc}{\eta\lambda}\,. \qquad (6.16)$$

From this calculation we see that the required pulse must have $21/\eta$ photons to avoid being mistaken for a logical **0** (with a probability of 10^{-9}).

──────────────── ∞ ────────────────

———————————————— ∞ ————————————————

Example: Calculate the number of photons N required for a detector with a quantum efficiency of 50% to have an error probability of 10^{-12}.

Solution: The number of photons is found from

$$P(n = 0) \quad = \quad \exp\left(-\frac{\eta\lambda E}{hc}\right) < 10^{-12} \tag{6.17}$$

$$-\frac{\eta\lambda E}{hc} \quad = \quad -27.6$$

$$E \quad \geq \quad \frac{hc}{\eta\lambda}(27.6) = \left(\frac{27.6}{0.5}\right)\left(\frac{hc}{\lambda}\right) = 55.3\frac{hc}{\lambda}$$

We recognize that hc/λ is the energy of a photon and that dividing E by the energy per photon will give the number of photons. Also we know that the number of photons must be an integer, so

$$N \quad = \quad \frac{E}{\frac{hc}{\lambda}} \geq 56 \text{ photons} \tag{6.18}$$

———————————————— ∞ ————————————————

Consider our case of the previous example where $21/\eta$ photons are required for a BER of 10^{-9}. If the number of logical **1**s and **0**s in the data stream is equal in an average sense and each bit period is T_B seconds long, then the average power P_{av} in the detected light must be

$$P_{\text{av}} = \frac{21(hc/\eta\lambda)}{2T_B} \tag{6.19}$$

to achieve the probability of error of 10^{-9}. Since this power level was derived assuming only shot noise, which is due to the quantum nature of light or charge, the resulting calculation is the *quantum limit* of the detection process (for the particular value of error probability under consideration).

To generalize our results to any desired value of probability, we will represent the probability as the *bit-error rate* or *BER*. We can then rewrite Eq. 6.15 on the facing page to be

$$\exp\left(-\frac{\eta\lambda E}{hc}\right) \leq \text{BER}. \tag{6.20}$$

Solving for the energy E gives

$$E \geq \frac{hc}{\eta\lambda}\ln\left(\frac{1}{\text{BER}}\right). \tag{6.21}$$

The minimum average power will be

$$P_{\text{av}} = \frac{E_{min}}{2T_B}. \tag{6.22}$$

———————————————— ∞ ————————————————

Example: Calculate the minimum power required to achieve a BER of 10^{-12} for the detector of the previous example with a 50% quantum efficiency at a data rate of 10 Mb·s^{-1}. The wavelength is 1300 nm.

Solution: The bit period T_B is given by

$$T_B = \frac{1}{\text{DR}} = \frac{1}{10 \times 10^6} = 10^{-7} \text{ s}. \tag{6.23}$$

The minimum power $P_{\min}$ is

$$\begin{aligned} P_{\min} &= \frac{E_{\min}}{2T_B} = \frac{1}{2T_B} \frac{hc}{\eta \lambda} \ln\left(\frac{1}{\text{BER}}\right) \\ &= \frac{(6.63 \times 10^{-34})(3.0 \times 10^8)}{2(10^{-7})(0.5)(1300 \times 10^{-9})} \ln(10^{12}) \\ &= 4.43 \times 10^{-11} \text{ W} = 4.43 \text{ nW}. \end{aligned} \tag{6.24}$$

∞

6.2.2   The Avalanche Photodiode

Physical Principles

The *avalanche photodiode* [10] dopes the p and n material higher (denoted by the p^+ and the n^+ in Fig. 6.6) and incorporates a narrow region of p type material between the intrinsic region and the n^+ region. As shown in the figure, the electric field in this region is larger than in the remainder of the depletion region due to the fact that most of the applied reverse bias is dropped across this region. (This particular structure is called a *reach-through avalanche photodiode* [RAPD].) In this high-field region, the field accelerates the charge carriers to velocities high enough to create more hole-electron pairs through collisions (*impact ionization*). (Field strengths in excess of 3×10^5 V/cm are required.) In the structure shown, the light enters through the p^+ region and is (ideally) absorbed in the i region. The generated carriers separate and drift across the i region. When the electrons enter the p region, they are accelerated and impact other atoms, creating more carriers. These carriers are then accelerated and, in turn, create more carriers, leading to the *avalanche effect*.

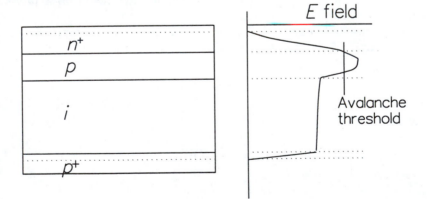

Figure 6.6 Structure of an avalanche photodiode.

The efficiency of a hole or electron creating a new hole-electron pair is expressed by the *hole ionization rate* or the *electron ionization rate*. To minimize noise generation, it has been found desirable to have as large a difference as possible in these ionization rates (i.e., it is desired to have only one type of carrier responsible for most of the avalanche process). Silicon exhibits this property with an electron ionization rate that is one hundred times larger than the hole ionization rate. Other materials such as Ge, GaAs and InGaAs have closer ratios ($\sim$ 5–10) thereby ensuring noisy operation in avalanche devices.

The net effect of the avalanche process is a multiplication of the current at the output terminals. The *multiplication factor M* is given by

$$M = \frac{I_M}{I} , \tag{6.25}$$

where I_M is the output current with multiplication and I is the output current without multiplication. Since the amount of multiplication is controlled by the reverse bias (as shown in Fig. 6.7), the latter current is measured under low bias conditions. It should be emphasized that the instantaneous amount of amplification or multiplication is a random value. The value M is the average amount of multiplication.

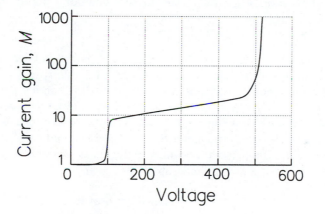

Figure 6.7 Multiplication vs. reverse bias voltage.

Responsivity of an APD

The sensitivity of an APD is also described by the *responsivity* of the detector. In a fashion similar to the pin diode, the responsivity of the avalanche photodiode is given by

$$\mathcal{R}_{\mathrm{APD}} \;=\; \frac{\eta q \lambda}{hc} M = \mathcal{R}_0 M , \tag{6.26}$$

where $\mathcal{R}_0$ is the responsivity at unity multiplication.

6.2.3   Detector Noise

In characterizing the noise performance of an optical receiver, we will use the *signal-to-noise ratio*. The signal is the signal power delivered to a hypothetical matched resistor by the signal current. The noise is characterized by the noise power delivered to the same hypothetical resistor. The signal-to-noise ratio can then be written as

$$\frac{S}{N} = \frac{P_{\text{signal}}}{P_{\text{noise}}} = \frac{<i_s^2> R/4}{<i_N^2> R/4} = \frac{<i_s^2>}{<i_N^2>}. \tag{6.27}$$

We note that the arbitrary resistor cancels out from the numerator and the denominator, so the signal-to-noise ratio is independent of this resistor, and we need to calculate only the mean-square currents.

There are two noise mechanisms associated with photodiodes, shot noise and thermal noise.

Shot Noise

The *shot noise* can be associated with either the quantization of charge into multiples of q or, equivalently, with the quantization of light energy into photons. The arrival of photons or, equivalently, the generation of charge carriers is characterized by Poisson statistics. The mean-square noise current $<i_N^2>$ associated with this noise source for a photodiode is given by [11]

$$<i_N^2> = 2qIB, \tag{6.28}$$

where I is the average output current of the device and B is the bandwidth of the electronics accepting the noise (e.g., a preamplifier, a noise meter). The interpretation of the equation is that there is this much noise power (delivered to a 1 ohm load) in the frequency region extending from $f_0 - (B/2)$ to $f_0 + (B/2)$, where f_0 is the center frequency of the passband of the output device. Note that the ac noise power centered within the band of frequencies is dependent on the dc value of output current. Note also that, since $<i_N^2>$ is independent of the central frequency, the noise is *white noise* (i.e., the noise has a uniform frequency distribution).

The dc current out of a pin diode is made up of three components:

$$I = I_L + I_{\text{background}} + I_{\text{dark}}, \tag{6.29}$$

where I_L is the load current out due to the incident light, $I_{\text{background}}$ is the current due to background illumination sources (assumed to be zero in fiber links), and I_{dark} is the *dark current* of the device. The dark current is the output of the device with no input illumination and is primarily due to thermal generation of charge carriers in the depletion region and surface leakage currents due to surface defects near the edges of the semiconductor.

Typically, the dark-current density is $10^{-6} - 10^{-7}$ A/cm^2 in silicon devices, is $10^{-4} - 10^{-6}$ A/cm^2 in InGaAs, and is 10^{-3} A/cm^2 in Ge. The high dark current of the long-wavelength detectors causes their noise characteristics to be inferior to the short-wavelength silicon-based devices.

--------------------------------∞--------------------------------

Example: A detector has a responsivity of 0.5 A/W at a wavelength of interest and a dark current of 1 nA. Calculate the mean-square noise current and the RMS noise current due to shot noise if the noise bandwidth is 50 MHz and the incident power is 100 μW.

Solution:

$$<i_N^2>$$
$$= 2qIB = 2q(I_L + I_{\text{dark}})B = 2q(\mathcal{R}P_i + I_{\text{dark}})B \tag{6.30}$$

$$
\begin{aligned}
&= 2(1.6 \times 10^{-19}) \left((0.5)(100 \times 10^{-6}) + 10^{-9} \right) \\
&\quad \times (50 \times 10^{6}) \\
&= 8 \times 10^{-16} \ \text{A}^2 \, .
\end{aligned}
$$

The RMS noise current is, by definition, the square root of the mean-square noise current.

$$
\sqrt{< i_N^2 >} = \sqrt{8 \times 10^{-16}} = 2.83 \times 10^{-8} \ \text{A} \, . \tag{6.31}
$$

Note: In this example the dark current turned out to be negligible compared to the signal current.

———————————————— ∞ ————————————————

Excess Shot Noise in APDs

In the avalanche photodiode, the avalanche process contributes noise above that previously described. In this device, the mean-square noise current is [11]

$$
< i_N^2 >= 2qIM^2BF(M) \, , \tag{6.32}
$$

where M is the average current gain and $F(M)$ is the *excess noise factor* that is the measure of the extra noise added by the avalanche process. The function $F(M)$ depends on the detector material, the shape of the E field in the device, and relative *ionization rates* of the carriers. (The ionization rate is a measure of the ability of the hole or the electron to generate other carrier pairs in the avalanche process.) If k is the ratio of the electron ionization rate to the hole ionization rate (or the inverse, since we require $k \leq 1$), the excess noise factor $F(M)$ can be given by [10]

$$
F(M) = kM + (1 - k)\left(1 + \frac{1}{M} \right) \, . \tag{6.33}
$$

Typical ranges of values of k are found in Table 6.2. The lowest excess noise occurs for the lowest value of k (i.e., we want the avalanche process to be predominately caused by either holes or electrons). An approximate form of $F(M)$ is

$$
F(M) \approx M^x \, , \tag{6.34}
$$

where typical ranges in x are shown in Table 6.2. This excess noise factor causes some unusual effects in the application of this device, as we shall see in considering the signal-to-noise ratio. In using Eq. 6.32, the current I is the dc signal current (*without gain*) plus the bulk dark effect current. Any surface dark currents are *not* amplified by the avalanche process and are negligible for reasonably large values of M.

Material	k	x
Silicon	0.02–0.04	0.3 – 0.5
Germanium	0.7–1.0	1.0
InGaAs	0.3–0.5	0.5 – 0.8

Table 6.2 Typical range of values of k and x in the expressions for the excess noise factor $F(M)$.

Thermal Noise

The second noise source of importance in photodiodes is *thermal noise*. Any resistive load or device with an associated resistance will produce a thermal-noise mean-square current given by

$$< i_N^2 > = \frac{4kTB}{R},$$ (6.35)

where k is Boltzmann's constant, T is the noise temperature of the device, B is the electronic bandwidth into which the noise is delivered, and R is the value of the resistor or input resistance. This noise is independent of the optical signal.

Having displayed the equations for the mean-square noise currents, we proceed to the consideration of the signal-to-noise ratio for the device operating into a simple load resistor.

6.2.4   Detector Signal-to-Noise Analysis

We now consider the signal-to-noise analysis of a detector loaded by a simple resistor of value R_L. We neglect, for now, the effects of any preamplifier in our effort to understand the ultimate limit placed on the system performance by the detector noise alone.

The general signal-to-noise ratio at the output of an arbitrary detector is given by

$$\frac{S}{N} = \frac{< i_s^2 > M^2}{2q(I_L + I_D)M^2 F(M)B + 2qI_{\text{surf}}B + 4kTB/R_L}.$$ (6.36)

The term in the numerator is the mean-square signal current out of the device. The gain M has a value of 1 if the device is a pin diode. The noise terms in the denominator are as follows:

- The first term in the denominator is the amplified shot noise due to the bulk currents-the signal-dependent current (I_L) (*before amplification*) and the dark current (I_D). The excess noise is accounted for by the $F(M)$ multiplier.

- The second term of the denominator is the shot noise due to the surface leakage current I_{surf}, if any.

- The third term of the denominator is the thermal noise due to the resistance of the load resistor R_L seen in Fig. 6.2 on page 163. (It is assumed that the load resistor is considerably smaller than the effective resistance of the reverse-biased diode and hence is the dominant thermal noise source.)

--------------------------------- ∞ ---------------------------------

Example: Consider an avalanche photodiode operating with a gain of 50 and an unamplified bulk dark current of 10 nA. The surface dark current is 1 nA. The unamplified responsivity of the device is 0.6 A/W. The excess noise factor for the device is $M^{0.4}$ and it is operating into a 50 Ω load at a noise temperature of 300K. The noise bandwidth is 10 MHz. Calculate the signal-to-noise ratio (in dB) of this detector when irradiated with 5 nW of light.

Solution: The signal current with a gain M of 1 is

$$i_s|_{M=1} = \mathcal{R}_0 P_i = (0.6)(5 \times 10^{-9}) = 3 \times 10^{-9} \text{ A} = I_L.$$ (6.37)

The excess noise factor $F(M)$ is estimated as

$$F(M) = M^{0.4} = (50)^{0.4} = 4.78.$$ (6.38)

The signal-to-noise ratio is expressed by

$$\frac{S}{N} = \frac{<i_s^2> M^2}{2q(I_L + I_D)M^2F(M)B + 2qI_{surf}B + 4kTB/R_L} . \tag{6.39}$$

We find the first part of the shot noise as

$$
\begin{aligned}
N_{shot} &= 2q(I_L + I_D)M^2F(M)B \\
&= 2(1.6 \times 10^{-19})(3 \times 10^{-9} + 10 \times 10^{-9})(50)^2(4.78) \\
&\quad \times (4.78)\left(10 \times 10^6\right) = 4.97 \times 10^{-16} \text{ A}^2 .
\end{aligned}
\tag{6.40}
$$

The shot noise due to the surface leakage current is

$$
\begin{aligned}
N_{shot\ surface} &= 2qI_{surf}B = 2(1.6 \times 10^{-19})(1 \times 10^{-9})(10 \times 10^6) \\
&= 2.6 \times 10^{-21} \text{ A}^2 .
\end{aligned}
\tag{6.41}
$$

The thermal noise is

$$
\begin{aligned}
N_{thermal} &= \frac{4kTB}{R_L} = \frac{4(1.38 \times 10^{-23})(300)(10 \times 10^6)}{50} \\
&= 3.312 \times 10^{-15} \text{ A}^2 .
\end{aligned}
\tag{6.42}
$$

Summing all of the noise source powers, we have

$$
\begin{aligned}
\sum Noise &= N_{shot} + N_{shot\ surface} + N_{thermal} \\
&= 3.81 \times 10^{-15} \text{ A}^2 .
\end{aligned}
\tag{6.43}
$$

Having found the total noise, we can now find the signal-to-noise ratio as

$$
\begin{aligned}
\frac{S}{N} &= \frac{(3 \times 10^{-9})^2(50)^2}{(3.81 \times 10^{-15})} \\
&= 5.91 \Rightarrow 7.71 \text{ dB} .
\end{aligned}
\tag{6.44}
$$

——————————— ∞ ———————————

When using the pin photodiode, $M = F(M) = 1$, and the dominant noise sources are usually the thermal noise source. The output current of the device for a constant input (such as a pulse, when present) is

$$i_s = I_L = \frac{\eta q \lambda P}{hc} = \mathcal{R}P , \tag{6.45}$$

where P is the incident optical power.

When using an ADP detector, the amplification factor will make the shot noise the dominant noise source. We note that while both the signal and noise terms increase as M^2, the excess noise factor will ensure that the noise will grow at a faster rate, $M^2F(M)$, than the signal. A plot of the signal-to-noise ratio vs. M for this device will reveal a maximum. The optimum value of M, M_{opt}, that gives the maximum signal-to-noise ratio is found by taking the derivative of Eq. 6.36 on the preceding page with respect to M and finding the value of M that will make this derivative equal zero. This evaluation gives

$$M_{opt}^{x+2} = \frac{2qI_{surf} + (4kT/R_L)}{xq(I_L + I_D)} \tag{6.46}$$

or

$$M_{\text{opt}} = \left(\frac{2qI_{\text{surf}} + (4kT/R_L)}{xq(I_L + I_D)} \right)^{\frac{1}{x+2}} .$$

(6.47)

Hence, operation of the APD in a low signal-to-noise environment requires adjustment of the gain to obtain optimum performance. The value of the optimum gain depends on the thermal noise, the signal level, the device dark current, and, through x, the ionization rates of the carriers. For silicon APDs, the optimum gain typically is in the range of 80 to 100, with improvements of 40× to 50× (16 to 17 dB) in the signal-to-noise ratio over the unamplified signal. Excessive noise in long-wavelength APDs has restricted their use. These materials suffer from nearly equal ionization rates for holes and electrons (i.e., $k \approx 1$ in Eq. 6.33 on page 173), compounded by relatively large dark currents.

───────────────────── ∞ ─────────────────────

Example: For the APD detector of the example on page 174, calculate the optimum gain that should be used.

Solution: The optimum gain is found from

$$
\begin{aligned}
M_{\text{opt}}^{x+2} &= \frac{2qI_{\text{surf}} + 4kT/R_L}{xq(I_L + I_D)} \\[2mm]
M_{\text{opt}}^{2.4} &= \frac{2(1.6 \times 10^{-19})(1 \times 10^{-9})}{0.4(1.6 \times 10^{-19})(3 \times 10^{-9} + 10 \times 10^{-9})} \\[2mm]
&\quad + \frac{(4)(1.38 \times 10^{-23})(300/50)}{0.4(1.6 \times 10^{-19})(3 \times 10^{-9} + 10 \times 10^{-9})} \\[2mm]
&= 3.98 \times 10^5 , \\[2mm]
M_{\text{opt}} &= (3.98 \times 10^5)^{1/2.4} = 215.4 .
\end{aligned}
$$

(6.48)

───────────────────── ∞ ─────────────────────

6.2.5 Linearity

The linearity of the output current vs. optical input power of optical photodiodes is excellent with typical devices exhibiting over 6 decades of linear operation. Linearity of APDs is not quite as good. Fortunately, the requirements for maximum linearity exist for analog signals with a high signal-to-noise output—a regime of operation best met by pin diodes, not APDs.

6.2.6 Speed of Response

The speed of response of a photodiode is limited by three factors:

1. The first factor is the *transit time* of the carriers as they drift across the depletion region. Usually the devices are designed and biased so that the carriers reach their scattering-limited velocity in the material. For silicon, the scattering-limited velocities of holes and electrons are 8.4×10^4 and 4.4×10^4 m·s^{-1}, respectively. For a depletion width of 10 μm, the response time caused by this effect is ≈ 0.1 ns.

2. The second factor is the *diffusion time* which is the time that it takes carriers created in the p or n material (close to the depletion region boundary) to move by diffusion into the depletion region, where the drift process takes over. A relatively small fraction of the carriers are involved in this process, but, because the diffusion process is quite slow, it can be a limiting effect on the device speed. This is especially true at low values of reverse bias, where the drift electric field is comparatively low. One way to minimize this effect (by ensuring that most of the carriers are generated in the depletion region) is to make the total depletion region much larger than $1/\alpha$ (where α is the material absorption coefficient of the material at the wavelength of interest). This guarantees that most of the light will be absorbed in this region of the device. (The increase in the size of the depletion region will increase the transit time, however.)

3. The third factor is the RC time constant of the device and any associated circuitry. Typically, R is the input resistance of the preamplifier in parallel with the load resistance, and C is the device capacitance. Most of the device capacitance is the junction capacitance associated with the reverse-biased diode. This capacitance is given by

$$C = \frac{\epsilon_s A}{w}, \tag{6.49}$$

where ϵ_s is the (total) permittivity of the semiconductor material, A is the cross-sectional area of the detecting portion of the device, and w is the depth of the depletion region. As seen from this equation, efforts to increase the efficiency of the device by increasing w also reduce the capacitance of the device. This increase in w also increases the transit time of the device, however, causing a trade-off in device design. (A usual compromise is to make $w \approx 2/\alpha$.) Detectors typically have capacitance values of <1 pF. The bandwidth limitation due to this RC time constant effect is given by

$$f_{\max} = \frac{1}{2\pi RC}. \tag{6.50}$$

To summarize, the primary mechanism limiting speed of response in a well-designed pin diode, used in a low-resistance circuit, is the transit time across the depletion region. Silicon devices have response times on the order of 1 ns.

In APDs, the response is typically slower because the carriers must drift into the avalanche region and then the created carriers (of the opposite type) must drift back across the depletion region, approximately doubling the total transit time approximately. Additionally, there is a constant gain-bandwidth product constraint for an avalanche device, which results from giving the avalanche process time to occur. Typical values of this gain-bandwidth product can range up to 200 GHz.

6.2.7   Reliability

Reliability has not proved to be a major problem in photodiode operation. Based on accelerated temperature lifetime testing, the projected lifetime is approximately 10^8 hours—ample lifetime for an optical detector.

6.2.8   Temperature Sensitivity

The gain of the avalanche photodiode is quite temperature sensitive, as shown in Fig. 6.8. Because the gain is also voltage dependent, it is fairly easy to incorporate a temperature-compensating feedback circuit that senses the change in gain due to a temperature change and adjusts the reverse bias in a direction to cancel the change in gain. Such circuitry is easily implemented through the use of a controllable supply voltage. Such a supply is also useful as an automatic gain-control circuit, frequently used when the avalanche diode is exposed to optical power levels with a wide dynamic range. (This gain control is necessary because the output voltage level of a high-gain APD is easily saturated with a large optical-input signal.)

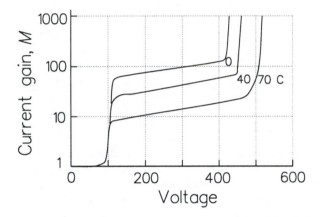

Figure 6.8 Temperature sensitivity of APD gain.

6.2.9   Analog Detector Analysis

Having considered how a simple detector (i.e., without any amplifier) responds to a digital signal, we now consider its performance for analog signals. As described in Chapter 5, an analog transmitter using intensity modulation operates with a small change in power about a dc power level. The received signal is

$$p(t) = P\left(1 + m(t)\right),\tag{6.51}$$

where $p(t)$ is the time-varying power, P is the received dc optical power with no modulation present, and $m(t)$ is the time-varying information, normalized such that $|m(t)| < 1$. The output current from a linear receiver is given by

$$i(t) = M\mathcal{R}_0 P\left(1 + m(t)\right) = I_L\left(1 + m(t)\right),\tag{6.52}$$

where $I_L = M\mathcal{R}_0 P$. The mean-square signal current is the mean-square value of the ac portion of the current and is given by

$$< i_s^2 > = I_L^2 < m^2(t) >,\tag{6.53}$$

where we have assumed that the time-varying portion of the output current is the only part of interest (or the only part of the output current that is passed by the ac electronics after the receiver). For the purpose of computation, a particular test signal is assumed for the time-varying signal. This test signal is assumed to be a cosine modulation at a frequency ω. The equation for this test signal is

$$p(t) = P\left(1 + m\cos(\omega t)\right), \tag{6.54}$$

where m is now a constant with a value between 0 and 1. The current from the detector into the load is

$$i_L(t) = M\mathcal{R}_0 P\left(1 + m\cos(\omega t)\right). \tag{6.55}$$

The mean-square signal current is given by

$$<i_s^2> = \frac{(mM\mathcal{R}_0 P)^2}{2} \tag{6.56}$$

since $<\cos^2(\omega t)> = 1/2$. (Note that we have included the gain term M^2 explicitly in this expression.) The detector noise is represented by

$$<i_N^2> = 2q(I_L + I_D)M^2 F(M)B + 2qI_s B + 4kTB/R_L \tag{6.57}$$

from the denominator of Eq. 6.36 on page 174. Assuming negligible surface leakage current ($I_s = 0$), the signal-to-noise ratio becomes

$$\begin{aligned}
\frac{S}{N} &= \frac{(1/2)(\mathcal{R}_0 mMP)^2}{2q(\mathcal{R}_0 P + I_D)M^2 F(M)B + 4kTB/R_L} \\
&= \frac{(1/2)(I_L Mm)^2}{2q(I_L + I_D)M^2 F(M)B + 4kTB/R_L}.
\end{aligned} \tag{6.58}$$

To simplify this expression, we now want to make various assumptions about which noise is dominant.

Photodiode Analog Response—Thermal Noise Dominant

For the pin diode, $M = 1$. For a small optical signal and small dark current in the detector, the thermal-noise term is the dominant noise source, so the signal-to-noise ratio becomes

$$\frac{S}{N} \approx \frac{(1/2)m^2 I_L^2}{4kTB/R_{eq}} = \frac{(1/2)m^2 \mathcal{R}_0^2 P^2}{4kTB/R_{eq}}. \tag{6.59}$$

Hence, for this case, we want to maximize the modulation index, to maximize the responsivity, to increase the signal power, to reduce the temperature (if possible), and to make the equivalent resistance as large as possible.

———————————— ∞ ————————————

Example: Consider an analog amplitude modulation system transmitting a 15 kHz tone with an 80% modulation index. A photodiode detector with a responsivity of 0.4 A/W operates into a 50 Ω load with a noise temperature of 400K. The bulk dark current is 1 nA; the surface dark current is negligible. The system bandwidth is 50 MHz.

(a) For an incident power of 1 μW, calculate the signal-to-noise ratio assuming that the signal and dark currents are so small that the thermal noise will be dominant.

Solution: For the case where the thermal noise is dominant, we have

$$\frac{S}{N} \approx \frac{(1/2)m^2\mathcal{R}_0^2 P^2}{4kTB/R_{eq}} = \frac{m^2\mathcal{R}_0^2 P^2 R_{eq}}{8kTB} \tag{6.60}$$

$$= \frac{\left[(0.8)(0.4)(1 \times 10^{-6})\right]^2 (50)}{(8)(1.38 \times 10^{-23})(400)(50 \times 10^6)}$$

$$= 2.32 \Rightarrow 3.65 \text{ dB}.$$

(b) Calculate the ratios of $< i_N^2 >_{\text{thermal}}$ to $< i_N^2 >_{\text{signal shot}}$ and $< i_N^2 >_{\text{dark shot}}$ to be certain that the thermal noise is dominant.

Solution: We find the ratio of the thermal noise to the signal-dependent shot noise to be

$$\frac{< i_N^2 >_{\text{thermal}}}{< i_N^2 >_{\text{signal shot}}}$$

$$= \frac{4kTB/R_L}{2q\mathcal{R}_0 P_i B} \tag{6.61}$$

$$= \frac{4(1.38 \times 10^{-23})(400)(50 \times 10^6)/50}{2(1.6 \times 10^{-19})(0.8)(1 \times 10^{-6})(50 \times 10^6)}$$

$$= \frac{4.42 \times 10^{-22}}{2.56 \times 10^{-31}} = 1.72 \times 10^3.$$

The ratio of the thermal noise to the shot noise due to the signal is

$$\frac{< i_N^2 >_{\text{thermal}}}{< i_N^2 >_{\text{dark shot}}}$$

$$= \frac{4kTB/R}{2qI_D B} \tag{6.62}$$

$$= \frac{4(1.38 \times 10^{-23})(400)(50 \times 10^6)/50}{2(1.6 \times 10^{-19})(1 \times 10^{-9})(50 \times 10^6)}$$

$$= \frac{2.21 \times 10^{-14}}{1.6 \times 10^{-20}} = 1.38 \times 10^6.$$

Hence, we have proven that the thermal noise, indeed, is dominant.

———————————————— ∞ ————————————————

Photodiode Analog Response—High-Illumination Case

For a large optical signal incident on the photodiode, the shot-noise term that is dependent on the average output current will be the dominant term. For this case

$$\frac{S}{N} \approx \frac{m^2 I_L}{4qB} = \frac{m^2 \mathcal{R}_0 P}{4qB}. \tag{6.63}$$

Since this latter case assumes that all thermal noise is negligible, that all device-related noise is also negligible, and that only the signal-dependent shot noise is left, this signal-to-noise is the theoretical limit, called the *quantum-limited signal-to-noise ratio*. This name results from the dominant noise term being caused by the quantization of electronic charge into integer multiples of q or, equivalently, the light being quantized into photons.

APD Analog Response

For the operation of an APD, one again finds the existence of an optimum value of M that will maximize the signal-to-noise ratio (Eq. 6.47 on page 176). This equation is valid for low levels of optical signal. For large values, the shot-noise term is again dominant. Since the shot noise increases faster than the signal (due to the excess noise factor), the S/N ratio of an avalanche photodiode receiver is *lower* than the S/N ratio of a pin diode receiver, and the APD should not be used in this regime of large-signal operation.

Summary for Analog Response

For low optical-power levels, an APD offers significant advantages. For small values of M, the signal can be amplified more than the noise, and the overall signal-to-noise level can be raised by increasing M (up to $M = M_{opt}$). Further increases in M have the effect of gradually decreasing the signal-to-noise ratio.

For higher optical-power levels, the advantage of APDs diminishes and, eventually, disappears (i.e., the value of M_{opt} eventually approaches a value of one for high incident power). The value of receiver power where the signal-to-noise ratio begins to decrease varies with the device and data rate and should be determined analytically or experimentally before committing to the more expensive APD.

6.2.10   Noise Equivalent Power

Another figure of merit used to describe the noise performance of a photodetector is the *noise equivalent power.*   For a device with negligible dark current, the signal-to-noise ratio at the output of the pin photodiode with a constant signal of P watts incident is

$$\frac{S}{N} = \frac{(\eta q \lambda P / hc)^2}{(2\eta q^2 \lambda P / hc)B + \dfrac{4kTB}{R}} . \tag{6.64}$$

For a given value of thermal noise, there will be a value of P (the optical power) that will make the signal-to-noise ratio equal to 1. The power level that makes the $S/N = 1$ is called the noise equivalent power (NEP) of the receiver. Usually, in a pin detector designed for fast response times, the thermal noise is the dominant noise source, producing an NEP of

$$\text{NEP} = \frac{hc\sqrt{< i_N^2 >_{\text{thermal}}}}{\eta q \lambda} . \tag{6.65}$$

Note that a value of receiver bandwidth, resistance values, and operating temperature should be specified with the NEP to allow for a meaningful comparison between detectors.

The signal-to-noise ratio for an APD can be written as

$$\frac{S}{N} = \frac{(\eta q \lambda / hc)^2 M^2 P^2}{(2\eta q^2 \lambda / hc) M^2 F(M) P B + \dfrac{4kTB}{R}} , \tag{6.66}$$

where we see that both the signal and the shot-noise terms have the M^2 amplification and that the shot-noise term is multiplied by the excess-noise factor term $F(M)$. For the simplified expression of $F(M)$ given in Eq. 6.34 on page 173, Eq. 6.47 on page 176 estimates the optimum gain.

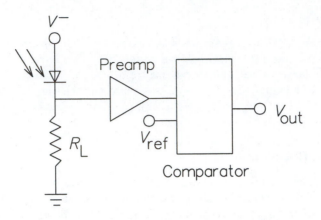

Figure 6.9 Block diagram of digital receiver.

6.3   Receiver Sensitivity and Bit-Error Rate

A digital receiver (Fig. 6.9) must take a weak optical signal and convert it into an electrical signal, decide whether the electrical output represents a logical **1** or **0** (using a comparator), and generate an electronically compatible voltage representative of the logic state. Since the output of the detector-preamplifier combination is contaminated by noise [8, 11, 12], the next step in our analysis is consideration of the value of the threshold voltage of the comparator (i.e., the *detection level*). This value will partially determine the *bit-error rate* (BER) of the receiver. By removing the amplification effects of the preamplifier, the threshold voltage can be converted to an equivalent threshold current at the diode output. We are concerned whether the instantaneous current (as corrupted by noise) is above or below this threshold.

Figure 6.10 on the next page illustrates the concept. The top curve is the noise-free output from a hypothetical detector. The value of threshold is shown on this curve as I_D. The noise-free signal in the top curve gives the correct logical values which represents perfect recovery of the transmitted signal. The middle curve shows a noise-corrupted received signal. For the threshold shown, the bottom curve is produced at the comparator output with one of the output bits in error. The number of errors in the recovered signal is a function of I_D, the threshold current.

While various assumptions can be made about the noise statistics of the output (with increasing accuracy bringing increasing computational complexity), we will use the simplest assumption to illustrate the design process. First is the assumption that the output currents of the detector are Gaussian random variables. The Gaussian probability distribution for the current is expressed by

$$p(i_N)\, di_N = \frac{1}{\sqrt{2\pi}\sigma} \exp\left(-\frac{i_N^2}{2\sigma^2}\right)\, di_N\,, \tag{6.67}$$

where $p(i_N)\, di_N$ is the probability of the output noise current being between the values i_N and $i_N + di_N$, and the standard deviation σ is a measure of the "width" of the probability distribution. As noted above, the mean-square noise current, predicted from consideration of the noise generation mechanism, is equal to the square of the standard deviation σ^2, i.e.,

$$\sigma^2 = <i_N^2>\,. \tag{6.68}$$

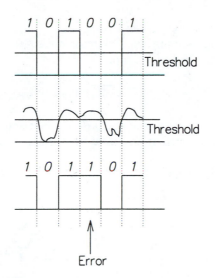

Figure 6.10 Recovery of signal. (Top.) Noise-free signal. (Middle.) Noise-corrupted signal. (Bottom.) Recovered signal with errors.

We will let the mean value of the receiver current with an incident logical **1** be represented by $\overline{i(1)}$ with a variance of σ_1. Similarly, we will denote the mean of the current when a **0** is received be denoted by $\overline{i(0)}$ with a variance of σ_0. (We will later assume $\overline{i(0)} = 0$ for simplicity, since this is usually true in a fiber link.)

If a logical **0** is sent, we note that an error will be made if the noise current is positive and of amplitude

$$i_N > I_D . \tag{6.69}$$

Similarly, if a given input is a logical **1**, we would have an incorrect output if i_N (the noise current contribution) were negative and met the condition

$$i_N < I_D . \tag{6.70}$$

The probability of making either one error or the other depends on the statistics of the data stream.

The total probability of making an error is the sum of the probability of calling a **1** a **0** (given that a **1** was sent), plus the probability of calling a **0** a **1** (given that a **0** was sent). Mathematically, this is written as

$$P_e = P(0|1)P(1) + P(1|0)P(0) , \tag{6.71}$$

where $P(0|1)$ is the probability of deciding that the output is a **0** when a **1** is sent, $P(1)$ is the probability that a **1** is sent, $P(1|0)$ is the probability of deciding that the output is a **1** when a **0** is sent, and $P(0)$ is the probability that a **0** is sent. If $P(1)$ is a and $P(0)$ is b, then the combined error probability P_e is

$$P_e = aP\left[i_N < -(1-k)\overline{i(1)}\right] + bP\left[i_N > k\overline{i(1)}\right] , \tag{6.72}$$

where $P(\cdot)$ is the probability of the event occurring.

We now make a simplifying assumption. For most messages we can assume that logical **1**s and **0**s are equally probable, so $a = b = 1/2$ and

$$P_e = \frac{1}{2}P\left[i_N < -(1-k)\overline{i(1)}\right] + \frac{1}{2}P\left[i_N > k\overline{i(1)}\right] . \tag{6.73}$$

From Fig. 6.11, we note that the error of Eq. 6.70 on the page before is the area underneath the tail of the upper probability distribution below the threshold, I_D, i.e.,

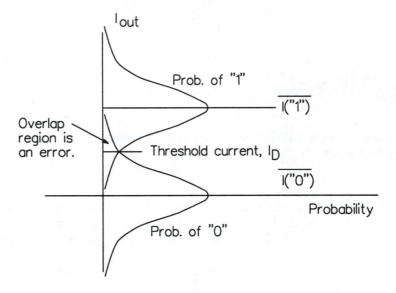

Figure 6.11 Probability density curves for a logical **1** and a logical **0**.

$$P(0|1) = \frac{1}{\sigma_1\sqrt{2\pi}} \int_{-\infty}^{I_D} \exp\left(-\frac{(I-\overline{i(1)})^2}{2\sigma_1^2}\right) dI = \frac{1}{2}\mathrm{erfc}\left(\frac{\overline{i(1)}-I_D}{\sigma_1\sqrt{2}}\right) , \tag{6.74}$$

where the complementary error function is defined as

$$\mathrm{erfc}\,(x) = \frac{2}{\sqrt{\pi}} \int_x^{\infty} e^{-y^2}\, dy . \tag{6.75}$$

The error, $P(1|0)$ is the area under the tail of the lower distribution above the threshold, I_D. It is found as

$$P(1|0) = \frac{1}{\sigma_0\sqrt{2\pi}} \int_{I_D}^{\infty} \exp\left(-\frac{(I_D-\overline{i(0)})^2}{2\sigma_0^2}\right) dI = \frac{1}{2}\mathrm{erfc}\left(\frac{I_D-\overline{i(0)}}{\sigma_0\sqrt{2}}\right) . \tag{6.76}$$

Hence, the probability of error (also called the *bit-error rate*, BER) is

$$P_e = \mathrm{BER} = \frac{1}{4}\left[\mathrm{erfc}\left(\frac{i(1)-I_D}{\sigma_1\sqrt{2}}\right) + \mathrm{erfc}\left(\frac{I_D-i(1)}{\sigma_0\sqrt{2}}\right)\right] . \tag{6.77}$$

The goal now is to select the value of I_D (i.e., set the threshold) to minimize the errors. It can be shown that the optimum value is the one that satisfies the relationship [13, 14]

$$\frac{\overline{i(1)} - I_D}{\sigma_1} = \frac{I_D - \overline{i(0)}}{\sigma_0} \tag{6.78}$$

or

$$I_D = \frac{\sigma_0 i(1) + \sigma_1 i(0)}{\sigma_0 + \sigma_1}. \tag{6.79}$$

We note that, in general, $\sigma_1 > \sigma_0$ due to the larger value of signal-dependent shot noise for a **1** than that of a **0**.

Using Eq. 6.78, we can define the Q parameter as

$$Q = \frac{\overline{i(1)} - I_D}{\sigma_1} = \frac{I_D - \overline{i(0)}}{\sigma_0}. \tag{6.80}$$

With the selection of optimal threshold setting, Eq. 6.77 on the preceding page is

$$\text{BER} = \frac{1}{2}\text{erfc}\left(\frac{Q}{\sqrt{2}}\right). \tag{6.81}$$

For $Q > 3$ a useful approximation to the right side of this equation is

$$\text{BER} = \frac{1}{2}\text{erfc}\left(\frac{Q}{\sqrt{2}}\right) \approx \frac{e^{-\frac{Q^2}{2}}}{Q\sqrt{2\pi}} \qquad \text{for } Q > 3. \tag{6.82}$$

This approximation is valid for $Q > 3$. A plot of this approximation is shown in Fig. 6.12 on the following page. The values of Q from the approximation for several benchmark values of BER are given in Table 6.3.

BER	Q exact	Q approx
10^{-3}	3.090	3.12
10^{-4}	3.719	3.73
10^{-5}	4.265	4.28
10^{-6}	4.753	4.76
10^{-7}	5.199	5.21
10^{-8}	5.614	5.62
10^{-9}	6.00	5.998
10^{-10}	6.36	6.361
10^{-11}	6.71	6.706
10^{-12}	7.034	7.04
10^{-13}	7.349	7.35
10^{-14}	7.651	7.65

Table 6.3 Values of Q-parameter for several values of BER (using both the exact formula and the approximate formula).

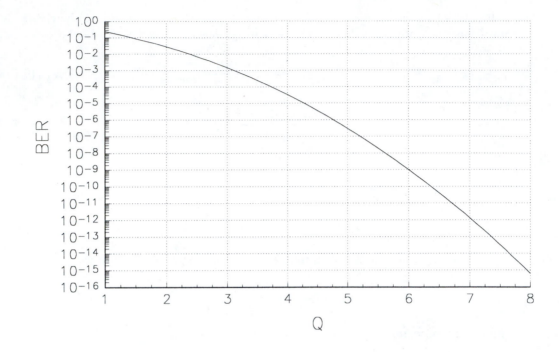

Figure 6.12 Plot of the approximation for BER vs. Q.

6.3.1   Simplified Model

We can now make another simplifying assumption that the probability distributions have equal widths (i.e., $\sigma_0 = \sigma_1 = \sigma$). For this assumption the optimum location for the threshold is $I_D = (\overline{i(0)} + \overline{i(0)})/2$ (i.e., the threshold is midway between the mean values of the curve). This assumption is correct for most pin-diode receivers where the noise is dominated by thermal noise. We also choose to let $\overline{i(0)} = 0$. The BER, then, is expressed by

$$\text{BER} \;=\; \frac{1}{2}\,\text{erfc}\left(\frac{\overline{i(1)}}{2\sigma\sqrt{2}}\right) = \frac{1}{2}\,\text{erfc}\left(\frac{\overline{i(1)}}{2\sqrt{2}\sqrt{<i_N^2>}}\right) \tag{6.83}$$

$$=\; \frac{1}{2}\,\text{erfc}\left(\frac{\sqrt{\frac{S}{N}}}{2\sqrt{2}}\right). \tag{6.84}$$

(Here, we have noted that $\overline{i(1)}/\sqrt{<i_N^2>}$ is $\sqrt{S/N}$ for our definition of S/N.) A plot of Eq. 6.84 is shown in Fig. 6.13 on the next page; the error probability is plotted vertically and the required $\sqrt{S/N}$ is plotted horizontally.

To use this curve, one begins with the desired error probability and determines the required signal-to-noise ratio. For example, a BER of 10^{-9} is a probability-of-error of 10^{-9} and is frequently used in modern digital communications. From the figure, we estimate the required ratio of $i_s/\sqrt{<i_N^2>}$ as 21.5 dB or a value of 11.89. From a knowledge of the limiting noise source in the detector (i.e., thermal noise for a pin diode or shot noise for an APD), we can calculate

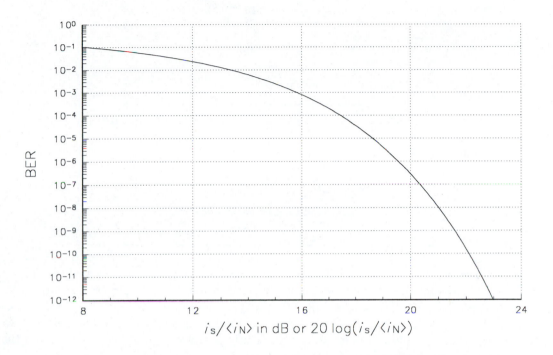

Figure 6.13 Dependence of bit error rate (error probability) on signal-to-noise ratio when $\sigma_0 = \sigma_1 = \sigma$ and $\overline{i(0)} = 0$. (Here, $i_s = \overline{i(1)}$.)

$\sqrt{< i_N^2 >}$. Knowing the required signal-to-noise ratio $\sqrt{(S/N)}$, we find the minimum RMS signal current from

$$< i_s >_{\min} = \sqrt{(S/N)_{\min}} \sqrt{< i_N^2 >}. \qquad (6.85)$$

The signal current is the product of the optical power and the responsivity $\mathcal{R}$ of the detector (i.e., $i_s = \mathcal{R}P$), giving

$$P_{\min} = \frac{\sqrt{(S/N)_{\min}} \sqrt{< i_N^2 >}}{\mathcal{R}}. \qquad (6.86)$$

This optical power must be available at the detector to achieve the desired error rate. We can then use this value to investigate possible trade-offs between the fiber losses, connector losses, coupling efficiency, and source power levels to complete the design.

———————————————— ∞ ————————————————

Example: Calculate the minimum power required to maintain a bit error of 10^{-6} achieved with a photodiode detector with a responsivity of 0.4 A/W. Assume that the signal-to-noise ratio is limited by thermal noise (with a 50 Ω load, a 400K noise temperature, and a 10 MHz noise bandwidth).

<u>Solution:</u> For BER$= 10^{-6}$, we find from Fig. 6.13,

$$20 \log \left(\frac{< i_s >}{< i_N >} \right) \approx 19.6 \text{ dB}. \qquad (6.87)$$

$$\frac{< i_s >}{< i_N >} = 10^{19.6/20} = 9.55 \,.$$

Since $< i_N >= \sqrt{< i_N^2 >}$,

$$
\begin{aligned}
< i_s > &= 9.55\sqrt{< i_N^2 >} = 9.55\sqrt{\frac{4kTB}{R}} \\
&= 9.55\sqrt{\frac{4(1.38 \times 10^{-23})(400)(10^7)}{50}} = 6.35 \times 10^{-7} \text{ A} = 635 \text{ nA}\,.
\end{aligned}
\tag{6.88}
$$

We then find $P_{\min}$ as

$$
\begin{aligned}
P_{\min} &= \frac{< i_s >}{\mathcal{R}_0} = \frac{635 \times 10^{-9}}{0.4} \\
&= 1.587 \times 10^{-6} \text{ W} = 1.587 \text{ } \mu\text{W}\,.
\end{aligned}
\tag{6.89}
$$

───────────────── ∞ ─────────────────

Recall, however, that various simplifying assumptions have been made in our analysis. The assumption of Gaussian noise reduced the complexity of the problem, but produces an $i_s/\sqrt{< i_N^2 >}$ value that is too low. We have also assumed that $\overline{i(0)} = 0$ and that a logical **1** and logical **0** are equally probable. (This latter assumption caused us to assume that the threshold was midway between $\overline{i(0)}$ and $\overline{i(1)}$.) Furthermore, we have implicitly assumed a noise-free amplifier and that there is negligible pulse spreading in the received pulses.

This last effect can lead to a form of noise termed *intersymbol interference*. As seen in Fig. 6.14 on the next page, any energy that spills out of the expected pulse position is lost to the receiver for use in detecting the pulse of interest. If large enough, this energy might trigger a false detection in the adjacent time slot. While this intersymbol interference may be partially corrected by the use of a properly designed equalization amplifier (at the expense of increased noise levels), the design of such a device is beyond the scope of this text. It has been found that the intersymbol interference is negligible if the bit spacing T_b is kept larger than 4 times the RMS pulse spreading of the fiber $\tau_{\rm rms}$ where the RMS pulse width is given (as in Chapter 5) by

$$\tau_{\rm rms} = \int t^2 h(t)\, dt - \int t\, h(t)\, dt\,, \tag{6.90}$$

where $h(t)$ is the pulse shape at the fiber output for an impulse input. If $h(t)$ is Gaussian (or approximated as Gaussian), then the RMS pulse width is equal to the pulse width measured at the $1/e$ points. Hence, we want

$$T_b \geq 4\tau_{\rm rms}\,. \tag{6.91}$$

6.4   Optical Receiver Noise and Sensitivity

The properties of the receiver are determined by the combination of the detector and amplifier rather than by the detector alone. In this section we will consider some typical receiver circuits and indicate some of the design considerations that must be made in applying the techniques to a fiber link. This section will follow the work by Kasper and Personick in Refs. [8], [12], [15] and [16].

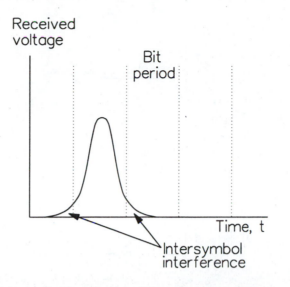

Figure 6.14 Example of intersymbol interference. (Note the energy spilling into adjacent bit intervals.)

The combination of the detector and the preamplifier is called the receiver *front-end*. Generally there are three common implementations (although there are many designs that do not fall into these three categories).

- The first is the *low-impedance front-end* (Fig. 6.15a). The detector operates into a low-impedance amplifier (usually with an input impedance of 50 Ω) through a coaxial cable. This design is popular because of the ready availability of low-impedance wideband RF amplifiers. To transfer the maximum power from the detector load resistor to the amplifier, the load resistor is chosen that is equal to the input resistance of the amplifier (e.g., a 50 Ω amplifier would call for a 50 Ω load resistor). This design does not provide much sensitivity (due to the small voltage developed across the input to the amplifier) and suffers from high noise (due to the $1/R$ dependence of the thermal noise) from the load resistance.

- The second design that we want to consider is the *high-impedance front-end* (Fig. 6.15b). A larger signal voltage (and less thermal noise) can be achieved by applying the detector output current to a load and amplifier combination with a high value of resistance. To achieve high input resistance, the amplifier uses an FET and the load resistor is made large; the parallel combination is numerically equal to the load resistor R_L. Several capacitances are in parallel with the combined load and amplifier resistance: the detector capacitance, the amplifier input capacitance, and parasitic capacitances. The total capacitance C_T is the sum of these individual capacitances. A current generator driving a parallel RC circuit has the characteristics of an integrator; hence, this type of front-end is also called an *integrating front-end*. The bandwidth of the combination is $1/2\pi R_L C_T$. For values of R_L in the 100s of kilohms to a few megohms and C_T values of a few picofarads or less, the bandwidth is quite low for high-data-rate systems. An equalization amplifier must be added to compensate for this poor bandwidth.

Figure 6.16 on page 191 shows one such equalizer. The preamp and detector are represented

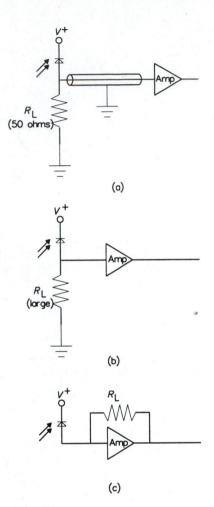

Figure 6.15 Block diagrams: (a) a low-impedance front-end, (b) a high-impedance front-end, and (c) a transimpedance front-end.

by an equivalent frequency-dependent voltage source $V_{\text{amp}}(\omega)$ and resistance R_{amp}. The transfer function of the circuit is

$$\frac{V_{\text{out}}(\omega)}{V_{\text{amp}}(\omega)} = \frac{R_2(1 + j\omega R_1 C)}{R_1 + R_{\text{amp}} + R_2 + j\omega R_1 C(R_{\text{amp}} + R_2)}. \tag{6.92}$$

If we choose R_1 and C such that

$$\frac{1}{R_1 C} = \frac{1}{R_L C_T}, \tag{6.93}$$

the zero of the equalizer will cancel the pole of the integrating front-end and the combined bandwidth will be larger than that of the front-end alone. The bandwidth of the combined

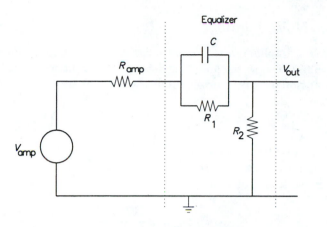

Figure 6.16 An RC equalizer for an integrating front-end.

circuit, f_{combined}, will be

$$f_{\text{combined}} = \frac{1}{R_1 C} \frac{R_1 + R_{\text{amp}} + R_2}{R_{\text{amp}} + R_2}. \tag{6.94}$$

Picking $R_1 \gg R_{\text{amp}} + R_2$ will ensure a higher bandwidth than the integrating front-end alone.

Another disadvantage of the high-impedance front-end, besides requiring an equalizer, is that the dynamic range of the receiver is limited. (This means that the range of optical power values, from the weakest to the strongest, is limited.) This is because the low-frequency components that are integrated can quickly saturate the preamplifier output.

- The third design is the *transimpedance front-end* (Fig. 6.15c). Here, a feedback resistor R_L connects the output and input of the preamp. Electronically, this circuit is a current-to-voltage convertor. The bandwidth of this amplifier is a factor of A (the amplifier gain) times $1/2\pi R_L C_T$ (i.e., the bandwidth is a factor of $A+1$ larger than the unequalized high-impedance amplifier bandwidth [8]). This wide bandwidth obviates the need for an equalization amplifier. Similarly, the low-frequency components are reduced by the same factor of $A+1$, lowering the possibility of electronically saturating the amplifier, and increasing the dynamic range of the front-end.

6.5 Amplifier Noise

So far, we have been working with the noise from a simple detector driving a load resistance. This analysis is useful to determine the fundamental noise limits of the detector, as the addition of an amplifying stage after the detector will only increase the noise. We now want to do some calculations to determine the signal-to-noise ratio with the added amplifier with gain G.

6.5.1 Amplifier Noise Figure

The noise added by an amplifier is a thermal noise source. Figure 6.17a shows a detector and amplifier combination. We show the amplifier noise as a current source across the output terminals. The total mean-square noise current from the three noise sources shown in Fig. 6.17b is

$$< i_N^2 >_{\text{total}} = G^2 < i_{\text{shot}}^2 > + \frac{4kTBG^2}{R_L} + < i_N^2 >_{\text{amp}} . \tag{6.95}$$

We now want to create an artificial noise source across the input which will generate the equivalent amount of noise. For reasons to be described shortly, we want to replace the $< i_N^2 >_{\text{amp}}$ term with an equivalent noise source of value

$$< i_N^2 >_{\text{amp}} = \frac{\dfrac{4kTBG^2}{R_L}}{F_n - 1} . \tag{6.96}$$

The quantity F_n is the *noise figure* [17] of the amplifier (sometimes called the *noise factor* of the amplifier). For the equality to hold, we want

$$F_n = 1 + \frac{< i_N^2 >_{\text{amp}} R_L}{4kTBG^2} . \tag{6.97}$$

If we move the artificial noise generator through the amplifier, we will need to divide the mean-square noise by G^2. (Figure 6.17c shows the artificial noise generator shifted to the amplifier input.) We can combine the noise generator of the load resistor and the artificial noise generator representing the amplifier noise by adding their mean-square noise currents,

$$\frac{4kTB}{R_L} + \frac{4kTB(F_n - 1)}{R_L} = \frac{4kTBF_n}{R_L} . \tag{6.98}$$

Figure 6.17d shows the combined noise generator. From this calculation we can see that the total thermal noise of the load resistor and the amplifier can be represented by multiplying the noise from the load resistor by F_n. This increase is the noise penalty imposed by the amplifier.

The noise figure of an amplifier is usually specified in units of dB; it must be converted to a numerical value for use in the formulas of interest. A typical noise figure for a good, low-noise amplifier is 3 dB (implying a total noise that is twice the noise of the load alone). Otherwise, the value might be 6 dB or larger. The value of F_n is determined by measuring the total noise output, subtracting the shot noise and thermal noise of the load resistor, and applying Eq. 6.97 to calculate F_n.

The signal-to-noise ratio for a pin-diode detector and amplifier is

$$\frac{S}{N} = \frac{G^2 \mathcal{R}^2 P^2}{2q(\mathcal{R}P + I_{\text{dark}})G^2 B + \dfrac{4kTBF_n G^2}{R_L}} . \tag{6.99}$$

The G^2 terms will cancel out from the numerator and denominator but have been left in the expression for completeness. Note the appearance of the F_n term in the thermal noise term in the denominator.

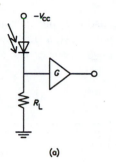

(a)

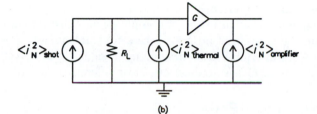

(b)

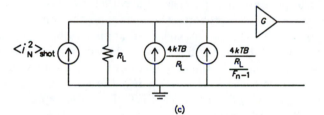

(c)

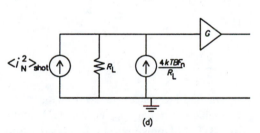

(d)

Note: All current generator labels show mean—square currents

Figure 6.17 (a) A detector-amplifier receiver. (b) Noise model of receiver with detector shot-noise source, load-resistor thermal-noise source, and amplifier-output-noise source shown. (c) Noise model of receiver with the output noise model of part (b) modeled by an equivalent noise generator at the input of the amplifier. (d) Noise model with resistor noise generator and equivalent amplifier noise generator combined into a single generator.

Considering an APD detector working with an amplifier, we have

$$\frac{S}{N} = \frac{G^2 M^2 \mathcal{R}_0^2 P^2}{2q(\mathcal{R}_0 P + I_{\text{bulk}})M^2 F(M)BG^2 + 2qI_{\text{surf}}BG^2 + \dfrac{4kTBF_n G^2}{R_L}} . \tag{6.100}$$

Notice that the surface-current shot noise and the thermal noise do *not* have the M^2 multiplier term. The signal-to-noise ratio will again have a maximum value at an optimum value of multiplication factor, Due to the excess noise factor of the avalanche process. This optimum value M_{opt} is found by solving the equation

$$M_{\text{opt}} = \left(\frac{2qI_{\text{surf}} + \dfrac{4kTF_n}{R_L}}{xq(\mathcal{R}_0 P + I_{\text{dark}})} \right)^{1/(x+2)} . \tag{6.101}$$

These expressions will allow us to calculate the signal-to-noise ratio for simple receivers. Now that we can accomplish this calculation, we need some signal-to-noise benchmarks to determine if we have enough signal-to-noise ratio for our application. Each analog application has its own required minimum signal-to-noise ratio for successful transmission. For example, the North American television standard requires a 35 dB minimum signal-to-noise ratio at the receiver for successful television reception. Other applications have different standards.

6.5.2   Noise in FET Front-Ends

The implementation of the preamplifiers [8] described earlier can use either FETs or bipolar transistors. FETs have superior noise properties compared to the bipolar transistor. The development of GaAs microwave FETs has enabled FET amplifiers to be realized for wideband high-data-rate receivers. Figure 6.18 on the facing page shows a representative common-source preamp. The principal sources of noise in the amplifier are thermal noise from the FET channel resistance, the thermal noise from the load resistor R_L, the shot noise due to the FET gate leakage current, and the electronic $1/f$ noise of the FET. The mean-square noise current of this amplifier has been shown [8, 18, 19] to be

$$<i^2_{\text{amp}}> = \frac{4kT}{R_L} I_2 B_R + 2q I_{\text{gate}} I_2 B_R + \frac{4kT\Gamma}{g_m} (2\pi C_T)^2 f_c I_f B_R^2 \tag{6.102}$$

$$+ \frac{4kT\Gamma}{g_m} (2\pi C_T)^2 I_3 B_R^3 .$$

Here, B_R is the bit rate, R_L is the load resistor (or feedback resistor for a transimpedance amplifier), I_{gate} is the FET gate leakage current, g_m is the FET transconductance, C_T is the total input capacitance, f_c is the FET $1/f$-noise corner frequency, and Γ is the FET channel-noise factor. (Table 6.4 on the next page gives the ranges of typical FET parameters for three FET types.)

The parameters I_2, I_3, and I_f are the *Personick integrals* [8, 18, 20]. (Be careful not to confuse them with currents.) They depend only on the pulse shape entering and leaving the fiber and the type of coding used to encode the data. Table 6.5 on page 196 gives values [8] of integrals for rectangular pulses entering the fiber and a particular shape (pulses having a raised cosine spectrum) that Personick used in his original analysis. Here NRZ (non-return-to-zero) coding is the usual on-off coding that follows the data; RZ (return-to-zero) coding ensures that there is a data transition during every bit period. This RZ coding is usually used to encode the clock information on the data stream. (Coding is discussed in more detail in Chapter 7.)

The FET channel-noise factor Γ is an FET parameter that describes the noise contribution from the channel resistance and the gate-induced noise; it also includes the effects that are due to these noise sources being correlated with each other [8].

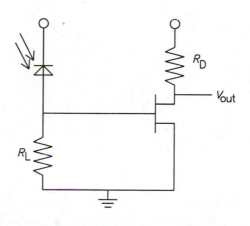

Figure 6.18 Typical common-source FET fiber receiver front-end.

	GaAs MESFET	Si MOSFET	Si JFET
g_m (mS)	15–50	20–40	5–10
C_{gs} (pF)	0.2–0.5	0.5–1.0	3–6
C_{gd} (pF)	0.01–0.05	0.05–0.1	0.5–1.0
Γ	1.1–1.75	1.5–3.0	0.7
I_{gate} (nA)	1–1,000	0	0.01–0.1
f_c (MHz)	10–100	1–10	<0.1

Table 6.4 Typical values of FET parameters. (From B.L. Kasper, "Receiver Design," in *Optical Fiber Telecommunications II*, S.E. Miller and I.P. Kaminow, eds., pp. 689–722, New York: Academic Press, 1988.)

The total capacitance C_T of the detector-preamp combination is given by

$$C_T = C_d + C_s + C_{gs} + C_{gd}, \tag{6.103}$$

where C_d is the detector capacitance, C_s is the stray capacitance, C_{gs} is the gate-to-source capacitance of the FET, and C_{gd} is the gate-to-drain capacitance of the FET. The stray capacitance is usually estimated, or a value is obtained from measured data. Integrated-circuit detector-preamp combinations have minimum stray capacitance.

The corner frequency f_c of the $1/f$ noise is an FET parameter defined as that frequency where the $1/f$ electronic noise of the device (dominant at low frequency operation) becomes equal to white thermal noise of the channel (characterized by Γ).

The first term of Eq. 6.103 on the preceding page is due to the thermal noise of the load resistor (or the feedback resistor). It can be minimized by making the resistor large. (This reduces the receiver dynamic range, however.) The second term is due to the shot noise associated with gate leakage current; it is minimized by choosing an FET with a low value of I_{gate}. The third term is due to the $1/f$ noise of the preamp; it is minimized by choosing a device with a low amount of $1/f$ noise (as indicated by a low value of f_c). The fourth term is due to the FET channel noise; it is minimized by choosing an FET with the maximum value of g_m/C_T^2 [8].

Frequently, the performance of FET preamps is expressed in terms of the FET short-circuit

	Coding	
	NRZ	RZ
I_1	0.548	0.500
I_2	0.562	0.403
I_3	0.0868	0.0361
I_f	0.184	0.0984

Table 6.5 Values of Personick integrals for RZ and NRZ (50% duty cycle) coding. Input pulses are rectangular; output pulses have raised cosine spectrum.

common-source gain-bandwidth product f_T. It can be shown that [8]

$$f_T = \frac{g_m}{2\pi(C_{gs} + C_{gd})}. \tag{6.104}$$

We are usually interested in optimizing the receiver performance at high bit rates. In this case, a well-designed receiver will be dominated by the fourth term in Eq. 6.103 on page 194, due to its B_R^3 dependence. In this case, the minimum noise current is [8]

$$<i_{\mathrm{amp}}^2>_{\mathrm{min}} \approx 32\pi kT \frac{\Gamma(C_d + C_s)}{f_T} I_3 B_R^3 \qquad \text{(for large } B_R). \tag{6.105}$$

The best results are obtained when an FET is chosen that has a maximum figure of merit $\mathrm{FOM_{FET}}$, of

$$\mathrm{FOM_{FET}} = \frac{f_T}{\Gamma(C_d + C_s)}. \tag{6.106}$$

Note that choice of the best FET depends on the capacitance of the optical detector. If we change the detector to one with a significantly different C_d, we should also change the preamp FET.

6.5.3 Noise in Bipolar Transistor Front-Ends

Bipolar preamplifiers [8] are also used in some fiber receiver front-ends. At low bit rates their noise is higher, but at high bit rates the noise is comparable to the FET noise. Figure 6.19 shows a representative common-emitter preamp using a bipolar junction transistor (BJT). The principal sources of noise in the amplifier are thermal noise from the load resistor R_L, the shot noise due to the base and collector bias currents (I_b and I_C), and the thermal noise from the base-spreading resistance $r_{bb'}$. The mean-square noise current of this amplifier has been shown [8, 18, 19] to be

$$<i_{\mathrm{amp}}^2> = \frac{4kT}{R_L} I_2 B_R + 2q I_b I_2 B_R + \frac{2q I_c}{g_m^2} (2\pi C_T)^2 I_3 B_R^3$$
$$+ 4kT r_{bb'} \left[2\pi \left(C_d + C_s\right)\right]^2 I_3 B_R^3. \tag{6.107}$$

The transconductance depends on the collector bias current as

$$g_m = \frac{I_c}{V_T}, \tag{6.108}$$

where

$$V_T = \frac{kT}{q}. \tag{6.109}$$

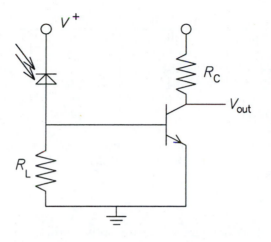

Figure 6.19 Typical common-emitter bipolar junction transistor receiver front-end.

The total capacitance for the BJT front-end is

$$C_T = C_d + C_s + C_{b'e} + C_{b'c}, \tag{6.110}$$

where $C_{b'e}$ and $C_{b'c}$ are capacitances from the small-signal hybrid-pi transistor model. The base-emitter capacitance $C_{b'e}$ is a function of the collector bias current I_c. It can be broken up into two components,

$$C_{b'e} = C_{je} + \frac{I_c}{2\pi V_T f_T}, \tag{6.111}$$

where C_{je} is the current-independent junction capacitance and the second term is a diffusion capacitance. In this second term, f_T is the short-circuit common-emitter gain-bandwidth product of the transistor (measured at high I_c to ensure that the diffusion term dominates). It can be shown [8, 19] that there exists an optimum collector bias current given by

$$I_{c\ opt} = 2\pi C_0 f_T V_T \psi(B_R), \tag{6.112}$$

where C_0 is the total capacitance at zero bias

$$C_0 = C_d + C_s + C_{b'c} + C_{je} \tag{6.113}$$

and $\psi(B_R)$ is given by

$$\psi(B_R) = \frac{1}{\sqrt{1 + \dfrac{I_2 f_T^2}{\beta I_3 B_R^2}}}, \tag{6.114}$$

with β being the transistor current gain,

$$\beta = \frac{I_c}{I_b}. \tag{6.115}$$

The total capacitance C_T can be written in terms of $\psi(B_R)$ as

$$C_T = C_0 \left[1 + \psi(B_R)\right]. \tag{6.116}$$

	GaAs MESFET	Si MOSFET	Si JFET
g_m (mS)	40	30	6
C_{gs} (pF)	0.38	0.8	4.0
C_{gd} (pF)	0.02	0.1	0.8
Γ	1.1	2.0	0.7
I_{gate} (nA)	2.0	0	0.05
f_c (MHz)	30	1.0	0

Table 6.6 Assumed FET parameters for noise and sensitivity calculations. (From B.L. Kasper, "Receiver Design," in *Optical Fiber Telecommunications II*, S.E. Miller and I.P. Kaminow, eds., pp. 689–722, New York: Academic Press, 1988.)

The mean-square amplifier-noise current at the optimum bias current can be written as [8]

$$< i^2_{\mathrm{amp}} >\big|_{\mathrm{optimum}\ I_c} \tag{6.117}$$

$$= \frac{4kT}{R_L}I_2B_R + \frac{4\pi kTC_0 f_T}{\beta}\psi(B_R)I_2B_R + \frac{4\pi kTC_0}{f_T}\frac{[1+\psi(B_R)]^2}{\psi(B_R)}I_3B_R^3$$

$$+ 4kTr_{bb'}\left[2\pi\left(C_d + C_s\right)\right]^2 I_3B_R^3 .$$

Again, we are usually concerned with design of a high-frequency receiver front-end. The bipolar transistor figure of merit $\mathrm{FOM_{BJT}}$ for this region of operation is [8]

$$\mathrm{FOM_{BJT}} = \frac{2f_T}{C_0 + \pi f_T r_{bb'}(C_d + C_s)} \qquad \text{(for high } B_R\text{)} \tag{6.118}$$

$$\approx \frac{2f_T}{C_0} \qquad \text{(for high } B_R \text{ and small } r_{bb'}\text{)}.$$

6.5.4   Comparison of Noise in FET and BJT Front-ends

Figure 6.20 on the next page shows a comparison [8] of the calculated noise for three FET front-ends and one BJT front-end as a function of bit rate. (See problems at end of chapter.) The FET designs use the data in Table 6.6. The sum of detector and stray capacitance $(C_d + C_s)$ is assumed to be 0.2 pF, and the load resistance R_L is assumed to be very large (to make its thermal noise negligible).

The noise for the silicon bipolar transistor is modeled from a device with the parameters of Table 6.7 on page 200. The optimum bias current is used to calculate the noise until the calculated optimum bias current falls below 0.1 mA. (The gain β starts to fall in value for bias currents below this value.) When the optimum bias current is computed to be below 0.1 mA, a bias current of 0.1 mA is assumed for calculating the noise current.

Figure 6.20 on the facing page shows that at low bit rates the bipolar transistor front-end is inferior to the FET front-end. At high bit rates this inferiority disappears. Among the FET front-ends, the silicon MOSFET is slightly advantageous at low bit rates, with the GaAs MESFET having slightly superior noise properties at high bit rates. (Silicon JFETs lose gain at bit rates above about 200 Mb·s^{-1} because of their relatively low gain-bandwidth product. These devices are, therefore, not suitable for high bit-rate designs.)

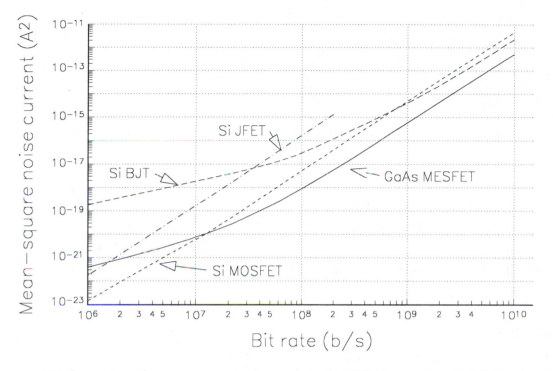

Figure 6.20 Comparison of mean-square noise current for typical FET front-ends and a BJT front-end. (Assumed device values are found in Tables 6.6 and 6.7.)

6.5.5   Sensitivity of Front-Ends

We now want to consider a calculation of the optical power required on the receiver to achieve a desired bit-error rate in the presence of both detector noise and amplifier noise [8]. We first consider the case of a pin-diode detector and then the more complicated case of an APD detector.

PIN Diode Sensitivity

The sensitivity of current pin-diode receivers is more than 20 dB from the theoretical minimum of the quantum limit. Hence, we neglect the signal-related shot noise. The total mean-square noise current is the sum of the amplifier mean-square noise current and the mean-square shot-noise current due to the dark current of the amplifier

$$< i_N^2 >_{\text{Total}} = < i_{\text{amp}}^2 > + 2qI_D I_2 B_R . \tag{6.119}$$

Here, the amplifier mean-square noise current is determined as in the prior sections, depending on the type of preamplifier used. Both expressions on the right side of this equation depend on the bit rate B_R. The required signal-to-noise ratio, S/N, for the desired bit-error rate is determined from Fig. 6.13 on page 187. The average optical power required to achieve this signal-to-noise

Parameter	Value
β	100
$r_{bb'}$	20 ohms
$C_{b'c}$	0.2 pF
C_{je}	0.8 pF
f_T	10 GHz

Table 6.7 Assumed BJT parameters for noise and sensitivity calculations. (From B.L. Kasper, "Receiver Design," in *Optical Fiber Telecommunications II*, S.E. Miller and I.P. Kaminow, eds., pp. 689–722, New York: Academic Press, 1988.)

ratio for the pin-diode receiver is [8]

$$P = \frac{hc}{q\lambda} \sqrt{\frac{S}{N}} \sqrt{< i_N^2 >_{\text{Total}}} \, . \tag{6.120}$$

The required optical power can be calculated and plotted as a function of the bit rate B_R, once the pin diode parameters and the amplifier type and parameters are known. (See the problems at the end of this chapter.)

APD Sensitivity

The sensitivity of an APD receiver is made more difficult to analyze because the gain of the APD is an additional variable. Also the APD produces excess noise which makes the model more difficult. In fact, there is an optimum value of diode gain M_{opt} which gives the best sensitivity. (This value of optimum gain depends on the device parameters, the preamplifier noise, and the bit rate.) At the optimum gain ($M = M_{\text{opt}}$), the APD noise is approximately (but not precisely) equal to the preamplifier noise [8]. The shot noise due to the dark current in an APD consists of two parts [19], one part due to the unamplified surface dark current I_{surf} and the other part due to the multiplied bulk dark current I_{bulk}, i.e,

$$< i^2 >_{\text{dark}} = 2q I_{\text{surf}} I_2 B_R + 2q I_{\text{bulk}} M^2 F(M) I_2 B_R \, . \tag{6.121}$$

Here $F(M)$ is the excess noise factor, as described earlier in this chapter. The surface dark current leaks around the edges of the device and does not engage in the avalanche process. The receiver sensitivity for an APD receiver is [8, 19]

$$P = \left(\frac{hc}{q\lambda} \right) Q \left[Q q B_R I_1 F(M) + \sqrt{\frac{< i^2 >_{\text{Total}}}{M^2} + 2q I_{\text{bulk}} F(M) B_R I_2} \right] , \tag{6.122}$$

where Q is the signal-to-noise ratio needed to achieve the required bit-error rate, I_1 and I_2 are Personick integrals described earlier, and

$$< i^2 >_{\text{Total}} = < i_{\text{amp}}^2 > + 2q I_{\text{surf}} I_2 B_R \, . \tag{6.123}$$

If the multiplied part of the dark current I_{bulk} is small enough that its noise is negligible (not a very good assumption for long-wavelength detectors), then the sensitivity is

$$P = \frac{hcQ}{q\lambda} \left(\frac{\sqrt{< i^2 >_{\text{Total}}}}{M} + q Q B_R I_1 F(M) \right) \quad \text{(for noise due to } I_{\text{bulk}} \text{ negligible)} \, . \tag{6.124}$$

For this same case, the optimum gain is [8, 19]

$$M_{\text{opt}} = \frac{1}{\sqrt{k}} \sqrt{\frac{\sqrt{<i^2>_{\text{Total}}}}{qI_1 B_R Q} - k + 1} \qquad \text{(for noise due to } I_{\text{bulk}} \text{ negligible)}, \qquad (6.125)$$

where k is the ionization ratio of the avalanche process, described earlier.

If the noise contribution from I_{bulk} is *not* negligible, M_{opt} is smaller than the value predicted by Eq. 6.125. The value of M_{opt} must be found graphically or numerically at each value of the bit rate by finding the value that minimizes the sensitivity of the receiver. Then the total noise and sensitivity can be calculated as a function of bit rate. A more detailed discussion can be found in Ref. [19].

6.5.6   Extinction Ratio Effects

The extinction ratio r (the ratio of the power transmitted for a logical **0** to the power transmitted for a logical **1**) indicates whether the optical source is turned off all the way when a logical **0** is transmitted. Not turning off the source causes a reduction in sensitivity (called a *sensitivity penalty*) [8, 19]. This penalty is incurred because there will be shot noise associated with the reception of a **0** and because not all of the received optical power is being modulated. For a pin-diode receiver with a nonzero extinction ratio, the power required to achieve a desired BER is multiplied by a factor of $(1 + r)/(1 - r)$. For the APD receiver, the extinction ratio affects the optimum APD gain (which must be found numerically) and reduces the required power in a complicated fashion. (See Refs. [8] and [19] for more details.)

6.6   Eye Pattern Analysis

The performance of a high-data-rate link can be measured from the *eye pattern* of the system. The eye-pattern measurement [13] is a time-domain measurement made with the simple setup shown in Fig. 6.21 on the next page. The eye pattern observed on the scope provides several pieces of meaningful information about the performance of digital data links. The pseudorandom generator produces a data stream with a pattern length and bit period controlled by the operator. Within the pseudorandom data stream, the bits are produced in a highly varied fashion, but the pattern is still predictable by the user (although this prediction is not worth the effort for long data streams). At the end of the data pattern, the pattern repeats itself. The received data is put into the vertical trace of the oscilloscope, while the data clock is used to trigger the oscilloscope. The resulting eye pattern (so-called for its resemblance to the human eye [see Fig. 6.22 on the following page]) is the superposition of output pulses from many data pulses. This pattern contains much easily observed information about the digital transmission characteristics of the optical link (see Fig. 6.21 on the next page).

- The horizontal width of the eye opening gives the optimum *sampling time interval* for the received signal to be sampled without error from intersymbol interference. The optimum sampling time is at the position of maximum eye opening.

- The vertical height of the eye opening is a measure of the *amplitude distortion* of the signal. As the upper 'limit of the frequency response of the system is reached, the vertical height of the eye opening will decrease and the eye will close.

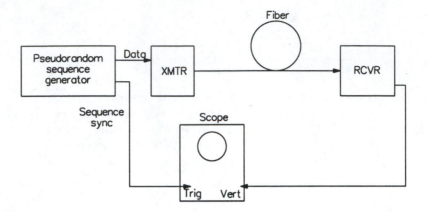

Figure 6.21 Experimental setup for observing the eye pattern of a fiber optic digital link.

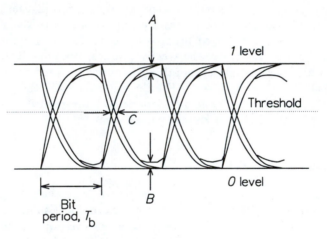

Figure 6.22 Representative eye pattern.

- The spacing of "A" on the figure indicates the amount of noise present when a logical **1** is sent; the spacing of "B" on the figure indicates the amount of noise present when a logical **0** is sent.

- The width of the threshold crossing ("C" on the figure) determines the *timing!jitter* (or *edge jitter*) of the system. The *jitter* is defined as

$$\text{Timing jitter (in \%)} = \frac{\Delta T}{T_b} \times 100\%, \tag{6.126}$$

where T_b is the bit spacing in the data stream. This jitter is an indication of the timing accuracy of the received pulse as modified by the receiving circuitry. (Meeting jitter specifications is receiving renewed attention in fiber links. To help in characterizing jitter performance, new instruments are becoming available to measure the jitter in a system.)

Characteristic	pin diodes			APDs	
	Silicon	Germanium	InGaAs	Silicon	Germanium
λ (μm)	0.4–1.1	0.5–1.8	1.0–1.5	0.4–1.1	0.5–1.65
Quantum efficiency	80%	50%	70%	80%	75%
Rise time (ns)	0.01	0.3	0.1	0.5	0.25
Bias voltage	15	6	10	170	40
Responsivity (A/W)	0.5	0.7	0.4	0.7	0.6
Gain	1	1	1	80-150	80-150

Table 6.8 Performance of representative detectors.

- If a long string of logical **1**s and **0**s is included in the data stream, the 10% to 90% rise (and fall) times can be measured from the rise (and fall) times of the eye.

- Eye patterns are also useful to identify the *performance penalty* associated with the introduction of a change to the link. The eye patterns with and without the desired changes are measured and compared. The ratio of the **1**-level voltage to the **0**-level voltage is expressed in dB. The difference in the dB values is the performance penalty associated with the change.

6.7 Summary

Table 6.8 summarizes the properties of pin diodes and APDs by presenting the properties of representative devices. (Individual devices can exhibit superior performance in some categories.) The silicon devices represent mature technology and operate close to theoretical limits in the short-wavelength region. At long wavelengths, the InGaAs detector exhibits the best characteristics, while germanium-based devices have fundamental difficulties with noise performance, construction of avalanche devices, and high dark current levels.

In the design of receivers, we have seen that the noise contributions of the preamplifier are important and that much design effort goes into the design of this amplifying stage. We have seen that, because of the capacitive nature of the detector, high-impedance preamplifiers, while providing the best sensitivity, integrate the current produced at the detector, necessitating the incorporation of equalization amplifiers. A frequently used alternative receiver is the transimpedance amplifier, which sacrifices some noise performance for increased dynamic range and simplicity of design and operation.

Figure 6.23 on the next page illustrates representative receiver sensitivities required for operation of a digital data link with a BER of 10^{-9}. We note that increased sensitivity is required at higher data rates and that silicon FET receivers are good up to approximately 70 Mb·s^{-1}. The APDs offer about 10 dB of increased sensitivity with disadvantages in requiring a high-voltage bias, increased cost, and the requirement for temperature compensation. We also note that the best detector/preamplifier combinations are approximately 10 dB from achieving quantum-limited detection. This 10 dB deficit indicates the degree of improvement in operation that might be expected with improved devices and preamplifiers. As a benchmark, we note that the pin diode requires -42 dBm at 100 Mb·s^{-1} while the APD requires -58 dBm.

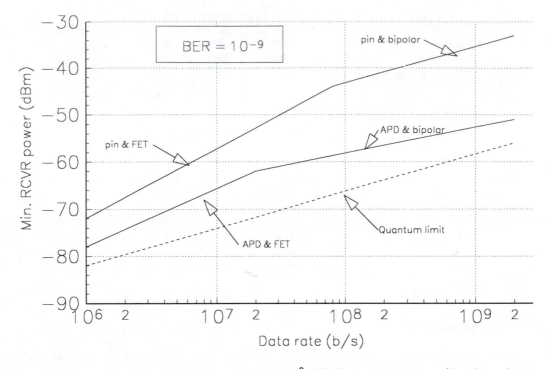

Figure 6.23 Representative sensitivities required for 10^{-9} BER for typical receivers (for silicon detectors and transistors).

6.8   Problems

1. (a) What are the energies (in joules and in eV) of photons having wavelengths of 820 nm and 1.3 μm?

    (b) What are the values of the free-space propagation constants for these wavelengths?

2. Calculate the cutoff wavelength of silicon ($E_g = 1.1$ eV) and germanium ($E_g = 0.785$ eV).

3. Consider the absorption coefficient for silicon as shown in Fig. 6.3 on page 165. Ignoring surface reflections, use a computer to plot the fraction of the incident power that is absorbed for depletion layer widths of 1, 5, 10, 20, and 50 μm over the wavelength range of 0.6 to 1.0 μm. Assume that $d = 20$ μm.

4. An avalanche photodiode has the following parameters:

Parameter	Value
dark current	1 nA
surface leakage current	1 nA
quantum efficiency	0.85
gain	100
excess noise factor	$M^{1/2}$
load resistance	$10^4 \Omega$

Consider a sinusoidal modulation on an 850 nm carrier with a modulation index m of 0.85. The average power level is −50 dBm and the detector operates at room temperature. Calculate the following S/N (in dB) if the bandwidth is 10 kHz and ...

(a) ... if the signal-dependent shot noise is the dominant term. (The mean-square signal current is given by $< i_s^2 >= (mM\mathcal{R}_0 P)^2/2$.)

(b) ... if the shot noise due to the dark current is the dominant term.

(c) ... if the shot noise due to the surface current is the dominant noise term.

(d) ... if the thermal noise is the dominant noise term.

(e) Which is really the dominant noise source? What will the actual S/N be?

5. Consider the avalanche device of the preceding problem.

(a) Use a computer to plot the S/N vs. M for values of M ranging from 1 to 100.

(b) Find the approximate value of M_{opt} and S/N at $M = M_{opt}$ from your graph.

(c) Calculate the values of M_{opt} and S/N at $M = M_{opt}$ from formulas and compare with your graphical results of the previous question.

6. Consider a silicon avalanche photodiode with parameters as given below, operating in a link with no intersymbol interference present.

Parameter	Value
$F(M)$	$M^{0.4}$
Responsivity (at $M = 1$)	0.3 A/W
Surface dark current	1 μA
Temperature	300 K
R_L	1 kΩ
Bulk dark current	1 nA
Bandwidth of receiver	10 MHz

(a) Calculate the dc optical power that must be incident on the detector to make the optimum gain of this APD have a value of 80.

(b) For a gain value gain of 80, calculate the ratio (in dB) of the mean-square noise current due to the shot noise caused by the bulk dark current to the mean-square noise current due to the thermal noise.

7. Consider a silicon photodiode operating at 850 nm ($\alpha = 10^3$ cm^{-1}).

(a) Calculate the area of the device if the capacitance is to be kept equal or less than 2 pF. (The relative permittivity of silicon is 11.7.)

(b) Calculate the maximum bandwidth of the detector when operating into 50 Ω load.

8. Consider a pin diode with the following properties at 920 nm–

Parameter	Value
Responsivity	0.5 A/W
Dark current	1.0 nA
Surface dark current	negligible
Operating temperature	300 K

This diode is irradiated with a constant 80 nW of optical power (at 920 nm). Find the signal-to-noise ratio (in dB) of the detector if it is operated into an equivalent load of 10 KΩ and a (noisefree) preamp with a bandwidth of 1 MHz.

9. Show that the relation between the Q-parameter and S/N is

$$Q = \frac{1}{2}\frac{S}{N}. \tag{6.127}$$

10. Using an iterative approach, show that $Q \approx 6.00$ for a BER of 10^{-9}.

11. Consider three FET front-ends having the parameters found in Table 6.6 on page 198. If $C_s + C_d = 0.2$ pF and $R_L = \infty$, plot (using a computer) the mean-square noise current of the front-end vs. the bit rate B_R for values of B_R falling between 1 Mb·s^{-1} and 10 Gb·s^{-1}. (Plot the graph on a log-log scale.) You may assume that $T = 300$ K and that NRZ coding is used.

12. Consider the bipolar transistor with parameter values given in Table 6.7 on page 200. It is to be used with a silicon pin detector with a device capacitance of 0.2 pF and a dark current of 1 nA. The stray capacitance is assumed to be 0.

   (a) Using a computer, calculate and plot the optimum bias current I_c as a function of bit rate B_R for values of B_R falling between 1 Mb·s^{-1} and 10 Gb·s^{-1}. (Plot the B_R axis on a logarithmic scale.)

   (b) For what values of B_R is the optimum bias current smaller than 0.1 mA?

   (c) For the range of values of B_R such that the optimum bias is greater than 0.1 mA, plot (using a computer) the mean-square noise current of the BJT front-end on a log-log scale, using the optimum bias current.

   (d) For the range of values of B_R such that the optimum bias is less than 0.1 mA, plot (using a computer, on the same figure as the previous part of this problem) the mean-square noise current of the BJT front-end, using a bias current of 0.1 mA. (Your result should be the same as the curve labelled "Si bipolar" in Fig. 6.20 on page 199.)

   (e) Plot the minimum power required to achieve a BER of 10^{-9} with thin pin diode and BJT transistor as function of bit rate B_R for the range of values mentioned in part (a).

13. Consider an InGaAs APD with the properties listed below, operating with a GaAs FET preamp with the properties listed in Table 6.6 on page 198 operating with an NRZ signal at a bit rate of 1 Gb·s^{-1}. Assume that the stray capacitance C_s is zero and that the noise temperature is 360K.

Parameter	Value
C_d	0.2 pF
I_{surf}	0 nA
I_{dark}	1 nA
k	0.3

   (a) Using a computer, plot the receiver sensitivity (i.e., the required optical power in dBm) for a BER of 10^{-9} ($Q = 6$) as a function of APD gain, M, for values of M from 1 to 50.

   (b) From your plot, estimate the optimum value of M.

   (c) From your plot, estimate the sensitivity of the receiver at the optimum gain.

References

1. T. P. Lee and T. Li, "Photodetectors," in *Optical Fiber Telecommunications* (S. E. Miller and A. G. Chynoweth, eds.), pp. 593–626, New York: Academic Press, 1979.

2. R. Plumb, "Detectors for fibre-optic communication systems," in *Optical Fibre Communication Systems* (C. Sandbank, ed.), pp. 184–205, New York: Wiley, 1980.

3. R. Smith, "Photodetectors for fiber transmission systems," *Proc. IEEE*, vol. 68, no. 10, pp. 1247–1252, 1980.

4. S. Personick, "Photodetectors for fiber systems," in *Fundamentals of Optical Fiber Communications* (M. F. Barnoski, ed.), pp. 257–293, New York: Academic Press, 1981.

5. M. Brain and T.-P. Lee, "Optical receivers for lightwave communication systems," *J. Lightwave Technology*, vol. LT–3, no. 6, pp. 1281–1300, 1985.

6. J. E. Bowers and Charles A. Burrus, Jr., "Ultrawide-band long-wavelength p-i-n photodetectors," *J. Lightwave Technology*, vol. LT–5, no. 10, pp. 1339–1350, 1987.

7. S. Forrest, "Optical detectors for lightwave communication," in *Optical Fiber Telecommunications II* (S. E. Miller and I. P. Kaminow, eds.), pp. 569–599, New York: Academic Press, 1988.

8. B. L. Kasper, "Receiver design," in *Optical Fiber Telecommunications II* (S. E. Miller and I. P. Kaminow, eds.), pp. 689–722, New York: Academic Press, 1988.

9. B. E. Saleh and M. C. Teich, *Fundamentals of Photonics.* New York: John Wiley and Sons, 1991.

10. B. L. Kasper and J. C. Campbell, "Multigigabit–per–second avalanche photodiode lightwave receivers," *J. Lightwave Technology*, vol. LT–5, no. 10, pp. 1351–1364, 1987.

11. A. Yariv, *Optical Electronics, Fourth Edition.* New York: Holt, Rinehart and Winston, 1991.

12. S. Personick, "Design of receivers and transmitters for fiber optic systems," in *Fundamentals of Optical Fiber Communications, Second Edition* (M. Barnoski, ed.), pp. 295–328, New York: Academic Press, 1981.

13. G. Keiser, *Optical Fiber Communications, Second Edition.* New York: McGraw-Hill, 1991.

14. G. P. Agrawal, *Fiber Optic Communication Systems.* New York: John Wiley & Sons, 1992.

15. S. Personick, "Receiver design for optical fiber systems," *Proc. IEEE*, vol. 65, no. 12, pp. 1670–1678, 1977.

16. S. D. Personick, "Receiver design," in *Optical Fiber Telecommunications* (S. E. Miller and A. G. Chynoweth, eds.), pp. 627–651, New York: Academic Press, 1979.

17. D. E. Meer, "Noise figures," *IEEE Trans. on Education*, vol. 32, no. 2, pp. 66–72, 1989.

18. R. G. Smith and S. D. Personick, "Receiver design for optical fiber communication systems," in *Topics in Applied Physics, Vol. 39* (H. Kressel, ed.), pp. 89–160, Berlin: Springer-Verlag, 1982.

19. T. V. Muoi, "Receiver design for high-speed optical-fiber systems," *J. Lightwave Technology*, vol. LT–2, no. 3, pp. 243–267, 1984.

20. S. Personick, "Receiver design for digital fiber optic communication systems," *Bell System Technical Journal*, vol. 52, pp. 843–874, 1973.

Chapter 7

Optical-Link Design

7.1   Introduction

Now that we have considered the building blocks of an optical link, we need a procedure to design a usable optical link to meet desired specifications. The procedure described in this chapter is iterative; that is, certain assumptions are made and the design is carried out based on those assumptions. The design is not finished at that point, however, as the designer must verify that it meets the objectives and represents an economical, as well as a technical, solution. If not, another pass through the design procedure is required. In particular, the assumptions need to be inspected to determine if changes might provide a simpler or cheaper alternative.

Techniques for optically amplifying the signal power to increase the length of the link are also discussed in this chapter. In particular, erbium-doped fiber amplifiers offer great capability.

At high data rates, dispersion can limit the transmission distance. This is especially true in fibers that were installed to operate at 1300 nm, but are now being upgraded to operate at 1550 nm so that optical amplifiers can be used. These fibers have a nontrivial amount of dispersion (on the order of 16 ps·nm^{-1}·km^{-1}). As the realization was made that dispersion can be either positive- or negative-valued in a fiber, depending on the fiber design, techniques of dispersion compensation have become popular where lengths of positive and negative valued fiber are alternated, making the total dispersion over the link length negligible. Special dispersion-compensating fibers have been designed.

Another technique for combatting dispersion effects is to use the nonlinearities in the fiber to cause a pulse compression that balances the pulse spreading of the dispersion. As long as the amplitudes of the waves are large enough to keep the nonlinear effect present, this technique has proven merit. Optical fiber amplifiers have been used to accomplish this maintenance of the amplitude level as described later in the chapter.

7.2   Data Coding

Before describing the design of a link, we want to first consider the data coding techniques [1] that can be used. In transmitting a digital signal, recovery of the data at the receiving end sometimes requires a sampling circuit that operates at the clock rate of the system. As seen in our discussion of the eye diagram in Chapter 5, timing information is required to sample the signal when the S/N ratio is maximized and to maintain proper pulse spacings. Low rate links

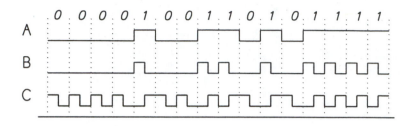

Figure 7.1 Examples of NRZ and RZ codes.

use simple free-running clocks that require periodic resynchronization for long-term operation. The period between resynchronization can be extended by the use of crystal-controlled clocks. Many links choose to encode the clock into the data stream for recovery at the receiver. While such encoding requires increased date rates in the fiber, the advantages of being able to recover the clock can frequently outweigh this disadvantage. Another alternative is to transmit the clock signal over a separate channel. Several standard techniques have evolved for the data encoding that will be described next. The various codes are represented in Fig. 7.1 and Table 7.1.

- The *nonreturn-to-zero (NRZ) code* is a code that is not required to return to the logical **0** level during the bit period. A string of logical **1**s will keep the output at the **1** level for the duration of the data string. The average (DC) output will vary with the signal content. While simple to generate and decode, this code has no clock-encoding capabilities or error detection or correction capability. This code requires the minimum bandwidth.

- The *return-to-zero (RZ) code* causes the output level to change from high to low (or low to high) within each bit period. (After encoding, we will call the period of the code that contains one bit of original data information the *bit period*. The rate of transmitting the original data will still be called the *data rate*; the rate of transmitting the encoded data will be the *baud rate*. The channel capacity required will be equal to the baud rate.) There are several variations of the RZ code. Curve B in Fig. 7.1 uses the first half of the bit period to represent the level of the bit; the second half of the bit period is always the **0** level. We see that two bauds (or code intervals) are required to represent a bit. Again, the DC level depends on the message content, since a string of **1**s would have a different average value than a string of **0**s. No clocking information is transmitted.

- The *Manchester code* is an RZ code. As shown in Fig. 7.1c, the signal makes a transition in the center of the bit period. The transition is downward for a **0** and upward for a **1**. Since the average of each bit period is constant, the DC value of the output is constant. The important feature of the Manchester code is that it encodes the clock by the transition in the center of the bit period. This transition is used at the receiver to recover the clock. (Integrated circuits are available that automatically encode and decode a data stream using the Manchester code.)

Many fiber-optic transmitter and receiver modules that are commercially available for the transmission of digital data require a specific coding of the signal (usually RZ). Inspection of the data sheets will reveal these requirements.

The increased bandwidth capability of a fiber-optic link allows increased freedom for the designer to incorporate coding for other purposes. Such coding usually inserts redundant bits

Code	Description	Baud rate	Clock encoded?	dc level?
Nonreturn-to-zero (NRZ)	1= High during entire bit period 0= Low during entire bit period	2x data rate	No	Yes
Return-to-zero (RZ)	1= Momentarily high 0= Low during entire bit period	2x data rate	No	No
Manchester	1= Positive transition 0= Negative transition	2x data rate	Yes	No

Table 7.1 Some simple digital communications codes.

Type	Short λ	Long λ
LED	< 150 Mb$\cdot$s$^{-1}\cdot$km	< 1.5 Gb$\cdot$s$^{-1}\cdot$km
Laser	< 2.5 Gb$\cdot$s$^{-1}\cdot$km	< 25 Gb$\cdot$s$^{-1}\cdot$km

Table 7.2 Source data rate-distance performance limits.

into the data stream (thereby increasing the required channel capacity). This extra data can be used to encode the clock, check for errors in the data stream, or to correct errors at the receiver. These techniques frequently use *block codes*. In an $mBnB$ code for example, a block (or word) of m bits is encoded into n bits ($m < n$). The encoded bits are transmitted by an NRZ or RZ code over the channel (with an increased channel-capacity requirement of n/m times the required capacity for the uncoded data). Integrated circuits are becoming available to generate various codes, primarily for error detection and correction.

Other types of data encoding are also possible with fiber links, including pulse position modulation (PPM), frequency-shift keying (FSK), or phase-shift keying (PSK). The simpler level-shifting methods are the most frequently used.

7.3 Source Selection

The starting point for a link design is choosing the operating wavelength, the type of source (i.e., laser or LED), and whether a single-mode or multimode fiber is required. In a link design, one usually knows (or estimates) the data rate required to meet the objectives. From this data rate and an estimate of the distance, one chooses the wavelength, the type of source, and the fiber type.

We begin knowing that a silica-based fiber operating with an LED source in the 800 to 900 nm region has a data rate-distance product of about 150 Mb$\cdot$s$^{-1}\cdot$km (due to the spectral dispersion). The same fiber operating with a laser source in the same region of the spectrum has a product of approximately 2.5 Gb$\cdot$s$^{-1}\cdot$km. In the region near 1300 nm, an LED can achieve a product of 1.5 Gb$\cdot$s$^{-1}\cdot$km and a laser can achieve products in excess of 25 Gb$\cdot$s$^{-1}\cdot$km. These benchmarks are summarized in Table 7.2. From these products, the decision is tentatively made whether to work with the lower-cost short-wavelength sources, or, if higher performance is required, to use the long-wavelength sources. A tentative choice of whether to use a laser or LED source can be made at this time as well, but this decision will be refined later.

The choice of fiber type involves the decision to use either multimode or single-mode fiber, and, if multimode, whether to use graded-index or step-index profiles. This choice is dependent on the allowable dispersion and the difficulty in coupling the optical power into the fiber. If an LED is chosen, then the obvious choice of fiber is a multimode fiber because the coupling losses into a single-mode fiber are too severe. (A long-wavelength LED can sometimes be used with a single-mode fiber for short-distance links.) For a laser source, either a multimode or single-mode fiber can be used. The choice depends on the required data rate, as losses in both types of fiber can be made quite low.

7.4   Power Budget

With the choice of the wavelength region and a tentative choice of fiber type made, one then proceeds to compute the power levels required at various locations in the circuit, as follows. From the data rate desired, combined with the desired bit-error rate (BER), we assume either a pin-diode detector or an APD and find the required detector power from sensitivity curves, such as Fig. 6.23 on page 204. The required receiver power can also be calculated from models, such as those used to derive expressions for the signal-to-noise ratio of the receiver in concert with calculations (or plots) that relate the error rate to the signal-to-noise ratio. As an initial estimate, we would probably begin by assuming a pin detector for its lower cost and simplicity of circuitry, unless we suspect that the application in mind is going to require an APD. The choice of preamplifier (i.e., whether it is a low-impedance receiver, an integrating front-end receiver, or a transimpedance receiver) depends on the data rate and the amount of noise that can be tolerated. From this, we obtain the receiver power P_R necessary to achieve the required performance.

Usually, we then choose a tentative source, based on considerations of the preceding section. With this choice of source, we know the power P_T available to be coupled into the fiber. The ratio of P_T/P_R, expressed in dB, is the amount of acceptable loss that can be incurred and still meet the specifications. This is expressed by the following equation,

$$\text{Losses(dB)} + l_M = 10 \log \left(\frac{P_T}{P_R} \right) , \tag{7.1}$$

where l_M is the *system margin*.

These losses can be allocated in any desired fashion by the system designer. Generally, the probable losses will be as follows:

- The source-to-fiber coupling loss l_T (dB) at the transmitter.

- The connector insertion loss l_C for each pair of connectors (or the splice insertion loss l_S for splices). If there are n connectors or splices, then the total losses will be nl_C (or nl_S), assuming that all losses are equal.

- The fiber-to-receiver loss l_R. In a well-designed link with a fairly sizable detector, this loss is usually negligible. Often the detector size is kept to a minimum to allow fast response speeds (since decreased size corresponds to decreased capacitance of the detector).

- allowance for device aging effects (especially for the reduction in laser power over time) and future splicing requirements l_A.

- fiber losses, expressed as the product of the losses per kilometer α times the link length L.

Equation 7.1 on the facing page can then be written as

$$10 \log \left(\frac{P_T}{P_R} \right) = \sum \text{losses} + l_M = \alpha L + l_T + n l_S + l_R + l_A + l_M \, . \tag{7.2}$$

After we solve Eq. 7.2 for the system margin, we find

$$l_M = P_T(\text{dBm}) - P_R(\text{dBm}) - \alpha L - l_T - n l_s - l_R - l_A \, . \tag{7.3}$$

(We can also use dBμ, if desired.) A positive system margin ensures proper operation of the circuit; a negative value indicates that insufficient power will reach the detector to achieve the BER. To extend the link length, we can either insert an opto-electronic repeater or an optical amplifier.

———————————————— ∞ ————————————————

Example: Let us design a system that operates at 830 nm at a data rate of 100 Mb·s^{-1}. We assume a BER of 10^{-9} so that we can use Fig. 6.23 on page 204.

Solution: Since this is a design example, we are free to select any components desired, constrained only by the availability of components with the parameters selected. This availability is determined by familiarity with the data sheets of a wide range of components that comes from experience with working in the lab or field with these components. Students are advised to check with your instructor for access to data sheets for representative components.

For the purpose of this problem, we will not calculate the problem from Eq. 7.3; instead, we will compute the problem step-by-step to illustrate the physical power levels required.

Assuming a pin-diode detector for simplicity, we find Fig. 6.23 on page 204 that we would require a receiver power level of approximately $P_R(\text{dBm}) = -40$ dBm to achieve this error rate at 100 Mb·s^{-1} (for a pin diode operating with a bipolar transistor amplifier).

For the source we will select an LED that produces 2 mW ($P_T(\text{dBm}) = 3$ dBm) in a spot that is 225 μm in diameter. (This is representative of a commercial LED that incorporates an internal microlens to produce the spot.)

For the fiber, we will assume that we will use a parabolic graded-index fiber ($g = 2$) with a 50 μm core and a 125 μm outer diameter that has a numerical aperture of 0.25.

We note that the effective source radius ($r_s = 112.5$ μm) is larger than the fiber radius ($a = 25.0$ μm), so we use Eq. 5.53 on page 152 to calculate the coupling efficiency.

$$\eta = [\text{NA}(0)^2] \left(\frac{a}{r_s} \right)^2 \left(\frac{g}{g+2} \right) = (0.25)^2 \left(\frac{25}{112.5} \right)^2 \left(\frac{2}{2+2} \right) \tag{7.4}$$

$$= 0.001543 = 0.1543\% \Rightarrow L_T = 28.1 \text{ dB} \, .$$

We note a considerable loss is incurred in coupling the light from the source into the fiber. This loss is subtracted from the optical power to produce

$$P_{T \text{ fiber}} = P_T - L_T = 3 - 28.1 = -25.1 \text{ dBm} \, . \tag{7.5}$$

(We note that the microlens has expanded the beam too much and, ideally, should be weaker in order to raise the coupling efficiency.)

To compute the losses at the receiver, we will assume that the detector area is bigger than the fiber core. The only losses, then, are the Fresnel reflection losses at the fiber-air and air-detector interfaces (assuming that no index matching liquid is used). These losses are approximately 0.2 dB per interface for a total loss of 0.4 dB at the receiver. If we will include a representative 6 dB allowance to compensate for aging effects, the required power in the fiber at the receiver is, then,

$$P_{\text{R fiber}} = -40 + 0.4 + 6 = -33.6 \text{ dBm}. \tag{7.6}$$

As an initial estimate, we will assume that there are no splices or connectors in the link, to get a feel for the representative length of the link with different fibers. Once we have computed these distances, we will iterate and include the splice effects.

A representative loss of graded-index fiber cable might be 5 dB/km. The fiber losses are found from

$$\alpha L = P_{\text{T fiber}} - P_{\text{R fiber}} = -25.1 - (-33.6) = 8.5 \text{ dB}. \tag{7.7}$$

The length of the fiber is

$$L = \frac{P_{\text{T fiber}} - P_{\text{R fiber}}}{\alpha} = \frac{8.7}{5} = 1.74 \text{ km}. \tag{7.8}$$

We now need to allow for splices or connectors. For the 1.7 km distance calculated, three additional joints might be assumed as typical since fiber cables might typically be available in lengths up to 1 km. If we assume splices with a loss of 0.1 dB per splice, then the 1.7 km distance is not changed very much since we have

$$
\begin{aligned}
\alpha L + 3L_S &= P_{\text{T fiber}} - P_{\text{R fiber}} \\
\alpha L &= P_{\text{T fiber}} - P_{\text{R fiber}} - 3L_S \\
L &= \frac{P_{\text{T fiber}} - P_{\text{R fiber}} - 3L_S}{\alpha} = \frac{8.7 - 0.3}{5} = 1.68 \text{ km}.
\end{aligned} \tag{7.9}
$$

For connectors with a loss of 1 dB, we have

$$
\begin{aligned}
\alpha L + 3L_C &= P_{\text{T fiber}} - P_{\text{R fiber}} \\
\alpha L &= P_{\text{T fiber}} - P_{\text{R fiber}} - 3L_C \\
L &= \frac{P_{\text{T fiber}} - P_{\text{R fiber}} - 3L_C}{\alpha} = \frac{8.7 - 3.0}{5} = 0.94 \text{ km}.
\end{aligned} \tag{7.10}
$$

We note now that we are below 1 km and only *two* pairs of connectors are required. Redoing the calculation for two pairs of connectors, we have

$$
\begin{aligned}
\alpha L + 2L_C &= P_{\text{T fiber}} - P_{\text{R fiber}} \\
\alpha L &= P_{\text{T fiber}} - P_{\text{R fiber}} - 2L_C \\
L &= \frac{P_{\text{T fiber}} - P_{\text{R fiber}} - 2L_C}{\alpha} = \frac{8.7 - 2.0}{5} = 1.34 \text{ km}.
\end{aligned} \tag{7.11}
$$

So, we have a conundrum. The assumption of two pairs of connectors being needed leads to a length of 1.34 km, a length that requires three pairs of connectors. (Another way of describing this situation is that the additional loss of the third pair of connectors makes the link margin negative.) The result? If we are required to use connectors with 1 dB loss per connection, we can achieve 1 km of link length with two connectors.

∞

7.5 Dynamic Range

In Eq. 7.2 on page 213 we left a term for the system margin l_M. While adequate system margin should be built into the link, too much margin can cause problems with the dynamic range of the system. For example, too much optical power at the detector might saturate the receiver. A calculation of the dynamic range for system operation may be useful for systems where we know the tolerances on the component characteristics, found on the data sheets. Most data sheets will identify a "typical" set of values, a set of "maximum" values, and a set of "minimum" values. Using a "best case/worst case" set of calculations, we can see whether our link has sufficient dynamic range. From Eq. 7.2 on page 213, we can write the system margin l_M as

$$l_M = l_{TR} - l_{\text{system}} \, , \tag{7.12}$$

where l_{TR} is the ratio of the transmitter power to the required receiver power, expressed in dB (i.e., $10\log(P_T/P_R)$), and l_{system} is the summation of the system losses, given by

$$l_{\text{system}} = \alpha L + l_T + nl_s + l_R + l_A \, . \tag{7.13}$$

The *dynamic range* of the system is found by first computing the system margin with the maximum power ratio l_{TR} and the minimum system losses. Then, the system margin is calculated for the minimum power ratio and the maximum expected system loss. The two computations are summarized by

$$l_{\text{M max}} = l_{\text{TR max}} - l_{\text{system min}} \tag{7.14}$$

$$l_{\text{M min}} = l_{\text{TR min}} - l_{\text{system max}} \, . \tag{7.15}$$

The system dynamic range $DR(\text{dB})$ is given by the difference in these values. Mathematically, we have

$$DR(\text{dB}) = l_{\text{M max}} - l_{\text{M min}} \, . \tag{7.16}$$

The receiver must have an equivalent dynamic range in order for the system to work properly.

We are basically concerned with keeping the power at the receiver above the minimum detectable power of the detector $P_{\text{R min}}$ and below the maximum-rated power of the detector $P_{\text{R max}}$. We can find the power at the detector from Eq. 7.2 on page 213 if we calculate the power values in dBm (or dBμ). The power is

$$P_R(\text{dBm}) = P_T(\text{dBm}) - l_{\text{system}} \, . \tag{7.17}$$

The transmitter power is a linear function of the drive current in the device. For a specified drive current there will be device-to-device variations in output power. There are similar variations in the fiber losses.

For example, the HFBR-1501 transmitter from Hewlett Packard is specified to produce a maximum power at a location 0.5 m into the fiber of -8.4 dBm with a drive current of 60 mA and to produce a minimum power of -14.8 dBm at the same current. Hewlett Packard's HFBR-3500 fiber-optic cable is specified to have a maximum loss of 0.63 dB/m at 665 nm (the wavelength of the HFBR-1501 source) with a minimum specified loss of 0.3 dB/m. The HFBR-2500 receiver used as a companion to the prior components requires a power level below -21.0 dBm to properly register a logical **0**, and requires a power level between a minimum of -21.0 dBm and a maximum of -12.5 dBm to avoid having a logical **1** confused with a logical **0**. (We

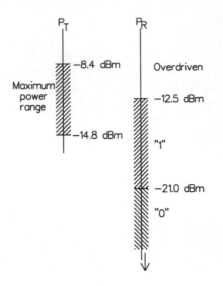

Figure 7.2 Location of power values for a Hewlett Packard data link.

will call these values $P_R(\mathbf{1})_{\mathrm{min}}$ and $P_R(\mathbf{1})_{\mathrm{max}}$, respectively.) Operating at or below the maximum power rating avoids overloading the detector. Figure 7.2 illustrates the locations of these power values.

If we assume that there are no connectors, no receiver losses, and no allowance for aging, then only the cable losses will contribute to the system losses. Since the optical power is specified at the output of a 0.5 m length of HFBR-3500 cable, we will have to represent the effective length of the link as $L - 0.5$ meters, where L is the total length of the link. The maximum/minimum conditions become

$$P_{\mathrm{R\,max}} = P_{\mathrm{T\,max}} - \alpha_{\mathrm{min}}(L - 0.5) \le P_R(\mathbf{1})_{\mathrm{max}} \tag{7.18}$$

$$P_{\mathrm{R\,min}} = P_{\mathrm{T\,min}} - \alpha_{\mathrm{max}}(L - 0.5) \ge P_R(\mathbf{1})_{\mathrm{min}}, \tag{7.19}$$

where the power levels are in dBm or dBμ. The first equation avoids overloading the receiver; the second equation ensures the minimum required power at the receiver.

————————————— ∞ —————————————

Example: Let's consider a 2 m link using the Hewlett Packard components just described. We can use Eq. 7.17 on the preceding page to calculate P_{max}.

Solution: We will assume that we transmit a logical **1**. We could be in danger of overdriving the receiver if we combined a source at maximum power output with a fiber that has minimum attenuation. We calculate the maximum transmitter power allowed in this case as

$$P_{\mathrm{T\,max}} = P_R(\mathbf{1})_{\mathrm{max}} + \alpha_{\mathrm{min}}(L - 0.5) \tag{7.20}$$

$$= -12.5 + (0.3)(2.0 - 0.5) = -12.05 \text{ dBm}.$$

Similarly, we could be in danger of underdriving the receiver when we combine the minimum transmitter power with a fiber having maximum attenuation. The expression for the minimum transmitter power allowed in this case is

$$P_{T\,min} = P_R(1)_{min} + \alpha_{max}(L - 0.5) \tag{7.21}$$
$$= -21.0 + (0.63)(2 - 0.5) = -20.1 \text{ dBm}.$$

Hence the maximum transmitter power for a 2 m link is -12.05 dBm (62.3 μW) and the minimum power is -20.1 dBm (9.77 μW). From these values we now want to compute the drive current of the LED source required to produce each value of power.

For the source specified, we can produce a maximum output of -8.4 dBm (144 μW) at the maximum drive current of 60 mA. The output power of any LED is known to be linearly dependent on the drive current,

$$\frac{P}{P_{max}} = \frac{I}{I_{max}}. \tag{7.22}$$

So, to produce the maximum allowed transmitter power of -12.05 dBm (62.3 μW), we can find the required drive current from the proportionality of

$$I = I_{max}\left(\frac{P}{P_{max}}\right) = (60 \times 10^{-3})\left(\frac{62.3}{144}\right) = 26.0 \text{ mA}. \tag{7.23}$$

The manufacturer recognizes that every device will not have the same output power at a given drive current, because of device-to-device variations. Hence the specifications indicate both the maximum power out at the maximum allowed drive current and the minimum power out at that current. The lowest rated output power at the 60 mA maximum-rated drive current is -14.8 dBm (33.1 μW). Hence, to produce the minimum output of 9.77 μW, we require a minimum drive current of

$$I = I_{max}\frac{P}{P_{max}} \tag{7.24}$$
$$I_{min} = (60 \times 10^{-3})\left(\frac{9.77}{33.1}\right) = 17.71 \text{ mA}.$$

(We note that Eq. 7.23 gave the *maximum* drive current.)

Thus, we conclude that a drive current between the values of 17.71 and 26.0 mA will ensure proper operation of the 2 m link if the source and receiver meet specifications and if the other sources of system losses (other than fiber loss) are truly negligible.

— ∞ —

If the same calculations are repeated for a 6 m spacing, the results are a maximum optical power of -10.85 dBm and a minimum optical power of -17.35 dBm, a narrower range of values. As the distance increases, the values approach each other until the difference goes to zero. The link can no longer operate for longer distances. The maximum length of the link is limited by the dynamic range of the receiver and the properties of the transmitter and fiber.

We can calculate this distance where we are dynamic-range limited by equating $P_{T\,max}$ with $P_{T\,min}$ in Eqs. 7.18 and 7.19 on the facing page. Combining these equations gives

$$P_R(1)_{max} + \alpha_{0\,min}(L_{max} - 0.5) = P_R(1)_{min} + \alpha_{max}(L_{max} - 0.5). \tag{7.25}$$

Solving for $L_{max} - 0.5$,

$$(L_{max} - 0.5) = \frac{P_R(1)_{max} - P_R(1)_{min}}{\alpha_{max} - \alpha_{min}} \tag{7.26}$$
$$= \frac{-12.5 + 21}{0.63 - 0.30} = 25.7 \text{ m}.$$

Hence, we find that these components have a dynamic-range-limited transmission distance of 25.7 m. From Eq. 7.26 on the page before we see that the way to increase this distance is to increase the receiver dynamic range (i.e., $P_R(1)_{\text{max}} - P_R(1)_{\text{min}}$) or to use a fiber with a tighter tolerance on the loss coefficient (i.e., $\alpha_{\text{max}} - \alpha_{\text{min}}$).

7.6   Timing Analysis

Once we have completed the power-budget analysis to ensure the proper maximum and minimum optical-power levels, the next step is to analyze the speed of the devices to ensure that the data-rate requirement can be met. The usual technique is to analyze the rise time of the system in terms of the rise times of the components. The *rise time of the system* t_{sys} is given by the following relationship [2],

$$\Delta t_{\text{sys}} = \sqrt{\sum_{I=1}^{N} \Delta t_i^2} \, , \tag{7.27}$$

where Δt_i is the rise time of each component in the system. Hence, we see that the system rise time is the square root of the sum of the squares of the system components. The four components of the system that can contribute to the system rise time are as follows:

- The rise time of the transmitting source Δt_S. This quantity is usually found in the spec sheet of the emitter or is measured with a pulsed input and a fast detector circuit.

- The rise time of the receiver Δt_R. This rise time can be measured directly in the time domain with a pulsed signal or can be calculated from the bandwidth as measured in the frequency domain. If B_{3dB} is the 3-dB frequency bandwidth of the receiver, the rise time can be calculated as

$$\Delta t_R = \frac{0.35}{B_{\text{3dB}}} \, . \tag{7.28}$$

- The material-dispersion time of the fiber Δt_{mat}. Equation 3.24 on page 45 gives the dispersion relation as

$$\Delta t_{\text{mat}} = -\frac{L}{c} \frac{\Delta \lambda}{\lambda} \left(\lambda^2 \frac{d^2 n}{d\lambda^2} \right) \, . \tag{7.29}$$

From this equation we see that the system rise time will be dependent on the length of the link (if this term is not negligible compared to the other terms).

- The modal-dispersion time of the fiber link Δt_{modal}. This term is dependent on many variables: the excitation conditions, the fiber construction, the fiber length, and the effect of splices on the modal distribution. For a step-index fiber with length L, the modal-dispersion delay is given by

$$\Delta t_{\text{modal}} = \frac{L(n_1 - n_2)}{c} \, . \tag{7.30}$$

The modal delay time for a graded-index fiber is a more complicated expression. The delay time is a function of the index profile g and often g is optimized to reduce the delay. For a parabolic-index fiber ($g = 2$), the delay is estimated as

$$\Delta t_{\text{modal}} = \frac{L}{c} \frac{\text{NA}(0)^2}{8 n_1^2} \, . \tag{7.31}$$

(In a single-mode fiber, of course, the modal delay is not present.)

In the above equations, we have assumed that there are no joints in the fiber, that the fiber was uniformly excited, and that modal equilibrium did not have a chance to occur. In an actual fiber link, any or all of these conditions might be violated. For these cases, empirical formulas have evolved which might be of use to the link designer. For example, the bandwidth $B_M(L)$ associated with a length L of fiber can be extrapolated from the bandwidth of a 1 km length of that fiber $B_M(1 \text{ km})$ by the expression [2]

$$B_M(L) = \frac{B_M(1 \text{ km})}{L^q}, \tag{7.32}$$

where q is an empirically fit parameter. For lengths where a steady-state modal equilibrium has been reached, the value of q is 0.5. For short distances where a steady-state modal equilibrium is *not* reached, the value of q is 1. For distances between these extremes, a reasonable estimate of q is 0.7. The relationship between the bandwidth of Eq. 7.32 and the rise time of the fiber, as required by Eq. 7.27 on the facing page, is given by

$$\Delta t_M = \frac{0.44}{B_M(L)}. \tag{7.33}$$

For the case where N sections of the same fiber are fastened together to form a long length, the total pulse broadening $\Delta t_M(N)$ can be estimated by [2]

$$\Delta t_M(N) = \left[\sum_{n=1}^{N} (\Delta t_n)^{(1/q)} \right]^q, \tag{7.34}$$

where Δt_n is the pulse spreading in the individual sections and q is the empirical value assigned as discussed in the previous paragraph.

—————————— ∞ ——————————

Example: Consider the 100 Mb·s^{-1} link previously described in the power-budget analysis. The postulated LED might have a rise time of 8 ns and a spectral width of 40 nm.

<u>Solution:</u> The pin diode might have a typical rise time of 10 ns.

For a silica fiber operating at 830 nm, the value of $\lambda^2(d^2n/d\lambda^2)$ is approximately 0.024 (from Fig. 3.8 on page 48). For a link distance of 2.5 km, the material-dispersion delay time is

$$\Delta t_{\text{mat}} = -\frac{L}{c} \frac{\Delta\lambda}{\lambda} \left(\lambda^2 \frac{d^2n}{d\lambda^2} \right) = -\left(\frac{2.5 \times 10^3}{3.0 \times 10^8} \right) \left(\frac{40}{830} \right) (0.024) \tag{7.35}$$

$$= -9.64 \times 10^{-9} \text{ s} = -9.64 \text{ ns}.$$

A typical intermodal dispersion for graded-index fibers is 3.5 ns/km. Hence, a 2.5 km link has $\Delta t_{\text{modal}} = 8.8$ ns. Calculating the system's rise time, we have

$$\Delta t_{\text{sys}} = \sqrt{(\Delta t_S)^2 + (\Delta t_R)^2 + (\Delta t_{\text{mat}})^2 + (\Delta t_{\text{modal}})^2} \tag{7.36}$$

$$= \sqrt{8^2 + 10^2 + 9.64^2 + 8.8^2} \text{ ns} = 18.3 \text{ ns}.$$

—————————— ∞ ——————————

We now must compare the system rise time with the required bit period T_B to achieve a bit rate B_R of 100 Mb·s^{-1} communications, where $T_B = 1/B_R$. The 100 Mb·s^{-1} bit rate of the previous example has a bit period of 10^{-8} s (or 10 ns). The requirement on the system rise time depends on how we choose to encode the data. Generally, the system rise time must be less than 70% of the bit period if NRZ (nonreturn-to-zero) coding is used, and it must be less than 35% of the bit period if RZ (return-to-zero) coding is used. These conditions are written as

$$\Delta t_{\text{sys}} \leq 0.7 T_B \quad \text{(NRZ coding)}$$
$$\Delta t_{\text{sys}} \leq 0.35 T_B \quad \text{(RZ coding)} . \tag{7.37}$$

∞

Example: The system rise time of the previous example was 18.3 ns. Using this value we can calculate the data rate that the system can support by inverting the previous equations:

Solution: For NRZ coding,

$$\Delta t_{\text{sys}} \quad \leq \quad 0.7 T_B \tag{7.38}$$
$$T_B \quad \geq \quad \frac{\Delta t_{\text{sys}}}{0.7}$$
$$B_R \quad \leq \quad \frac{0.7}{\Delta t_{\text{sys}}} \leq \frac{0.7}{18.5 \times 10^{-9}} \leq 38.3 \text{ Mb} \cdot \text{s}^{-1} .$$

For RZ coding, we have

$$B_R \quad = \quad \frac{0.35}{\Delta t_{\text{sys}}} = \frac{0.35}{18.3 \times 10^{-9}} = 19.1 \text{ Mb} \cdot \text{s}^{-1} . \tag{7.39}$$

Surprisingly, *neither* coding will support the desired 100 Mb·s^{-1} data rate. To consider possible solutions to this dilemma, we look at Eq. 7.36 on the preceding page. Each term of the sum must *individually* be below the desired speed of response for the system. The receiver speed and the material dispersion are too large; the modal dispersion contribution is small because the distance is so short. We need to use a faster detector. To reduce the material dispersion, inspection of Eq. 7.29 on page 218 reveals that one should reduce $\Delta\lambda$. Two methods of doing this would be

1. to use an LED with a longer wavelength (while keeping $\Delta\lambda$ constant), or
2. to use a laser source with its reduced value of $\Delta\lambda$. (The choice of a laser source would also allow increased distance in exchange for the increased cost and increased complexity.)

After choosing a new set of components, the power budget and timing analysis, need to be recalculated to ensure proper operation of the system.

∞

7.6.1 Dispersion-Limited Transmission Distance

The previous calculation illustrates an important result. In a fiber-optic system at long distances or high data rates, the system can be limited either by the losses (*attenuation-limited transmission*) or, assuming that the link is not limited by the source or detector speed, by the dispersion of the fiber (*dispersion-limited transmission*).

The calculation of the dispersion-limited transmission distances is easily accomplished. We ignore the rise times of the source and receiver since we want to find the distances that are limited only by the fiber dispersion; then we isolate each dispersion factor and consider it as the sole source of dispersion.

Material Dispersion-Limited Transmission

Consider material dispersion in a link using RZ coding. We require

$$\Delta t_{\mathrm{mat}} \leq 0.35 T_B \,. \tag{7.40}$$

Hence

$$0.35 T_B \;\geq\; \left(\frac{L_{\max}}{c}\right)\left(\frac{\Delta\lambda}{\lambda}\right)\left(\lambda^2 \frac{d^2 n}{dn^2}\right) \tag{7.41}$$

$$L_{\max} \;=\; (0.35 T_B c)\left(\frac{\lambda}{\Delta\lambda}\right)\left(\frac{1}{\left(\lambda^2 \dfrac{d^2 n}{d\lambda^2}\right)}\right)\,.$$

(The reader is encouraged to find the corresponding expression for NRZ coding.)

Modal Dispersion-Limited Transmission

For modal dispersion in a *step-index fiber*, we have

$$L_{\max} = \frac{0.35 c T_B}{n_1 - n_2} = \frac{0.35 c}{(n_1 - n_2) B_R} = \frac{0.7 c n_1}{\mathrm{NA}^2 B_R}\,. \tag{7.42}$$

For modal dispersion in a *graded-index fiber*, we have

$$L_{\max} = \frac{2.8 T_B c n_1^2}{[NA(0)]^2} = \frac{2.8 c n_1^2}{[NA(0)]^2 B_R}\,. \tag{7.43}$$

(Again, the reader is encouraged to find the corresponding expression for NRZ coding.)

These latter three equations are useful for estimating the dispersion-limited transmission distances when waveguide dispersion is not significant (i.e., the results are not valid near 1300 nm).

---------------------------------- ∞ ----------------------------------

Example: Calculate the modal-dispersion-limited transmission distance for a 50 Mb·s^{-1} data link using SI and GI fibers with $\Delta = 1\%$ and $n_1 = 1.45$. The coding is return-to-zero.

Solution: We begin with $n_1 - n_2 = \Delta n_1 = 0.01(1.45) = 0.0145$.

For the SI fiber,

$$L_{\max\,\mathrm{SI}} = \frac{0.35 c}{(n_1 - n_2)\, B_R} = \frac{(0.35)(3.0 \times 10^8)}{(0.0145)(50 \times 10^6)} = 1.448 \times 10^2 \text{ m}\,. \tag{7.44}$$

For the GI fiber,

$$L_{\max\,\mathrm{GI}} \;=\; \frac{2.8 c n_1^2}{[NA(0)]^2\, B_R} = \frac{2.8 c n_1^2}{2 n_1 (n_1 - n_2)\, B_R} = \frac{1.4 c n_1}{(n_1 - n_2)\, B_R} \tag{7.45}$$

$$=\; \frac{(1.4)(3.0 \times 10^8)(1.45)}{(0.0145)(50 \times 10^6)} = 840 \text{ m}\,.$$

The small transmission distances are caused by the use of a multimode fiber for a fairly high-data-rate signal instead of a single-mode fiber. Whether these distances are acceptable depends on the application.

---------------------------------- ∞ ----------------------------------

These limitations on the link length have become more important as achievable data rates have increased. With the advent of multi-gigabit-per-second links, dispersion-limited transmission has become more common. Techniques to lower the dispersion or to compensate for it have become increasingly popular. These dispersion compensation techniques are discussed later in this chapter.

Attenuation-Limited Transmission Length

For comparison purposes we frequently want to calculate the maximum link distance for a system limited only by the fiber attenuation. The formula for this is

$$L_{\max} = \frac{P_T(\text{dBm}) - P_R(\text{dBm})}{\alpha_{\text{fiber}}}. \tag{7.46}$$

(One could also use dBμ for the units of power.) Here P_T is the power of the transmitter (usually independent of data rate, except for very high data rates), P_R is the power that the receiver requires to maintain the bit-error rate or the signal-to-noise ratio, and α is the fiber-attenuation value. To overcome the limits of attenuation, we must generate a stronger signal. Two techniques are used, the opto-electronic repeater that receives and regenerates the digital pulses and the optical amplifier, to be discussed later in this chapter.

---------------------------------- ∞ ----------------------------------

Example: Consider a graded-index fiber with $n_1 = 1.45$ and $\Delta = 1\%$ and a loss of 1 dB/km. It is used with an 850 nm source that produces an output power (in a fiber) of -10 dBm. The source linewidth is 60 nm. The receiver is a pin-diode receiver that requires a power given by

$$P_R(\text{dBm}) = -65.0 + 20 \log \text{DR}'[\text{Mb} \cdot \text{s}^{-1}], \tag{7.47}$$

to maintain a BER of 1×10^{-9} where DR$'$ is the data rate in Mb·s^{-1}. The coding is RZ coding. Set up the equations to find ...

(a) ... the material-dispersion-limited distance,

Solution: The equation for the material-dispersion-limited distance is

$$L_{\max} = \frac{0.35c\lambda}{B_R \, \Delta\lambda \left(\lambda^2 \dfrac{d^2 n}{d\lambda^2} \right)}. \tag{7.48}$$

The value of $\lambda^2(d^2 n/d\lambda^2)$ is estimated to be 0.022 from Fig. 3.8 on page 48. Hence, we have

$$L_{\max} = \frac{(0.35)(3.0 \times 10^8)(850 \times 10^{-9})}{(B_R)(60 \times 10^{-9})(0.022)} = \frac{6.76 \times 10^{10}}{B_R}. \tag{7.49}$$

(b) ... the modal-dispersion-limited distance.

Solution: This distance is

$$
\begin{aligned}
L_{\max} &= \frac{2.8 c n_1^2}{[NA(0)]^2 \, B_R} = \frac{1.4 c n_1^2}{n_1^2 \, \Delta \, B_R} = \frac{1.4 c}{\Delta \, B_R} \\
&= \frac{1.4 (3.0 \times 10^8)}{(0.01)(B_R)} = \frac{4.14 \times 10^{10}}{B_R} .
\end{aligned} \tag{7.50}
$$

(c) ... the attenuation-limited distance as functions of the data rate.

Solution: This distance is

$$
\begin{aligned}
L_{\max} &= \frac{P_T(\mathrm{dBm}) - P_R(\mathrm{dBm})}{\alpha} = \frac{-10 - (-65.0 + 20 \log(B_R))}{1.0} \\
&= 55.0 - 20 \log(B_R) .
\end{aligned} \tag{7.51}
$$

(d) Plot the results for a data rate range extending from 1 kb·s^{-1} to 10 Mb·s^{-1}.

Solution: Figure 7.3 shows the plot of these curves. Note that for data rates below about 700 kb·s^{-1}, the link length is attenuation-limited. Above 700 kb·s^{-1}, the link is limited by the modal dispersion. The material-dispersion limit is slightly longer than the modal-dispersion limit.

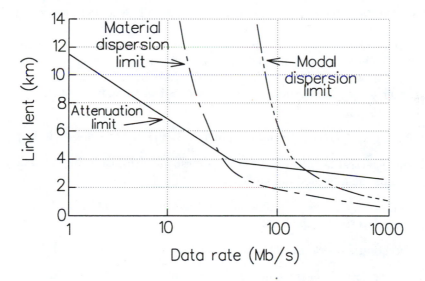

Figure 7.3 Maximum transmission distance vs. data rate as determined by the attenuation limit and dispersion limits.

7.7   Commercial Fiber-Optic Modules

For moderate data rates and moderate distance transmission, the link design reduces to the selection of commercially available transmitter and receiver modules. These devices are easily used. They provide the proper signal levels so that TTL or ECL input voltages will produce TTL or ECL output voltages. The user design is only in the selection of the proper modules and in the design of the electronic interface circuitry. The optical portion of the design has been optimized (the user hopes) by the manufacturer. Only a careful reading of the spec sheets and some experimentation (or a recalculation using our design procedure) will reveal if a particular module set can be successfully used.

7.8   Optical Amplifiers

Once the optical power in the link reaches the minimum detectable power, we need to find a way to increase the power (if we are not yet at the link destination). Two techniques have been used, repeaters and optical amplifiers.

An electronic *repeater* consists of an optical detector, signal-recovery circuits, and an optical source. The optical detector transforms the optical signal into an electrical one. The signal-recovery circuits recover the clock from the data and detect the data stream in electrical form. The recovered data are retransmitted by the optical source. Many long-distance links use this technique, but it suffers the disadvantage of requiring a full electronic recovery system at each repeater.

An alternative approach is to amplify the optical signal without ever forming an electronic equivalent. This technique has potential economic advantages and is conceptually simple [3]. Both semiconductor amplifiers and fiber amplifiers are being studied, with recent advances in fiber amplifiers showing great potential. The *ideal* optical amplifier would have the following properties:

- provide high gain (power gains on the order of 30 dB or more are desirable),

- have a wide spectral bandwidth (to allow several wavelengths to be transmitted simultaneously through the fiber and amplifier),

- provide uniform gain over amplifier spectral width (to maintain the relative strengths of the different spectral components),

- allow bidirectional operation (to allow bidirectional use of the optical fiber),

- add minimum noise from the amplifier,

- demonstrate lack of interference between multiple wavelengths within the spectral band (i.e., no crosstalk between spectral channels),

- provide a gain that is independent of the signal polarization (to maintain constant gain as the polarization of the signal varies due to environmental factors along the length of the single-mode fiber),

- have a low insertion loss when the amplifier is inserted into a fiber link (to avoid using significant amounts of the amplifier's gain just to overcome the insertion loss),

- have a gain that is applicable over a wide range of input power levels (i.e., a gain that does not saturate at high values of input power),

- use an amplifier pump source that is small and compact (on the order of the size of a diode laser), and

- have a good conversion efficiency (at converting the pump power into amplifier gain.

The addition of amplifiers to a data link brings its own problems, however. The amplifiers add a different type of noise to the signal (due to their broadband spontaneous emissions). The spontaneous emissions that are guided by the amplifier fiber increase as they proceed, resulting in *amplified spontaneous emission* (or ASE) noise. Figure 7.4 on the following page shows an example of the optical power entering and leaving an amplifier. At the output of the amplifier, the narrowband signal is joined by the broadband ASE noise. The added noise of an amplifier is represented by the noise figure F, defined as

$$F = \frac{\text{SNR}_{\text{in}}}{\text{SNR}_{\text{out}}},\qquad(7.52)$$

where SNR_{in} and SNR_{out} are the values of the signal-to-noise ratios at the input and output of the amplifier. (The SNRs are calculated by hypothesizing that the receiver is located at either position.)

Another problem is that the amplifier gain can be saturated by the introduction of too strong a signal. The gain coefficient g of a material (and, hence, the overall gain of a given length of the material) can be modeled [4] by

$$g = \frac{g_0}{1 + \dfrac{P}{P_{\text{sat}}}},\qquad(7.53)$$

where g_0 is the unsaturated gain coefficient (whose value is determined by the material and the amplifier pump power), P is the strength of the optical wave being amplified, and P_{sat} is an amplifier parameter called the *amplifier saturation power*. The overall power gain of the amplifier is modeled as [4]

$$G = G_0 e^{\frac{(1-G)P_{\text{in}}}{P_{\text{sat}}}},\qquad(7.54)$$

where G_0 is the unsaturated gain of the amplifier, P_{in} is the input power, and P_{sat} is the amplifier saturation power. This nonlinear equation needs to be solved for G, once G_0, P_{in}, and P_{sat} are known or specified. This gain saturation means that the noise will reduce the amplifier gain below the value that it might have had without the presence of the ASE noise. The ASE noise robs the amplifier of part of its potential signal gain; the stronger the noise, the more gain it robs. The *gain saturation* is also detrimental because of nonlinear effects that can increase the crosstalk in a link carrying more than one optical signal.

Hence, the dynamic ranges of the signal and the noise in the link become important; too strong a signal will saturate the amplifier and too weak a signal will be lost in the noise. The presence of the noise leads to an interesting trade-off. The noise becomes larger as the gain grows larger. The least noise is generated when the amplification is distributed over the entire length of the fiber (rather than occurring at discrete locations within the link). The optimum gain for this case is the gain required to just compensate for the fiber losses [5]. In other words, the pump energy is used to make the fiber transparent (i.e., zero losses) to the signal. The problem is that we do not know how to make long lengths of weakly amplifying fiber efficiently, nor do

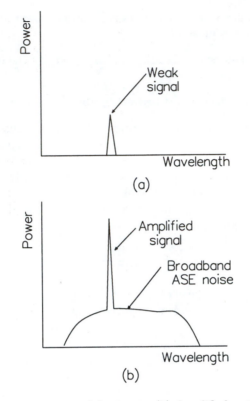

Figure 7.4 (a) Weak optical signal at amplifier input. (b) Amplified optical signal at amplifier output with amplified spontaneous emission (ASE) noise.

we know how to maintain a uniform pump-power level over the entire length of the fiber (since we can introduce the pump only at the transmitter and receiver end of the fiber). The second best solution is to have a lot of small-gain amplifiers spaced closely together. This is uneconomic, since the amplifiers are currently quite expensive. To keep the total amplifier costs down, we have to space our amplifiers widely apart and accept the noise penalty.

Optical amplifiers and isolators also present problems in the use of optical time-domain reflectometers (see Chapter 3). Design efforts have begun [6] to allow the amplifiers to be bypassed for measurement purposes.

Besides being used as in-line amplifiers to replace repeaters, optical amplifiers can also be useful as preamplifiers (placed immediately before the receiver as in Fig. 7.5a, postamplifiers (placed immediately after the laser source as in Fig. 7.5b, or to compensate for splitting losses when an optical signal passes through a 1×N splitter [7]. When used as a preamplifier, the important parameters are the gain and the noise figure. Gains in excess of 30 dB have been demonstrated with noise figures as low as 3–5 dB [8]. The sensitivity of the receiver with the preamplifier can be improved as much as 15–20 dB over the detector without the preamplifier. Postamplifiers emphasize the power out of the device (as well as reasonable size and conversion efficiencies. Power outputs as high as +23 dBm have been achieved in fiber postamplifiers [8].

In-line amplifiers are measured by the span length (i.e., the distance between amplifiers) and the noise that they add. These devices are discussed in detail in the following material.

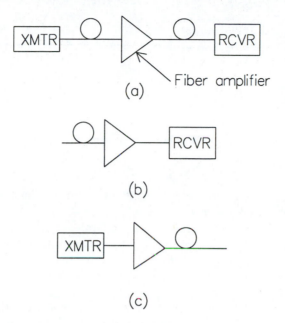

Figure 7.5 Optical amplifier used as (a) an in-line amplifier, (b) a preamplifier (before a receiver), and (c) a postamplifier (after the sources).

7.8.1   Erbium-Doped Fiber Amplifiers

An optical fiber doped with erbium ions has become a potential all-optical amplifier, called an *erbium-doped fiber amplifier* [9–12], operating in the 1520 to 1550 nm window of the fiber (near the fiber-optic attenuation minimum). Diode-laser pumps operating at 950 or 1480 nm are the pump source, with most amplifiers using a 1480-nm diode-laser source for a pump. Gains of 30 to 40 dB with output power levels of 1 mW can be achieved in these fiber amplifiers with lengths of 10s of meters [13]. The pump power required for such gains is a few 10s of mW, a level that is achievable with modern sources. Figure 7.6 on the next page illustrates the coupling of the pump power into the fiber amplifier. (When the pump enters the amplifier at the same end as the signal, the beams are *copropagating*. An alternative geometry is possible where the pump laser is introduced at the opposite end of the fiber amplifier and propagates in the opposite direction from that of the signal. The beams are then *counterpropagating*.) The insertion loss when the fiber amplifier is spliced into the link is low, typically less than 1 dB. Among the potential disadvantages of these amplifiers are that they are limited to operation near 1550 nm (although they are moderately broadband in the vicinity of that wavelength), that they require a non-trivial amount of pump power (40 to 50 mW) at 1480 nm, that they require lengths of fiber that exceed several meters (i.e., shorter lengths do not give as much gain), and that they add noise to the optical signal [14, 15]. More complicated amplifiers can be realized by cascading

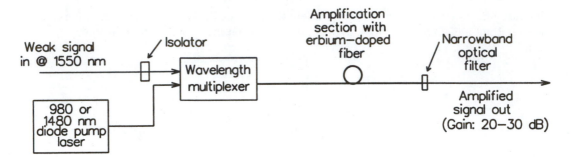

Figure 7.6 Erbium fiber laser geometry.

amplifier stages [8]; the incorporation of optical filters and other devices allows tailoring of the overall multi-stage amplifier parameters.

Investigations to produce fiber amplifiers that operate at 1300 nm have centered on the praseodymium-doped fibers [16]. One problem is that silica is not a good host for these ions; another host must be used. Efforts have been focused on fluoride glasses to serve as the host material. Gains of 20 to 30 dB have been demonstrated with these materials with pump-laser powers ranging from 100 to 300 mW.

The interaction between the signal beam, the amplifying medium, the pump beam, and the amplified spontaneous emission are described by models [4, 17] that are beyond the scope of this text. We can, however, perform some simple analyses based on the results of those models.

7.8.2 Cascaded Amplifiers

A typical application of the amplifiers in a long-distance high-data-rate link is shown in Fig. 7.7a, where the fiber amplifiers act as in-line repeaters. The amplifiers serve to boost the signal strength after attenuation from traversing the fiber between the repeaters. The signal is successively amplified at each amplifier, as shown in Fig. 7.7b. Unfortunately, the process cannot be carried out indefinitely, as the amplifiers all add noise to the signal.

The primary noise that is added to the signal is due to light from spontaneous emissions occurring within the fiber amplifier. A portion of the spontaneous emitted light is in the same direction as the signal and is amplified the remaining length of the fiber. (Generally, the co-propagating pump geometry will introduce less noise than the counterpropagating pump beam.) This added light, called *amplified spontaneous emission* or ASE, is characterized as having a wide spectrum compared to the signal as seen in Fig. 7.4b. The spectral width of this light can be reduced by including a narrow optical filter centered on the signal wavelength. (This filter will increase the insertion loss of the amplifier, however.)

To perform our analysis we will follow the work of Giles and Desurvire [4]. We want to consider a string of N amplifiers. To make the analysis tractable, we will assume that the amplifiers are equally spaced with an equal multiplicative fiber transmission loss of L between amplifiers. Each amplifier can have a different gain, in general, of G_i. (The size of the gain is controlled by the amount of pump power and the length of the amplifying fiber. We note, in passing, that, for a given fiber and pump strength, there is an optimum fiber length to maximize the gain.) Both L and G_i are numerical values; they are *not* in dB. We will assume that each amplifier will amplify

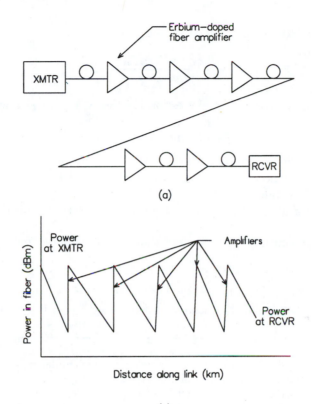

Figure 7.7 (a) Example of system using erbium-doped fiber amplifiers. (b) Power levels in the fiber.

the signal and the noise from the previous amplifiers and will also add its own ASE power to the output. We assume that the output of each amplifier has an optical filter of bandwidth B_0, centered on the signal wavelength, to reduce the broadband ASE noise. At the output of the i-th amplifier, the output signal power $P_{\mathrm{s,i,out}}$ is

$$P_{\mathrm{s,i,out}} = G_i P_{\mathrm{s,i,in}} = G_i L P_{\mathrm{s},i-1,\mathrm{out}}, \qquad (7.55)$$

where $P_{\mathrm{s,i,in}}$ is the optical signal power entering the i-th amplifier, and $P_{\mathrm{s},i-1,\mathrm{out}}$ is the signal power out of the previous amplifier. The total ASE power out of the i-th amplifier $P_{\mathrm{ASE,i,out}}$ is given by [4]

$$
\begin{aligned}
P_{\mathrm{ASE,i,out}} &= G_i P_{\mathrm{ASE,i,in}} + b P_{\mathrm{ASE,i}}(G_i, n_{\mathrm{sp}}) \qquad (7.56) \\
&= G_i L P_{\mathrm{ASE},i-1,\mathrm{out}} + b P_{\mathrm{ASE,i}}(G_i, n_{\mathrm{sp}}),
\end{aligned}
$$

where $P_{\mathrm{ASE,i,in}}$ is the total ASE power from previous amplifiers at the input to the i-th amplifier, $b = B_o/\Delta\nu$ (with B_o being the the bandwidth [in Hz] of the optical bandpass filter placed after each amplifier and $\Delta\nu$ being the spectral width of the amplifying medium), and $P_{\mathrm{ASE,i}}$ is the portion of the forward-propagating ASE power that is captured and guided down the fiber. This latter power is a function of the gain of the amplifier and n_{sp}, the ratio of the ASE power at the output of the amplifier $P_{ASE}(l)$ and $2h\nu\,\Delta\nu\,(G-1)$, i.e.,

$$n_{sp} = \frac{P_{ASE,out}}{2h\nu\,\Delta\nu\,(G-1)}.\tag{7.57}$$

This denominator is used because the expression for the ASE optical power P_{ASE} in a single mode is [14]

$$P_{ASE} = \mu h\nu\,\Delta\nu\,(G-1),\tag{7.58}$$

where μ is a measure of the population inversion efficiency within the amplifying medium (with a maximum value of 1). Hence, we see that n_{sp} is a normalized ASE power. In Eq. 7.57 on the preceding page, the initial values are $P_{s,0,out} = P_s$, where P_s is the signal power in the fiber at the transmitter and $P_{ASE,0,out} = 0$ (i.e., there is no ASE power present at the transmitter end of the link).

Adding these power contributions, we find that the total power out of the i-th amplifier, $P_{total,i,out}$, is

$$\begin{aligned} P_{total,i,out} &= LG_i P_{total,i-1,out} + 2n_{sp}(G_i - 1)h\nu B_o \\ &= LG_i \left(P_{s,i-1,out} + P_{ASE,I-1,out}\right) + 2n_{sp}\left(G_i - 1\right)h\nu B_o. \end{aligned}\tag{7.59}$$

∞

Example: As an example of the analysis of a cascaded chain of amplifiers, we consider the case where the total output power of each amplifier is the same (i.e., $P_{total,i,out} = P_{total,in} = P_{s,0}$). We will use the following values: $P_{sat} = 8$ mW, $P_{s,0} = 9$ mW, $G_0 = 35$ dB, $LG_0 = 3$, and $n_{sp} = 1.3$. We will also assume that the optical filter has a bandwidth of $B_o = 126 \times 10^9$ Hz (or 1 nm), that the optical amplifier has a spectral width of 25 nm ($\Delta\nu = 3.10 \times 10^{12}$), and that the signal wavelength is 1545 nm.

(a) Find the value of the gain G_i required for each amplifier.

Solution: The value of gain is found from Eq. 7.60 as

$$G_i = \frac{P_{s,0} + 2n_{sp}h\nu B_o}{2n_{sp}h\nu B_o + LP_{s,0}}.\tag{7.60}$$

For the case where $P_{s,0} \gg 2n_{sp}h\nu B_o$, the equation reduces to

$$G \approx \frac{1}{L}.\tag{7.61}$$

This gain is what is required to just balance the attenuation incurred in the fiber between stages, as we would intuitively expect. We are given that $G_0 = 35$ dB or $G_0 = 3.16 \times 10^3$, so

$$L = \frac{LG_0}{G_0} = \frac{3}{3.16 \times 10^3} = 9.49 \times 10^{-4} \Rightarrow 36.2 \text{ dB}.\tag{7.62}$$

The gain of each stage, then, is

$$G_i \approx \frac{1}{L} = \frac{1}{9.49 \times 10^{-4}} = 1.049 \times 10^3 \Rightarrow 30.2 \text{ dB}.\tag{7.63}$$

(b) Find the value of P_{sat} required to achieve this gain.

Solution: To find the saturation power at each stage, we solve Eq. 7.54 on page 225 for P_{sat} to find

$$P_{\text{sat,i}} = \frac{(1-L)P_{s,0}}{\ln(LG_0)} = \frac{(1-9.49 \times 10^{-4})(9 \times 10^{-3})}{\ln(3)} \qquad (7.64)$$
$$= 8.19 \times 10^{-3} \text{ W} = 8.19 \text{ mW}.$$

(c) Calculate and plot $P_{s,i,out}$, $P_{total,i,out}$, and $P_{ASE,i,out}$ for each of 100 stages. These powers are, respectively, the signal power out of each stage, the total power out of each stage, and the ASE power out of each stage.

Solution: The equation for $P_{total,i,out}$ is easy; it is

$$P_{\text{total},i,out} = P_{s,0,\text{in}} = 9 \times 10^{-3} \text{ W} = 9 \text{ mW}. \qquad (7.65)$$

The ASE power is found from Eq. 7.57 on the preceding page as

$$bP_{\text{ASE,i,out}} = i(2n_{\text{sp}})(G_i - 1)h\nu B_o. \qquad (7.66)$$

The signal power is obtained by subtracting the power $P_{s,i,out}$ from the total power $P_{total,i,out}$, i.e.,

$$P_{\text{s,i,out}} = P_{\text{total,i,out}} - P_{\text{ASE,i,out}}. \qquad (7.67)$$

Using an iterative approach, we can find and plot these power values at each stage. (One way to perform the iterative calculations is to use a spreadsheet program.) Figure 7.8 on the following page shows a plot of the resulting power levels in the link.

Looking at the figure, we note the following:

1. As required by our desired goal of keeping the total power at the amplifier outputs constant, the plot of $P_{\text{total,i,out}}$ is flat.

2. After the first amplifier, the ASE power begins to grow due to the subsequent amplification of the attenuated ASE power from the prior amplifier(s) and the addition of a fixed amount of ASE power by each amplifier.

3. The signal power steadily falls as it traverses the chain of amplifiers. The increasing ASE power is using an increasing portion of the gain and the signal uses the declining remainder of the gain. (The signal is decreasing much slower than it would without the amplifiers, however.)

4. Eventually, the ASE power would equal or exceed the signal power, if more stages were added.

—————————————— ∞ ——————————————

Receiver Signal-to-Noise Ratio

We have seen from our example that, as a signal passes through the amplifier chain, the signal power decreases and the ASE power grows. Eventually, we suspect, the signal will be lost in the noise; where this occurs depends on the detection of light at a receiver. The hypothetical receiver can be placed at the output of any amplifier (or anywhere else in the link).

When we detect the light that is the combination of the signal and the ASE, several noise sources are possible [14]:

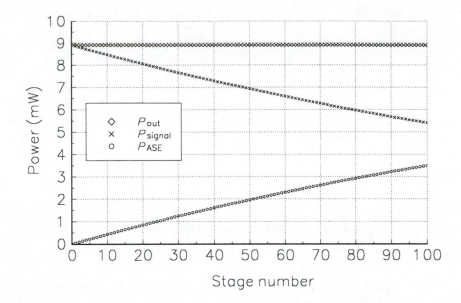

Figure 7.8 Power levels in a sample 100-amplifier fiber link (see example for link parameter values). The powers shown are the total power out of each amplifier, the signal power out of each amplifier, and the power in amplified stimulated emission at the output of each amplifier. The link is designed to keep the total power out of each stage the same.

- shot noise due to the signal,

- thermal noise associated with the detector load resistance,

- noise due the mixing (or "beating") of the signal and the ASE (which produces frequencies within the electronic bandwidth of the receiver), and

- noise due to the mixing of different ASE components with each other (which can also produce frequencies within the receiver bandwidth).

The latter two sources are new noises introduced by the ASE light.

The mean-square noise currents associated with these latter noise sources are [4]

$$< I_{\text{sig}-\text{ASE}}^2 > \quad = \quad 2 I_{\text{ASE}} I_{\text{sig}} \frac{B_e}{B_o} \tag{7.68}$$

$$< I_{\text{ASE}-\text{ASE}}^2 > \quad = \quad I_{\text{ASE}}^2 \frac{B_e}{B_o} , \tag{7.69}$$

where B_e is the electrical bandwidth of the receiver. (The electrical bandwidth can be estimated as one-half the data rate.) The currents in these expressions are found from

$$I_{\text{ASE}} \quad = \quad \mathcal{R} P_{\text{ASE}} \tag{7.70}$$

$$I_{\text{sig}} \quad = \quad \mathcal{R} P_{\text{sig}} , \tag{7.71}$$

where $\mathcal{R}$ is the responsivity of the detector at the ASE wavelengths (assumed constant over the band of ASE wavelengths) and signal wavelength, P_{ASE} is the ASE power in the optical bandwidth B_o at the receiver, and P_{sig} is the signal power at the receiver.

If we define the ratio of the optical filter bandwidth to the electrical receiver bandwidth as

$$R_B = \frac{B_o}{B_e} \tag{7.72}$$

and the ratio of the ASE power to the signal power as

$$R_{\text{ASE}} = \frac{P_{\text{ASE}}}{P_{\text{sig}}}, \tag{7.73}$$

then the value of the Q-parameter (defined in the previous chapter) required to meet the link's BER requirement is [4]

$$Q = \frac{I_{\text{sig}}}{\sqrt{I_{\text{N1}}^2} + \sqrt{I_{\text{N0}}^2}}, \tag{7.74}$$

where I_{N1}^2 (I_{N0}^2) is the noise in the receiver when a logical **1** (logical **0**) is received. Assuming Gaussian noise and that only the ASE-signal and ASE-ASE beat noises are important, then [4]

$$Q = \frac{2\sqrt{R_B}}{R_{\text{ASE}} + \sqrt{4R_{\text{ASE}} + R_{\text{ASE}}^2}}. \tag{7.75}$$

We now want to illustrate the application of this formula.

------------------------------ ∞ ------------------------------

Example: Consider a link that uses a cascaded series of optical amplifiers with the properties of the previous example. The data rate is to be 2.5 Gb·s^{-1} and the BER is to be 10^{-9}. We want to see if we will have enough signal power (compared to the ASE noise) to achieve the desired BER after traversing 100 amplifiers and the fiber between them.

<u>Solution:</u> The electrical bandwidth is estimated as

$$B_e \approx \frac{B_R}{2} = \frac{2.5 \times 10^9}{2} = 1.25 \times 10^9, \tag{7.76}$$

and the ratio R_B is

$$R_B = \frac{B_o}{B_e} = \frac{126 \times 10^9}{1.25 \times 10^9} = 100.8. \tag{7.77}$$

From Table 6.3 on page 185, we find that a BER of 10^{-9} requires a Q of 6.0, so

$$Q = \frac{2\sqrt{R_B}}{R_{\text{ASE}} + \sqrt{4R_{\text{ASE}} + R_{\text{ASE}}^2}} \tag{7.78}$$

$$6.0 = \frac{2\sqrt{100.8}}{R_{\text{ASE}} + \sqrt{4R_{\text{ASE}} + R_{\text{ASE}}^2}}$$

$$R_{\text{ASE}} = 1.046.$$

Figure 7.9 on the following page shows a plot of R_{ASE} for the cascaded amplifiers of the previous problem.

We find that the value of R_{ASE} stays below 70% after 100 stages. We could add appreciably more stages before we would achieve the ratio of 104%, where the BER would reach a value of 10^{-9}.

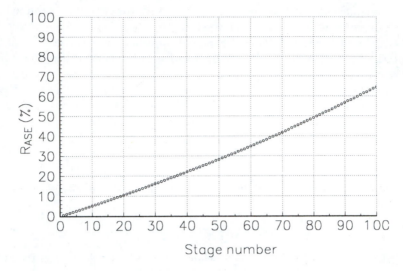

Figure 7.9 Ratio of P_{ASE} to $P_{\text{out,i,signal}}$ for cascaded amplifiers in a fiber-optic link (see example for link specifications). This ratio is R_{ASE}.

───────────── ∞ ─────────────

From these examples (and problems at the end of the chapter), we find that each cascaded amplifier allows the signal power to be increased. The penalty imposed for the amplification is an addition of ASE noise. Eventually, the ASE-to-signal power ratio is appreciable and exceeds the upper limit required by the BER and data rate of the link.

7.8.3 Cascaded Amplifiers: Other Effects

An additional effect of N cascaded amplifiers will be an accompanying decrease in the bandwidth of the channel. Any element with a bandpass response will contribute to this effect; hence, the amplifier spectral response and the optical filter response contribute to the bandwidth reduction. If the responses are mathematically modeled, the net effect is that the value for B_o that is used in the calculations must be replaced by an effective bandwidth for the cascaded units. This effective bandwidth will reduce the noise contributions from the ASE noise of previous stages; only amplifiers later in the chain contribute the most noise [4]. In addition, nonuniform spectral response complicates the spectrum of the ASE noise; some parts are attenuated more heavily than other parts. Complicated models based on measured spectral response can be used [17].

Also, large amplifier gains call for the use of optical isolators after the amplifiers to keep reflections from being amplified and upsetting the data link. These optical isolators do not allow bidirectional links and do not allow optical time domain reflectometers (discussed in Chapter 3) to operate without a special apparatus to subvert the isolators, when desired.

Long strings of cascaded amplifiers accentuate the effects on nonlinearities in the fiber [3, 18, 19]. These effects, such as four-wave mixing, cross-phase modulation, and self-phase modulation (discussed in Chapter 3) can be negligible over short distances but they can accumulate over the longer distance that optical amplifiers make possible. In addition, the amplified stimulated emission (ASE) may become large enough to interact with the signal through these interactions. The reader is referred to the literature for detailed descriptions of these effects in long-distance links.

For multi-channel links, cascaded amplifiers accentuate the effects of the wavelength-dependent gain of these amplifiers. Signals operating on the stronger gain regions will grow ever-stronger; eventually, they will rob the gain from the other signals and the output will be a wide variety of signal strengths. Efforts can be made to flatten the amplifier gain, to predistort the signals to compensate, or to incorporate filter elements into the amplifier chains.

Several demonstration projects have been done showing long-distance transmission [6, 20–22] and in the transmission of *solitons* (as described in a later section of this chapter). (Solitons are pulses of moderate power that induce nonlinear interactions in the fiber to counter the dispersion in the fiber, allowing ultra-low dispersion transmission. Multi-gigabit-per-second signals have been sent through millions of kilometers of [simulated] fiber, using solitons.) Work is proceeding to incorporate fiber amplifiers in undersea telecommunication cable systems [18, 19].

7.8.4 Semiconductor Optical Amplifiers

Semiconductor optical amplifiers (or *semiconductor laser amplifiers*) can also be used to amplify the light [23, 24]. These devices are laser devices that are biased just below oscillation. They have gain and can amplify a wave passing through the medium, but they do not have enough gain to oscillate. The insertion losses of these amplifiers tend to be 6 dB or more when they are placed in a fiber link. Another problem that presents itself in multichannel systems is that amplitude modulation in one channel can modulate the gain characteristics for other channels, causing crosstalk. Nonlinearities in the gain medium can also affect the amplification process [13]. Due to the increased popularity of the erbium-doped fiber amplifiers, few applications of diode amplifiers are available.

Other optical amplifiers include the *Raman fiber amplifier* that uses the Raman nonlinear effect discussed earlier in Chapter 3 on fiber properties. Here, light at one wavelength co-propagates (or counter-propagates) with the signal beam, transferring its energy to the signal beam by the Raman process. These amplifiers [24, 25] are under investigation, but have not yet proven to be commercially viable.

7.9   Dispersion Compensation

We have seen that for low data-rate links, the signal power can limit the length of the link. In some cases optical amplifiers offer an opportunity to concatenate fiber to allow longer link lengths. Even in these links and in high-data-rate single-mode links, the dispersion can limit the operating link length. One solution is to operate in the 1300-nm low-dispersion region; however, practical commercial amplifiers operate only at 1550 nm. To operate at this wavelength, dispersion-shifted fibers have been produced that offer low dispersion at this wavelength. Dispersion-shifted fiber can greatly increase the dispersion-limited distance since the dispersion-shifted fiber can have dispersions that are typically a factor of one-tenth of the dispersion of conventional fibers (typically, about 2 ps·nm^{-1}·km^{-1} for a dispersion-shifted fiber to 20 ps·nm^{-1}·km^{-1} for a conventional fiber). However, many millions of kilometers of conventional fiber has been installed with a dispersion minimum at 1300 nm. Such dispersion can limit the link distance in 1550-nm links operating at high data rates, as has been described in our discussion of dispersion-limited propagation distances. One goal is to be able to upgrade installed links from 1300-nm operation to 1550 nm without incurring a dispersion penalty [3, 18].

Techniques that allow this upgrading of the 1300-nm links are called *dispersion compensation* techniques (also called *dispersion management*). The goal is to add elements to the link that reverse the effects of the pulse spreading due to dispersion and to shorten the pulse length. (This process is also called *equalization*.) These dispersion-compensation techniques include:

- Adding optical components to the link to compensate for the pulse spreading including

  - dispersion compensating (i.e., dispersion-reversing) fiber lengths [26, 27],

  - specialized devices with a dispersion-reversing effect For example, an alternative technique to achieve high values of dispersion in a small device is to propagate the light in one of the higher order modes of a fiber. Using optical devices called *mode convertors*, the ordinary LP_{01} mode can be converted to a higher order mode (e.g., the LP_{11} mode which travels with negative-valued dispersions that can be as large as several hundred $ps \cdot km^{-1} \cdot nm^{-1}$). After traversing a short length of specially made fiber that supports this mode, another mode convertor changes the light back into the LP_{00} mode for propagation through the conventional single-mode fiber.

  - specialized optical amplifiers [28],

  - optical phase conjugation devices [29], and

  - optical gratings and filters [30, 31] (as described at the end of Chapter 4).

  Among the optical components are

- Predistorting the emitted pulse to compensate for the dispersion effects.

- Electronically processing the pulse after detection at the receiver to compress the pulse.

We will consider only the technique of adding dispersion-canceling fiber lengths; the other techniques are still in the research phase and it is not clear which will be commercially usable. (We described the use of fiber gratings to cancel fiber dispersion in Chapter 4.)

7.9.1 Dispersion-Compensating Fiber

In using *dispersion-compensating fiber*, the idea is to alternate lengths of fiber with positive-valued dispersion with lengths of negative-valued dispersion. Representative dispersion plots are shown in Fig. 7.10 on the next page. If a fiber has a representative dispersion of, say, $+20$ $ps \cdot nm^{-1} \cdot km^{-1}$ at the wavelength of interest, a dispersion compensating fiber with, say, a dispersion of -100 $ps \cdot nm^{-1} \cdot km^{-1}$ can be used to reverse the dispersion. The high value of dispersion is chosen to minimize the lengths of compensating fiber that must be used. This high-dispersion fiber also has relatively high losses, so a fiber amplifier is also included in the method to compensate for the combined losses of the compensating fiber and the conventional fiber. Figure 7.11 on page 238 represents the physical layout and the dispersion map as one traverses a typical link. Other techniques use other high-dispersion devices to achieve the same benefit.

The important parameters for the dispersion-compensating fiber or device are the dispersion, the insertion loss, and the bandwidth (i.e., the frequency or wavelength span over which it offers the quoted dispersion value). Another parameter of interest is the wavelength slope of the

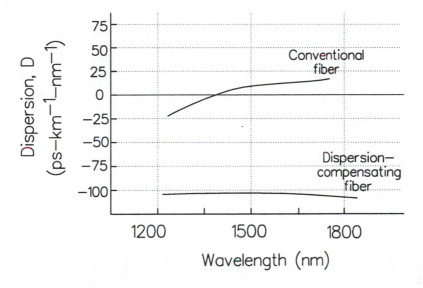

Figure 7.10 Plot of dispersion vs. wavelength for representative conventional unshifted fiber, dispersion-shifted fiber, and dispersion-compensating fiber.

dispersion (i.e., $dD/d\lambda$ in units of ps·km^{-1}·nm^{-2}) for use in fibers supporting several wavelengths simultaneously (i.e., WDM systems).

One figure of merit for a dispersion-compensating fiber is the ratio of dispersion, D, of the fiber (in ps·km^{-1}·nm^{-1}) to the loss α (in dB/km), i.e.,

$$\text{FOM} = \frac{D}{\alpha}. \tag{7.79}$$

Values as high as 300 ps·nm^{-1}·dB^{-1} have been achieved in fibers with losses as low as 0.32 dB/km and dispersions in the range of of 80 ps·km^{-1}·nm^{-1}.

In recent years it has become important that the nonlinearities of the fiber are important since dispersion-compensating fibers usually combine high values of nonlinear coefficient (more than 1.5 times larger, due to a highly doped core) with extra-small values of mode-field diameter, raising their vulnerability to nonlinearities. The effective area of these fibers can be in the 20–30 μm^2 range compared to approximately 50 μm^2 for a typical single-mode fiber. Most dispersion-compensating fibers are designed taking into account both their dispersion properties and their nonlinear effects to predict the propagation of the signal through the fiber. An alternative figure of merit [32] for dispersion-compensating fibers that encompasses nonlinear effects has been proposed as

$$F_{\text{nl}} = \frac{A'_{\text{eff}}}{A_{\text{eff}}} \frac{\alpha'_p}{\alpha_p} \frac{(e^{\alpha_p L} - 1)^2}{e^{\alpha_p L}} \frac{e^{\alpha'_p L'}}{(e^{\alpha'_p L'} - 1)^2} \tag{7.80}$$

where A'_{eff}, α'_p, L', and D' are the effective area, the loss coefficient, the length, and the dispersion of the dispersion-compensating fiber, respectively. The unprimed variables are for the segments of conventional fiber in the link. (The reader should be able to prove that $D'L' = LD$.) The goal is to manipulate the fiber variables to optimize this figure of merit which includes the nonlinearities.

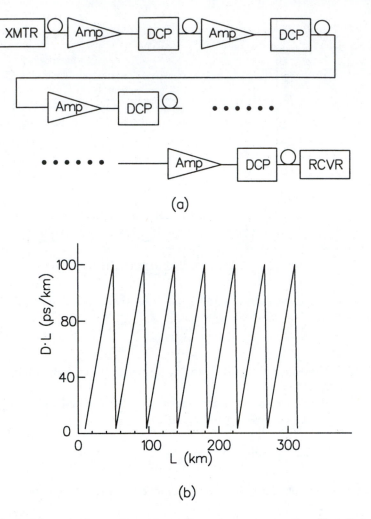

Figure 7.11 (a) Physical layout of a representative dispersion-compensating fiber link and (b) map of accumulated dispersion vs. distance along the link.

Many compensation devices are optical filters that have a resonant response (such as the fiber grating devices described in Chapter 4); their dispersion compensating abilities may be confined to a small region in the vicinity of the resonant wavelength. For these devices, the wavelength region must also be specified. Some of these filters have a periodic frequency response (and an almost-periodic wavelength response), offering the possibility of multiple channels at different wavelengths; others do not have a periodic response, perhaps limiting their use in multichannel systems.)

7.10   Solitons

We learned in an the earlier discussion of nonlinearities that self-phase modulation can cause an apparent frequency chirp during the rising and falling edges of an optical pulse. In fact, it is possible for the nonlinear effects of the modulation to cancel the effects of dispersion. In this case, the tendency of the pulse to spread wider in time can be canceled by the pulse compression effects of the nonlinear interaction. In this case, the pulse retains its initial shape as it traverses the fiber waveguide; such unchanging pulses are called *solitons*. (The historical background of solitons is found in Refs. [33–39].) Of course, fiber losses eventually cause the nonlinear interaction to become negligible; hence, we use periodically spaced optical amplifiers to restore the wave amplitude to keep the nonlinear interaction going. The treatment of solitons that follows is based on that of of Agrawal [24, 25].

In general, the dispersion can be a positive value (*normal dispersion*) or a negative value (*anomalous dispersion*). It is possible to find a soliton pulse shape for the anomalous dispersion case (*bright solitons*); the normal dispersion case provides solutions that are decreases in brightness from a bright background (so-called *dark solitons* [24, 25]). We will consider only bright solitons in our discussion.

Neglecting, for now, the fiber losses, solitons are solutions to the following propagation equation [24, 25],

$$\frac{\partial A}{\partial z} + \beta_1 \frac{\partial A}{\partial t} + \frac{i}{2}\beta_2 \frac{\partial^2 A}{\partial t^2} - \frac{1}{6}\beta_3 \frac{\partial^3 A}{\partial t^3} = i\gamma|A|^2 A \tag{7.81}$$

where $A(z,t)$ is the envelope of the pulse, $\beta_1 = d\beta/d\omega = 1/v_g$ (v_g is the group velocity), $\beta_2 = d^2\beta/d\omega^2$, $\beta_3 = d^3\beta/d\omega^3$, and γ is the nonlinearity parameter given by

$$\gamma = \frac{2\pi n_2}{\lambda A_{\text{eff}}} \tag{7.82}$$

where n_2 is the nonlinear-index coefficient and A_{eff} is the effective area, as defined earlier in our discussion of nonlinearities. The parameters, β_2 and γ, describe the effects of the dispersion and the nonlinearities, respectively. (This propagation equation is good for pulse widths of a few picoseconds or longer; additional higher-order nonlinearities need to be added to describe the behavior of ultra-short pulses.) In addition, as long as the operating wavelength is *not* the same as the zero-dispersion wavelength, we can assume that $\beta_3 = 0$. A useful expression for β_2 relates it to the dispersion D as

$$\beta_2 = -\frac{D\lambda^2}{2\pi c} . \tag{7.83}$$

Using the normalized variables,

$$\tau = \frac{t - \beta_1 z}{T_0} \tag{7.84}$$

$$\xi = \frac{z}{L_D} \tag{7.85}$$

$$U = \frac{A}{\sqrt{P_0}} \tag{7.86}$$

where T_0 is the pulse width, P_0 is the peak power of the pulse, and L_D is the *dispersion length*, defined as

$$L_D = \frac{T_0^2}{|\beta_2|} , \tag{7.87}$$

we can convert the propagation equation to

$$i\frac{\partial U}{\partial \xi} + \frac{1}{2}\frac{\partial^2 U}{\partial \tau^2} + N^2|U|^2 U = 0. \tag{7.88}$$

(The second term has a minus sign for the case of normal dispersion.) The parameter N is defined as

$$N^2 = \gamma P_0 L_D = \frac{\gamma P_0 T_0^2}{|\beta_2|}. \tag{7.89}$$

It will turn out that the equation has physically realizable solutions only for integer values of N. Equation 7.88 is the *nonlinear Schrodinger equation*; we can use it to describe the propagation of pulses in a lossless, dispersive, nonlinear medium.

When the excitation is

$$U(0, \tau) = \text{sech}(\tau), \tag{7.90}$$

the propagating pulse will keep its initial shape if $N = 1$. This shape is called the *fundamental soliton*. For integer values of N not equal to one, the wave will change its shape as it propagates but will be periodic (i.e., it will return to its original shape periodically as it propagates through the fiber). The *period of the soliton*, z_0, is

$$z_0 = \frac{\pi}{2}L_D = \frac{\pi T_0^2}{2|\beta_2|}. \tag{7.91}$$

The equation for the fundamental soliton (including its phase) is

$$U(\xi, \tau) = \text{sech}(\tau)e^{i\xi/2}. \tag{7.92}$$

The shape of the soliton pulse is quite stable; minor perturbations in the beam shape will be resolved and the wave will regain its soliton shape after a short duration. Hence, although the shape and amplitude of the pulse are theoretically known, we can have considerable leeway in experimentally exciting the soliton in practice. (For example, we could excite the fiber with a Gaussian approximation $[U(0, \tau) = e^{-\tau^2/2}]$ to the sech function and it would eventually become the fundamental soliton after propagation.) This dynamic adaptation of the pulse shape comes at the expense of some pulse energy, however.

When the fiber losses are added to the nonlinear Shrodinger equation, it becomes [24, 25]

$$i\frac{\partial U}{\partial \xi} + \frac{1}{2}\frac{\partial^2 U}{\partial \tau^2} + N^2|U|^2 U = -i\Gamma U \tag{7.93}$$

where the losses are included in the term,

$$\Gamma = \frac{\alpha_p L_D}{2} = \frac{\alpha_p T_0^2}{2|\beta_2|} = \frac{\alpha_p z_0}{\pi}. \tag{7.94}$$

Here, α_p is the fiber loss in m^{-1}. When $\Gamma \ll 1$, the approximate solution is

$$U(\xi, \tau) = \text{sech}(\tau e^{-2\Gamma\xi}) \exp\{(i/8\Gamma)[1 - \exp(-4\Gamma\xi)]\}. \tag{7.95}$$

This soliton pulse has a width T that expands as it propagates according to the relation,

$$T = T_0 e^{2\Gamma\xi} = T_0 e^{\alpha_p z}, \tag{7.96}$$

where T_0 is the width of the pulse at the fiber input. Obviously, an exponential increase in pulse width cannot occur indefinitely; the solution is valid only for $\Gamma\xi \ll 1$ [24, 25]. For values beyond this limit, the actual broadening is less than that predicted by the exponential increase. It must be noted that the soliton-pulse broadening that occurs with the losses is still much less than the broadening that occurs with nonsoliton pulses. Hence, even with fiber loss, solitons are *still* advantageous and are envisioned to allow the transmission of optical pulses over distance of many 10,000s of kilometers without electronic regeneration of the pulses.

To achieve long-distance transmission, links using soliton pulses must include optical amplifiers to recover the original pulse peak power in order to maintain the nonlinear interaction. In nonsoliton links, amplifier spacings can be on the order of 50 to 100 km between amplifiers. In soliton systems, amplifiers must be closer, typically 10 to 30 km apart, due to the energy loss that occurs as the soliton readjusts its width back to that of the fundamental soliton in the fiber after the amplifier. This energy is lost into a dispersive wave that accompanies the soliton. These short propagation distances between amplifiers do not allow significant deviation from the fundamental soliton shape, hence little energy is diverted into this dispersive wave. Simulations show that the spacing between amplifiers, L_A, should meet the criterion that $L_A \ll z_0$ to provide a good design. In conventional dispersive fibers, z_0 is typically 100 km for data rates above 1 Gb/s; the resulting amplifier spacing of 10 km or less is too low to be practical. In dispersion-shifted fibers, however, typical values of z_0 can be 500 to 1000 km, allowing satisfactory amplifier spacings of 30 to 50 km.

7.10.1   Soliton Link Design

We now want to consider some of the trade-offs that must be made in designing an optical data link that uses solitons. The input excitation for the fundamental soliton in terms of the unnormalized variables is [24, 25]

$$A(0, t) = \sqrt{P_0}\,\mathrm{sech}(t/T_0) \tag{7.97}$$

where

$$P_0 = \frac{|\beta_2|}{\gamma T_0^2}\,. \tag{7.98}$$

The width parameter, T_0, is related to the full-width half-maximum width of the soliton by

$$T_0 = \frac{T_{\mathrm{FWHM}}}{2\ln(1 + \sqrt{2})} = 0.567 T_{\mathrm{FWHM}}\,. \tag{7.99}$$

The energy, E_p, in the pulse is

$$E_p = \int_{-\infty}^{+\infty} |A(0, t)|^2 dt = 2P_0 T_0\,. \tag{7.100}$$

Soliton Spacing

Soliton links utilize an RZ code format; in addition, the soliton typically occupies only a small fraction of the bit period. In fact, the fundamental soliton solution that we are using requires that there be no other pulses in the vicinity; otherwise, the solution is not valid. Hence, we have a requirement that the soliton pulses not be *too* close together; this will limit the maximum data rate that can be achieved. We can define a normalized separation parameter q_0 by [24, 25]

$$q_0 = \frac{T_B}{2T_0} = \frac{1}{2B_R T_0} = \frac{0.88}{B_R T_{\text{FWHM}}}, \tag{7.101}$$

where T_B is the bit period $(= 1/B_R)$ and B_R is the bit rate. The criterion for avoiding having the solitons too close is

$$\frac{\pi}{2} e^{q_0} L_D \gg L, \tag{7.102}$$

where L_D is the dispersion length $(= T_0^2/|\beta_2|)$ and L is the total transmission distance. The criterion becomes an equation identifying some trade-offs [24, 25],

$$B_R^2 L \ll \frac{\pi e^{q_0}}{8q_0^2 |\beta_2|}. \tag{7.103}$$

This equation allows us to see the effects on the bandwidth and/or distance as we choose q_0.

————————————— ∞ —————————————

Example: Suppose that $\beta_2 = -2 \text{ ps}^2 \cdot \text{km}^{-1}$ and $q_0 = 10$.

(a) Find the allowed total link length, L, for a 10 Gb·s^{-1} soliton data link.

Solution: We note that $\beta_2 = -2 \times 10^{-24} \text{ s}^2 \cdot \text{km}^{-1}$. We begin by finding the total transmission distance, L, from

$$B_R^2 L \ll \frac{\pi e^{q_0}}{8q_0^2 |\beta_2|} = \frac{\pi e^{10}}{(8)(10)^2 |-2 \times 10^{-24}|} = 13.7 \text{ (Tb/s)}^2 \cdot \text{km} \tag{7.104}$$

$$L \ll \frac{B_R^2 L}{B_R^2} = \frac{13.7}{(0.01)^2} = 1.377 \times 10^5 \text{ km}.$$

So, we see that a total link length of about 1,377 km could be accommodated.

(b) Find the pulse width (full-width half-maximum) for this link.

Solution: The soliton pulse width is

$$T_{\text{FWHM}} = \frac{0.88}{q_0 B_R} = \frac{0.88}{(10)(10 \times 10^9)} = 8.8 \text{ ps}. \tag{7.105}$$

(c) Find the fraction of the bit period that is occupied by the soliton.

Solution: The fraction occupied is

$$\frac{T_{\text{FWHM}}}{T_B} = T_{\text{FWHM}} B_R = (8.8 \times 10^{-12})(10 \times 10^9) = 8.8\%. \tag{7.106}$$

Hence, we have found that the soliton represents less than 10% of the bit period.

————————————— ∞ —————————————

Amplifier Spacing

The amplifier spacing, L_A, should be as large as possible for economic reasons. However, the best performance occurs will short spacings; the ideal limit is when the amplification is continuously distributed along the length of the fiber. If the amplifier power gain is G, we can write the amplitude of the wave at the amplifier output as

$$U'(L,\tau) = \sqrt{G_0}\, U(L_A,\tau) \qquad (7.107)$$

where G_0 is the amplifier gain that just balances the fiber losses such that

$$G_0 = \frac{1}{e^{-\alpha_p L_A}} \,. \qquad (7.108)$$

Simulations show [24, 25] that the solitons will keep their shape and be successfully propagated over long distances if $L_A \ll z_0$. Since $z_0 = \pi L_D/2$, we can consider an equivalent condition, $L_A \ll L_D$. This provides a trade-off condition relating the amplifier spacing, the bit rate, and the soliton separation of

$$B_R^2 L_A < \frac{1}{4q_0^2|\beta_2|} \,. \qquad (7.109)$$

∞

Example: (a) Consider a soliton system with $q_0 = 10$ operating with a conventional fiber having a dispersion of $\beta_2 = -20$ ps$^2\cdot$km^{-1}. Find the amplifier spacing for a link operating at 5 Gb/s.

Solution: We begin with the trade-off equation,

$$B_R^2 L_A \quad < \quad \frac{1}{4q_0^2|\beta_2|} = \frac{1}{(4)(10^2)(20 \times 10^{-24})} \qquad (7.110)$$
$$< \quad 1.25 \times 10^{20} \text{ (b/s)}^2 \cdot \text{km} \,.$$

We find the distance between amplifiers as

$$L_A < \frac{B_R^2 L_A}{B_R^2} = \frac{1.25 \times 10^{20}}{(5 \times 10^9)^2} = 5.00 \text{ km} \,. \qquad (7.111)$$

We note that this distance is too small to be practical (due to our choice of a conventional-dispersion fiber).

(b) Find the amplifier spacing for a link operating at 5 Gb/s if a dispersion-shifted fiber with $\beta_2 = -2$ ps$^2\cdot$km^{-1} is used.

Solution: We again begin with the trade-off equation,

$$B_R^2 L_A \quad < \quad \frac{1}{4q_0^2|\beta_2|} = \frac{1}{(4)(10^2)(2 \times 10^{-24})} \qquad (7.112)$$
$$< \quad 1.25 \times 10^{21} \text{ (b/s)}^2 \cdot \text{km} \,.$$

We find the distance between amplifiers as

$$L_A < \frac{B_R^2 L_A}{B_R^2} = \frac{1.25 \times 10^{21}}{(5 \times 10^9)^2} = 50 \text{ km} \,. \qquad (7.113)$$

We note that this distance is now practical due to our choice of a dispersion-shifted fiber with lower dispersion value at the 1550-nm region.

∞

Source Chirp

It is important to note that the input soliton pulse should have a hyperbolic-secant shape (or close to it) but also must be free of frequency chirp. (Significant amounts of chirp will upset the balance between the dispersion and the self-phase modulation.) Commercial lasers are beginning to appear that can generate the fundamental soliton without significant chirp.

Gordon-Haus Limit

Another effect of the amplifiers in a soliton link is due to the spontaneous emission of the amplifiers. The light added by this spontaneous emission also causes a time jitter in the output pulse [40] of the amplifier. If the arrival times of the solitons at the receiver is modeled as being Gaussian shaped (centered around the expected arrival time), the variance of the distribution (in ps) is [41]

$$\sigma_t'^2 = 4138 n_{\rm sp} F(G) \frac{\alpha_p'}{A_{\rm eff}'} \frac{D}{\tau} L'^3 , \qquad (7.114)$$

where $F(G) = (G-1)^2/(G \ln G)$, α_p' is the fiber power attenuation in units of km^{-1}, $A_{\rm eff}'$ is the effective area as defined in our discussion of nonlinear effects in units of μm^2, D is the fiber dispersion in units of ps·nm^{-1}·km^{-1}, τ' is pulse width in ps, and L' is the link length in units of megameters. This timing jitter presents a limit to the distance that the link can maintain the desired error performance. This limit is called the *Gordon-Haus limit* [42].

———————————————— ∞ ————————————————

Example: Calculate the Gordon-Haus timing jitter limit, $\sigma'_t{}^2$, for a soliton length of 50 ps. Assume that $D = 1$ ps·nm^{-1}·km^{-1} $A_{\rm eff} = 35$ μm^2, $\alpha_p' = 0.0576$ km^{-1} (corresponding to $\alpha = 0.25$ dB/km), $F(G) = 1.24$ (corresponding to a span between amplifiers of $L_A = 25$ km), and $n_{\rm sp} = 1.5$. Let the total link length be 9000 km.

Solution: The variance of the jitter is

$$\begin{aligned}
\sigma_t'^2 &= 4138 n_{\rm sp} F(G) \frac{\alpha_p'}{A_{\rm eff}'} \frac{D}{\tau} L'^3 & (7.115) \\
&= (4136)(1.5)(1.24) \left(\frac{0.0576}{35} \right) \left(\frac{1}{50} \right) (9)^3 = 184.7 \ {\rm ps}^2 , \\
\sigma_t' &= \sqrt{\sigma_t'^2} = \sqrt{184.7} \ {\rm ps} = 13.59 \ {\rm ps} .
\end{aligned}$$

We note that this jitter is an appreciable fraction of the soliton width. (Can you find the gain of the amplifiers from the given information to prove that the value of $F(G)$ is as stated?)

———————————————— ∞ ————————————————

Assuming that there are M amplifiers in the chain, that the jitter added by each is independent, and that the root-mean-square timing difference at the receiver should not exceed some fraction, f_j, of the bit period, it can be shown [24, 25] that the following trade-off condition results

$$(B_R L)^3 < \frac{0.1372 T_{\rm FWHM} B_R f_j^2 A_{\rm eff}}{h \alpha_p n_2 D} \qquad (7.116)$$

or

$$B_R L < \sqrt[3]{\frac{\pi f_j^2}{q_0 \lambda h \alpha_p \gamma D}} . \qquad (7.117)$$

where h is Planck's constant.

—————————————— ∞ ——————————————

Example: Consider a soliton communication system operating at 1550 nm with $q_0 = 10$, $\alpha = 0.2$ dB/km, $\gamma = 10$ (W·km)$^{-1}$, a dispersion given by $D = 2$ ps·km^{-1}·nm^{-1} (typical of a dispersion-shifted fiber), and a jitter parameter $f_j = 20\%$. Find the total transmission distance for a bit rate of $B_R = 5$ Gb/s. We will keep α_p and D in units of km. The quantities λ and D will be kept in units of nanometers. We will express D in terms of seconds.

Solution: We begin by converting α into α_p by

$$\alpha_p = \frac{\alpha[\text{dB/km}]}{4.34 \times 10^3} = \frac{0.2}{4.34 \times 10^3} \tag{7.118}$$

$$= 4.6 \times 10^{-5} \text{ m}^{-1} = 0.046 \text{ km}^{-1}.$$

We observe that $D = 2$ ps·km^{-1}·nm^{-1} is the same as $D = 2 \times 10^{-12}$ s·km^{-1}·nm^{-1}. The bit-rate-distance product is found from

$$B_R L < \sqrt[3]{\frac{\pi f_i^2}{q_0 \lambda h \alpha_p \gamma D}} \tag{7.119}$$

$$= \sqrt[3]{\frac{(\pi)(0.2)^2}{(10)(1550)(6.63 \times 10^{-34})(0.046)(10)(2 \times 10^{-12})}}$$

$$= 2.37 \times 10^{13} \text{ b} \cdot \text{s}^{-1} \cdot \text{km} = 2,370 \text{ Gb} \cdot \text{s}^{-1} \cdot \text{km}.$$

Hence,

$$L = \frac{B_R L}{B_R} = \frac{2.36 \times 10^{13}}{5 \times 10^9} = 4.72 \times 10^3 \text{ km}. \tag{7.120}$$

The link length needs to be 4720 km or shorter to avoid time jitter effects (i.e., the Gordon-Haus limit). So, we see that, even with dispersion-shifted fibers having low values of D, the link length can be limited by the timing jitter. Larger values of dispersion in the fiber will affect the distance as $D^{-1/3}$.

—————————————— ∞ ——————————————

The solution to the Gordon-Haus limit is the inclusion of optical filters after the amplifiers [43–45]. The filter widths B_o should be broader than the soliton (roughly ten times broader); however, the filters "recenter" the frequency content of the soliton back towards the desired center frequency. The undesirable wider deviations in frequency are trimmed down by the filter. A cascade of many filters centered on the same frequency has the undesirable property of having a narrow linewidth. This problem was solved by slightly offsetting the center frequency of the filters. This concept has been called *sliding filters*. The filters have the additional benefit of reducing the added ASE noise of the amplifiers, as described in the amplifier section of this chapter.

Single-channel data rates of up to 15 Gb/s have been achieved over distances of 25,000 km with a BER of $\times 10^{-10}$ [46] with this technique. Since the filters used have periodic wavelength response, multiple channels can be accommodated by choosing a wavelength separation equal to the period of the filters.

Experimental Results

Soliton experiments have been performed which demonstrate the concept. Both long-distance direct fiber links and recirculating-loop experiments have been performed. (The latter experiments save money and fiber by setting up a loop using a fiber splitter. The loop contains sections of fiber and one or more optical amplifiers. A soliton pulse is excited and enters the loop through the couple. At every transit of the loop a portion of the pulse is transmitted out of the loop through the coupler; the rest of the pulse traverses the loop again.) High-data-rate pulses (at multi-gigabits per second rates) have been successfully transmitted over tens of thousands of kilometers. The reader is advised to peruse the recent research literature for state-of-the-art results, as new experimental benchmarks are continually being set in this fast-developing area.

7.10.2 Multichannel Soliton Links

Upon comparing the performance of a soliton link with amplifiers spacing on the order of 30 km with a traditional dispersion-shifted fiber link with an amplifier spacing of about 70 to 100 km, we see that the traditional link is superior. However, in traditional link only one wavelength is at the dispersion-free wavelength; other channels will have increasing amount of dispersion as their wavelengths lie away from the dispersion-free wavelength. Soliton links, on the other hand, offer the promise of being able to carry many wavelengths simultaneously without having the effects of pulse spread due to dispersion.

Allowing several solitons to be present in the fiber, each at a different wavelength, raises the question of soliton collisions. Due to material dispersion, the soliton pulses at the different wavelengths will travel at slightly different velocities. Solitons at the faster velocities will overtake and pass through solitons at the slower velocities; hence, solitons from one channel will periodically collide with the solitons in another channel. Solitons have the property that they can collide and that the solitons will reform after the collision process. Due to the cross-phase modulation nonlinear effect, however, there are time shifts induced in the carriers. These effects are especially pronounced within the fiber amplifiers. If the collision takes place within the amplifier, the solitons reform with a random time shift that has been induced by the interaction. These time shifts result in a *time jitter* in the arrival times of the solitons at the receiver. Such time jitters can cause an unacceptably large power penalty in the performance of the link.

The time jitter induced within the amplifiers can be reduced by ensuring that the distance that the solitons overlap is much longer than the spacing between the amplifiers. In this way the portion of the distance that the solitons overlap within the amplifier fiber will be small compared to the amplifier-fiber length. We can define the *collision length* of a link as the distance over which one soliton passes through the width (the full-width at half-maximum) of the second soliton. This collision length, $L_{\text{collision}}$, is given by [24, 25]

$$L_{\text{collision}} = \frac{2T_{\text{FWHM}}}{D\,\Delta\lambda},\tag{7.121}$$

where D is the fiber dispersion and $\Delta\lambda$ is the wavelength separation of the two solitons.

——————————— ∞ ———————————

Example: Find the collision length of two solitons separated by 5 nm if the fiber dispersion is 1 ps·km^{-1}·nm^{-1}. The soliton widths are 70 ps.

Solution: The collision length is

$$L_{\text{collision}} = \frac{2T_{\text{FWHM}}}{D\,\Delta\lambda} = \frac{(2)(70 \times 10^{-12})}{(1 \times 10^{-12})(5)} \text{ km} = 28.0 \text{ km}. \qquad (7.122)$$

We want to keep the amplifier spacing below 28 km if we are using this link for a multichannel system.

———————— ∞ ————————

We want to ensure that the collision length is greater than the amplifier spacing, say $L_{\text{collision}} > 2L_A$. This can be ensured by requiring that the wavelength separation of the channels be

$$\Delta\lambda < \frac{T_{\text{FWHM}}}{DL_A}. \qquad (7.123)$$

The spectral width of the soliton, $\Delta\nu_{\text{FWHM}}$ is related to the temporal width by $\Delta\nu_{\text{FWHM}} = 0.34/T_{\text{FWHM}}$. If we assume that the channel separation is at least three times $\Delta\nu_{\text{FWHM}}$, we have [24, 25]

$$\Delta\lambda > \frac{\lambda^2}{cT_{\text{FWHM}}}. \qquad (7.124)$$

The combination of relations gives

$$L_A < \frac{cT_{\text{FWHM}}^2}{D\lambda^2} \qquad (7.125)$$

or

$$B_R^2 L_A < \frac{0.12}{q_0^2 |\beta_2|}. \qquad (7.126)$$

This latter relationship is another bit rate-amplifier spacing trade-off relationship for multichannel soliton systems.

———————— ∞ ————————

Example: Consider a multichannel soliton link with $q_0 = 10$ and $\beta_2 = -2 \text{ ps}^2 \cdot \text{km}^{-1}$.

(a) Find the $B_R^2 L_A$ product.

Solution: The product is found from

$$\begin{aligned}
B_R^2 L_A &< \frac{0.12}{q_0^2 |\beta_2|} = \frac{0.12}{(10^2)(|-2 \times 10^{-24}|)} \qquad (7.127)\\
&= 6 \times 10^{20} \text{ b/s}^2 \cdot \text{km}^{-1} = 600 \text{ (Gb/s)}^2 \cdot \text{km}^{-1}.
\end{aligned}$$

(b) Find the allowable amplifier spacing for a 5 Gb/s link.

Solution: We find the amplifier spacing as

$$L_A < \frac{B_R^2 L_A}{B_R^2} = \frac{600}{5^2} = 24 \text{ km}. \qquad (7.128)$$

———————— ∞ ————————

Hence, we see that the amplifier spacing should be less than 24 km to ensure that the jitter due to soliton collisions inside the amplifier is within acceptable limits.

Solitons offer promise of being able to propagate light pulses long distances without being dispersion limited. We have seen the link requires optical amplifiers to keep the optical amplitude large enough to ensure that the required nonlinear phase modulation occurs. These amplifiers are closer together than the distances in conventional amplitude-limited links. The use of solitons also requires a source that can generate pulses with a shape that is close to the desired soliton pulse shape. Such sources are becoming commercially available. Currently, soliton links are being developed in research laboratories; they have not been commercially fielded. The dispersion-compensating techniques, described earlier, also offer an opportunity to minimize or eliminate the dispersion. It is too early to tell which technology will be the choice to overcome dispersion in fiber links at 1550 nm [18, 19].

7.11 Summary

In this chapter, we have considered the methodology used in the design of a fiber-optic data link. Since a link can be either attenuation-limited or dispersion-limited, calculations of the power budget and the system rise time are required to assure that each meets the performance requirements of the link (as usually expressed by the desired data rate, BER, and approximate link distance). Additionally, we have seen that the channel capacity required to meet the data rate performance depends on the type of data encoding used for either clock encoding or error correction (or both).

Optical amplifiers, especially erbium-doped fiber amplifiers, allow us to amplify the signal without detecting it and electronically regenerating it, as is done in repeaters. Although the signal is amplified, we found that noise is also generated within the amplifier and subsequently amplified by the following amplifiers. The signal-to-noise ratio gradually decreases as the signal proceeds through a chain of concatenated amplifiers until the bit-error rate requirement can no longer be met. These amplifier techniques are proving increasingly popular in long-haul high-data-rate systems.

A variety of techniques are being investigated to try to reverse the pulse-spreading effects of dispersion. These pulse compression techniques can overcome the dispersion-limited distance much as the use of optical amplifiers allows us to extend the attenuation-limited distances in a fiber link. We discussed the concatenation of optical fibers with reverse signs of their dispersion. Long lengths of fiber with modest values of positive dispersion can be compensated for with the intermittent addition of short lengths of large-valued negative dispersion fiber. The techniques using other optical or electronic dispersion cancellation are also being intensely explored.

Solitons offer the opportunity to transmit optical pulses over long distances at high data rates. By incorporating periodic optical amplifiers it is possible to adjust the amplitude and shape of the pulse to have the nonlinear phase interactions due to the shape of the pulse cancel the dispersion effects, resulting in a pulse that is constant amplitude and shape over the entire transmission distance. This work offers the potential of having a communications medium where attenuation and dispersion are totally compensated, potential providing transmission media without power or dispersion limitations as addressed in the early sections of this chapter. The "trick" of using solitons is to introduce a pulse close to a soliton shape and to provide amplifiers at a distance spacing that is economical. Our trade-off analyses have indicated some of the compromises that must be made to accommodate soliton propagation. Generally, the amplifiers must be spaced closer than in a conventional fiber link and dispersion-shifted fiber is required.

7.12   Problems

1. Consider a 50/125 step-index fiber with $n_1 = 1.45$ operating with a source having a fractional spectral linewidth of 1.5%. Calculate the approximate bandwidth-distance product at 820 nm of this fiber if the length-normalized pulse spreading due to material dispersion is $t_M/L = 1.5$ ps/km (and is dominant).

2. A short-range system is to operate at 700 Mb·s^{-1} at a 1500 nm wavelength. Choose a source and detector for the system and explain your choices.

3. A system operates at 1550 nm with the following components:

   - LED: −13 dBm coupled into a fiber pigtail (of negligible length).
   - Pin diode: −45 dBm incident power is required in the detector's connectorized pigtail to achieve link requirements.
   - Fiber: 1.3 dB/km losses. The fiber is available in any length up to 1 km maximum.
   - Connectors: 2 dB loss.

   Calculate the maximum length of the link as determined by attenuation, allowing 6 dB margin for aging effects, etc.

4. Consider a system operating at a 1300 nm wavelength at a 100 Mb·s^{-1} data rate with the following components:

   - Transmitter: InGaAs LED producing 50 μW into a fiber pigtail. The spectral width is 40 nm.
   - Connectors: 1 dB losses.
   - Fiber:
     - Losses = 1.5 dB/km.
     - Intermodal bandwidth-distance product (measured in 1 km sample) = 800 MHz·km.
     - Mode mixing—moderate mode mixing observed ($q = 0.7$)
     - Lengths—any length is available.
   - Receiver:
     - InGaAs pin photodiode with a connectorized pigtail
     - Sensitivity required for a BER = 10^{-9} is given by
     $$P_R(\text{dBm}) = (11.5 \log B_R) - 60.5 \,,$$
     where P_R is in dBm and B_R is the data bit rate in Mb·s^{-1}.
     - Allowance for aging, etc.: 6 dB.
     - Coding: NRZ.

   (a) Calculate the maximum link distance L assuming that the link is attenuation-limited.

   (b) Calculate the maximum link distance L assuming that the link is dispersion-limited.

5. A receiver has the following performance:

   - $P_R < -30$ dBm is a logical 0.

- -30 dBm $< P_R < -15$ dBm is a logical **1**.
- $P_R > -15$ dBm is not allowed due to detector saturation.

This receiver is to be used with a fiber that has an attenuation specified as being between a minimum of 3 dB/km and a maximum of 4 dB/km. If the source has a power into the fiber of -3 dBm, calculate the dynamic range (in dB) of a 2 km link (neglecting all other losses except fiber losses).

6. Consider a link that uses a source with a rise time of 1 ns and a receiver with a rise time of 2 ns operating at a data rate of 50 Mb·s^{-1} with RZ coding.

    (a) If the link has moderate mode conversion ($q = 0.7$), find the maximum distance between the source and receiver for a modal-dispersion-limited fiber (i.e., material dispersion is negligible). Assume that the measured bandwidth of a 1 km sample of this fiber is 400 MHz.

    (b) If a single-mode fiber is used and the source operates at a wavelength of 900 nm with a spectral linewidth of 3 nm, calculate the dispersion-limited distance between this source and this receiver.

7. In the example of a cascaded chain of equally spaced fiber amplifiers on page 230, we assumed that the *total* output power of each amplifier was constant. Consider the case where the *signal* power out of the amplifiers is kept constant. The output signal power is

$$P_{s,i,out} = P_{s,0,in} , \tag{7.129}$$

and, hence, $LG_i = 1$ and

$$P_{total,i,out} = P_{s,0.in} + 2in_{sp}(G_i - 1)\frac{hc}{\lambda}B_o . \tag{7.130}$$

The following parameters apply: $LG_0 = 3$, $G_0 = 35$ dB, $P_{s,0,in} = 5$ mW, $n_{sp} = 1.3$, $\lambda = 1.545$ nm, and $B_o = 1$ nm ($= 126$ GHz).

    (a) Show that the saturation power required at the i-th stage is [4]

$$P_{sat,i} = \frac{1-L}{\ln(LG_0)}\left[P_{s,0,in} + 2in_{sp}\left(\frac{1}{L}-1\right)\frac{hc}{\lambda}B_o\right] .$$

    (b) Using a computer, plot $P_{s,i,out}$, $P_{sat,i}$, and $P_{ASE,i,out}$ as a function of the stage number for 100 stages (similar to Fig. 7.8 on page 232).

8. Consider the prior problem, where the output signal power of each amplifier is kept constant.

    (a) Using Table 6.3 on page 185, find the value of R_{ASE} required to achieve a BER of 10^{-14} for a data rate of 2.5 Gb·s^{-1}.

    (b) Using your results of the previous problem, find the number of amplifiers that can be used in the link.

    (c) If the fiber loss is 0.5 dB·km^{-1}, calculate the distance between amplifiers (using the values of the previous problem) and the end-to-end distance of the link.

9. In the example of a cascaded chain of equally spaced fiber amplifiers on page 230, we assumed that the *total* output power of each amplifier was constant. Consider the case where there is no attempt made to regulate the output power of the amplifiers, but the saturation power $P_{sat,i}$ of each amplifier is equal. The following parameters apply: $LG_0 = 3$, $G_0 = 35$ dB, $P_{sat} = 8$ mW, $P_{s,0,in} = 1$ mW, $n_{sp} = 1.3$, $\lambda = 1545$ nm, and $B_o = 1$ nm ($= 126$ GHz).

(a) Using a computer, find and plot the value of the saturated gain G_i at each stage of a 100-stage amplifier chain.

(b) Using a computer, plot $P_{s,i,out}$, $P_{total,i,out}$, and $P_{ASE,i,out}$ as a function of the stage number for 100 stages (similar to Fig. 7.8 on page 232).

10. An optical filter used after a fiber amplifier has a spectral response $f(\lambda)$ that is modeled by [4]

$$f(\lambda) \; = \; \cfrac{1}{1 + \left(\dfrac{\lambda - \lambda_c}{B_1}\right)^6} \, ,$$

where λ_c is the center wavelength of the passband and B_1 is the 3-dB spectral width of the filter.

(a) The frequency response of a cascade of N filters is equal to $[f(\lambda)]^N$. Using a computer, plot the spectral response of a cascade with N equal to 1, 2, 5, 10, 20, 50, and 100 if $B_1 = 1$ nm. (The horizontal axis should be $\lambda - \lambda_c$ and should extend from -1 nm to $+1$ nm. The vertical axis should be in dB (relative to the response at the center wavelength) and should extend from 0 to -20 dB.)

(b) The 3-dB spectral bandwidth after N stages is reduced by a factor of $(\ln(2)/N)^{1/6}$ [4]. Estimate (from your plot) and calculate the spectral bandwidth after 50 and 100 filters.

11. Consider an optical fiber link operating at 1550 nm. The fiber dispersion is that of conventional fiber optimized to have a zero-dispersion wavelength of 1300 nm. The dispersion in this fiber is $+80$ ps·km^{-1}·nm^{-1} at 1550 nm and the loss is 0.5 dB/km. Design a dispersion-compensation scheme using a fiber amplifier and a length of dispersion-compensating fiber. The distance between amplifiers is to be 50 km. You may assume that the dispersion in this fiber is $-100 +80$ ps·km^{-1}·nm^{-1}

(a) Find the length of the compensating fiber.

(b) Find the gain of the amplifier (in dB).

References

1. Hewlett-Packard, Palo Alto, CA, *Digital data communications with the HP fiber optic system*, 1978. Technical report.

2. G. Keiser, *Optical Fiber Communications, Second Edition.* New York: McGraw-Hill, 1991.

3. T. Li, "The impact of optical amplifiers on long-distance lightwave telecommunications," *Proc. IEEE*, vol. 81, no. 11, pp. 1568–1579, 1993.

4. C. R. Giles and E. Desurvire, "Propagation of signal and noise in concatenated erbium-doped fiber optical amplifiers," *J. Lightwave Technology*, vol. 9, no. 2, pp. 147–154, 1991.

5. K. Rottwitt, A. Bjarklev, J. H. Povlsen, O. Lumholt, and T. A. Rasmussen, "Fundamental design of a distributed Erbium-doped fiber amplifier for long-distance transmission," *J. Lightwave Technology*, vol. 10, no. 11, pp. 1544–1552, 1992.

6. K. Nakagawa, K. Aida, K. Aoyama, and K. Hohkawa, "Optical amplification in trunk transmission networks," *IEEE Lightwave Telecommunications Systems (LTS)*, vol. 3, no. 1, pp. 19–26, 1992.

7. R. Ramaswami and K. Liu, "Analysis of effective power budget in optical bus and star networks using erbium-doped fiber amplifiers," *J. Lightwave Technology*, vol. 11, no. 11, pp. 1863–1871, 1993.

8. J. Delavaux and J. Nagel, "Multi-stage erbium-doped fiber amplifier designs," *J. Lightwave Technology*, vol. 13, no. 5, pp. 703–720, 1995.

9. E. Desurvire, J. R. Simpson, and P. C. Becker, "High-gain erbium-doped traveling-wave fiber amplifier," *Optics Letters*, vol. 12, no. 11, p. 888, 1987.

10. R. J. Mears, J. M. Jauncey, and D. N. Payne, "Low-noise erbium-doped fiber amplifier operating at 1.54 μm," *Electronics Letters*, vol. 23, no. 19, p. 1206, 1987.

11. B. J. Ainslie, "A review of the fabrication and properties of erbium-doped fibers for optical amplifiers," *J. Lightwave Technology*, vol. 9, no. 2, pp. 220–227, 1991.

12. E. Desurvire, *Erbium-doped Fiber Amplifiers: Principles & Applications*. New York: John Wiley & Sons, 1994.

13. I. P. Kaminow, "Non-coherent photonic frequency-multiplexed access networks," *IEEE Network*, pp. 4–12, 1989.

14. A. Yariv, *Optical Electronics, Fourth Edition*. New York: Holt, Rinehart and Winston, 1991.

15. A. Yariv, "Signal-to-noise considerations in fiber links with periodic of distributed optical amplification," *Optics Letters*, vol. 15, no. 19, p. 1064, 1990.

16. T. J. Whitley, "A review of recent system demonstrations incorporating 1.3-μm praseodymium-doped fluoride fiber amplifiers," *J. Lightwave Technology*, vol. 13, no. 5, p. 760, 1995.

17. C. R. Giles and E. Desurvire, "Modeling erbium-doped fiber amplifiers," *J. Lightwave Technology*, vol. 9, no. 2, pp. 271–283, 1991.

18. J. Thiennot, F. Pirio, and Jean–Baptiste Thomine, "Optical undersea cable systems trends," *Proc. IEEE*, vol. 81, no. 11, pp. 1610–1623, 1993.

19. H. Taga, N. Edagawa, S. Yamamoto, and S. Akiba, "Recent progress in amplified undersea systems," *J. Lightwave Technology*, vol. 13, no. 5, pp. 829–840, 1995.

20. C. Giles, E. Desurvire, J. Talman, J. Simpson, and P. Becker, "2 Gbit/s signal amplification at $\lambda = 1.53$ μm in an erbium-doped single-mode fiber amplifier," *J. Lightwave Technology*, vol. 7, no. 4, p. 651, 1989.

21. P. Urqhart and T. Whitley, "Long-span fiber amplifiers," *Applied Optics*, vol. 29, no. 24, p. 2503, 1990.

22. S. Saito, T. Imai, and T. Ito, "An over 2200-km coherent transmission experiment at 2.5 Gb/s using erbium-doped-fiber in-line amplifiers," *J. Lightwave Technology*, vol. 9, no. 2, pp. 161–169, 1991.

23. M. O'Mahony, "Semiconductor laser optical amplifiers for use in future fiber systems," *J. Lightwave Technology*, vol. 6, no. 4, pp. 531–544, 1988.

24. G. P. Agrawal, *Fiber Optic Communication Systems*. New York: John Wiley & Sons, 1992.

25. G. P. Agrawal, *Nonlinear Fiber Optics: Second Edition*. San Diego: Academic Press, 1995.

26. Y. H. Cheng, "Optimal design for direct-detection system with optical amplifiers and dispersion compensators," *J. Lightwave Technology*, vol. 11, no. 9, pp. 1495–1499, 1993.

27. C. D. Poole, J. M. Wiesenfeld, D. DiGiovanni, and A. Vengsarkar, "Optical fiber-based dispersion compensation using higher order modes near cutoff," *J. Lightwave Technology*, vol. 12, no. 10, pp. 1746–1758, 1994.

28. R. Li, P. Kumar, and W. L. Kath, "Dispersion compensation with phase sensitive amplifiers," *J. Lightwave Technology*, vol. 12, no. 3, pp. 541–549, 1994.

29. S. Watanabe, T. Naito, and T. Chikama, "Compensation of chromatic dispersion in a single-mode fiber by optical phase conjugation," *IEEE Photonics Technology Letters*, vol. 5, pp. 92–95, 1993.

30. C. D. Poole, "Dispersion compensation in lightwave systems," in *Optical Fiber Conference '94*, *Tutorial Proceedings*, (Washington DC), pp. 123–144, Optical Society of America, 1994.

31. F. Oullette, J. Cliche, and S. Gagnon, "All-fiber devices for chromatic dispersion compensation based on chirped distributed resonant coupling," *J. Lightwave Technology*, vol. 12, no. 10, pp. 1728–1738, 1994.

32. F. Forghierhi, R. Tkach, A. Chraplyvy, and A. Vengsarkan, "Dispersion-compensating fiber: Is there merit in the figure of merit?," in *Optical Fiber Communications Conference, Vol. 2, 1996 Technical Digest*, (Washington DC), Optical Society of America, 1996.

33. L. F. Mollenauer and R. H. Stolen, "The soliton laser," *Optics Letters*, vol. 9, p. 13, 1984.

34. L. F. Mollenauer, J. P. Gordon, and M. N. Islam, "Soliton propagation in long fibers with periodically compensated loss," *IEEE J. Quantum Electronics*, vol. QE–22, no. 1, pp. 157–173, 1986.

35. L. F. Mollenauer and K. Smith, "Demonstration of soliton transmission over more than 4,000 km in fiber with loss periodically compensated by Raman gain," *Optics Letters*, vol. 13, pp. 675–677, 1988.

36. K. Smith and L. Mollenauer, "Experimental observation of adiabatic compression and expansion of soliton pulses over long fiber paths," *Optics Letters*, vol. 14, p. 751, 1989.

37. K. Smith and L. M. Mollenauer, "Experimental observation of soliton interaction over long fiber paths: discovery of a long-range interaction," *Optics Letters*, vol. 14, pp. 1284–1286, 1989.

38. A. R. Chraplyvy, "Limitations on lightwave communications imposed by optical-fiber nonlinearities," *J. Lightwave Technology*, vol. 8, no. 10, pp. 1548–1557, 1990.

39. A. Hasegawa and Y. Kodama, *Solitons in Optical Communications*. Oxford: Oxford University Press, 1995.

40. J. Gordon and H. Haus, "Random walk of coherently amplified solitons in optical fiber transmission," *Optics Letters*, vol. 11, p. 665, 1986.

41. L. Mollenauer, "Solitons in ultralong distance transmissions," in *OFC '95 Tutorial Sessions*, (Washington DC), pp. 97–143, Optical Society of America, 1995.

42. H. A. Haus, "Optical fiber solitons, their properties and uses," *Proc. IEEE*, vol. 81, no. 7, pp. 970–983, 1993.

43. A. Mecozzi, J. D. Moore, H. A. Haus, and Y. Lai, "Soliton transmission control," *Optics Letters*, vol. 16, p. 1841, 1991.

44. Y. Kodama and A. Hasegawa, "Generation of asymptotically stable optical solitons and suppression of the Gordon-Haus effect," *Optics Letters*, vol. 17, p. 31, 1992.

45. L. F. Mollenauer, J. P. Gordon, and S. G. Evangelides, "The sliding-frequency guiding filter: an improved form of soliton jitter control," *Optics Letters*, vol. 17, no. 22, p. 1575, 1992.

46. L. F. Mollenauer, P. V. Mamyshev, and M. J. Neubelt, "Measurements of timing jitter in filter-guided soliton transmission at 10 Gb/s and achievement of 375 Gb/s-Mm error free, at 12.5 and 15 Gb/s," *Optics Letters*, vol. 19, no. 10, p. 704, 1994.

Chapter 8

Single-Wavelength Fiber-Optic Networks

8.1 Introduction

In this chapter, we consider some of the fundamentals of fiber networks operating at a single wavelength. In the next chapter on wavelength-division multiplexing (WDM), we consider some networks that operate with several wavelengths simultaneously.

In recent years several emerging trends have been challenging the telecommunications industry:

- Telecommunication volume and requirements demand an ever increasing bandwidth. Table 8.1 shows some representative services offered by telecommunications companies to meet these expanding requirements.

  In the computer world, the data rate of local area networks has also been constantly increasing. Table 8.2 on the following page indicates the increasing data rates of some present and proposed standard computer interconnections.

- Telecommunications systems now carry a mixture of analog signals (e.g., voice, TV channels) and digital data (e.g., computer interchanges). While the communications trunk system is thoroughly digitized, the user connections are typically analog (i.e., a twisted pair of wires connected to the microphone and speaker in the handset). Worldwide efforts are being made to provide total digital service throughout the network as the proportion of digital traffic increases steadily.

- The telecommunications industry is increasingly global in scope. Worldwide standards need to be agreed upon so that national telecommunications systems can be interconnected.

Data transmission rates	Application
1200/2,400/9,600/14,400/28,800 b/s	modems
128 kb/s	ISDN (2 64-kb/s channels and 1 16-[or 14]-kb/s channel)
1.544 Mb/s	T1 data transmission
44.783 Mb/s	T3 data transmission

Table 8.1 Data rates of various services.

Data rate	Network
9600 b/s	RS232 printer interface
4 or 16 Mb/s	Token Ring Network
10 Mb/s	Ethernet
50 Mb/s	coax Hyperchannel
100 Mb/s	fiber Hyperchannel
100 Mb/s	FDDI, fast Ethernet
600 Mb/s	GaAs FDDI (proposed)
800 Mb/s	HSC channel (used in Cray supercomputers)

Table 8.2 Data rates of representative computer networks.

- Because of the advent of personal computers and the increasing popularity of workstations, there is currently a trend away from centralized computers to a confederation of "peer" computers (or servers with clients) and their peripherals. Along with this trend comes the need for intercommunication of data between the elements of the confederation.

- New trends toward video conferencing and picture transmission are arising that will tax the current bandwidth capabilities of communications networks. Digital picture transmission in real time is a voracious consumer of bandwidth.

- Also, increased bandwidth is required for token passing (or random access strategies), greater complexity in packet headers, use of cyclic error-correcting (CRC) codes, greater complexity of network control overhead as network use increases, higher security requirements, and requirements for graceful network degradation.

Standards need to be set to allow interconnection of units on a network. Users are unwilling to commit major capital resources to a network unless assured that the units will intercommunicate and will communicate with networks of other users.

Networks can be divided into the following (roughly defined) categories:

1. *Local Area Networks (LANs)* interconnect users within a department, a building, or a campus.

2. *Metropolitan Area Networks (MANs)* interconnect users within a city or the metropolitan area surrounding a city.

3. *Wide Area Networks (WANs)* can interconnect users within a large geographic area (e.g., within a country).

The wide bandwidth offered by optical fibers is helping to lead the way into innovative data networks [1–3] that will be capable of carrying the mixture of traffic demanded by the users including video signals, voice, and high-speed data. Multiplexing techniques [4] can include time-division multiplexing (TDM), frequency-division multiplexing (FDM), code-division multiple-access multiplexing (CDMA) [5–8], and wavelength-division multiplexing (WDM) [9]. (The latter topic is considered in more detail in the next chapter.) The protocols that are used to control access to the network can include [10] central control (frequently with controlled switches [11]), token passing [12], and contention arbitration schemes similar to the carrier-sense multiple access/collision detection (CSMA/CD) scheme used in Ethernet [13, 14].

In this chapter we want to introduce some of the topologies that are possible for fiber-optic networks and to consider the trade-offs that network designers must analyze to meet their goals. This is again done through the power budget analysis. This analysis shows that the star configuration has some advantages over the linear data bus for applications with many users. We then introduce two new standards that are evolving for fiber-optic networks, the Fiber Distributed Data Interface (FDDI) for local area networks and the Synchronous Optical NETwork (SONET) for interoperability of equipment on wide area networks. Finally, we consider the combination of ATM (Asynchronous Transfer Mode) technology with the SONET transmission structure.

8.2   Network Topologies

Several topologies of networks have become popular. These include the star network, the linear bus network, and the ring network (see Fig. 8.1 on the next page).

In the *star network* (Fig. 8.1a), the transmission lines are brought together at a common point and distributed among the other lines, and the receiver lines branch out from that point. The star has the disadvantage of requiring a central node which can disrupt the entire network if damaged. It has the advantage of making it relatively simple to add terminals to the network, especially if spare lines have been designed into the node at its inception.

In the *linear bus network* (Fig. 8.1b), the network takes the form of a backbone with individual stations receiving or adding data as required. In electrical form it is very easy to add stations by tapping into the line; in fiber form it is more difficult to splice in a station, as the network must be shut down.

In the *ring network* (Fig. 8.1c), the stations are arranged into a continuous circuit. Each station receives a message from its upstream neighbor and, if it is not the addressee, repeats the message to its downstream neighbor. The topology is characterized by its closure at the ends. It suffers from the disadvantage that damage to any station on the ring will disrupt the entire network unless bypass switches or network redundancy are built in.

Other topologies exist, as do combination topologies. Figure 8.2 on page 259, for example, shows a "leaf" topology that combines the linear bus and star networks. Such a configuration is useful for interconnecting isolated groups of users with fairly large distances between them.

8.3   Network Design Trade-Offs

We now want to consider some of the trade-offs that are needed to analyze networks, so we will compare the star network and the linear bus network.

8.3.1   Data-Bus Power Budget

The power budget [15, 16] for a data network is important because we have to ensure that we have enough power in the source to reach the farthest station without saturating the nearest station. (This is the dynamic range of the source.) We will also find that, everything else being equal, the star network has an inherent lower-loss advantage over the linear bus network as the number of stations grows large. While we shall not consider the use of fiber amplifiers in networks, work is emerging [17] that shows that these amplifiers are particularly useful in improving the number of stations that can exist on a linear bus.

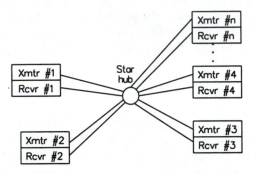

(a)

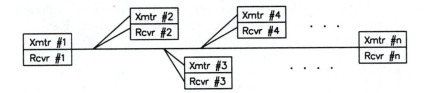

(b)

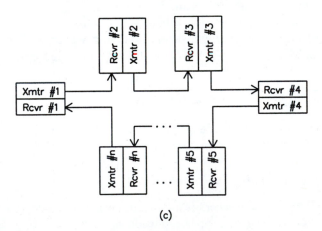

(c)

Figure 8.1 Typical network configurations. (a) A star network. (b) A linear bus network. (c) A ring network.

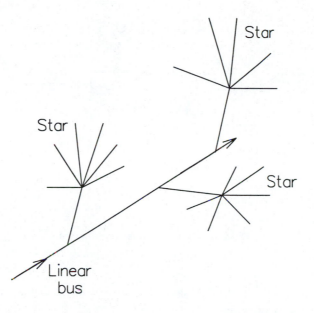

Figure 8.2 A combination of a linear bus network and star networks into a "leaf" topology.

Star Network Power Budget

We begin by calculating the power budget for a star network. Consider the power entering a fiber at the input to a coupler, P_F, and the power required by the receiver, P_R. We will assume an insertion loss, L_{insert}, a power splitting loss, $L_{\text{pwr split}}$, a connector loss, L_C, a system margin, L_M, and a fiber loss, α (dB/km). If L is the distance from the star coupler to each station, the path of the power from the transmitter to the receiver would be as follows:

- P_F in the fiber at the transmitter,

- a fiber loss of αL in going from the transmitter to the star,

- a loss of L_C as the light passes through the fiber/coupler connector pair to enter the star coupler,

- a loss of $L_{\text{pwr split}}$ as the power is divided among the output fibers of the star,

- a loss of L_C as the power passes through coupler/fiber connector pair leaving the star coupler,

- an additional loss of L_{insert} due to the insertion loss of the star coupler, and

- a fiber loss of αL in going to the receiver. (We assume ideal coupling into the receiver.)

Adding up all of these dB losses, the link power budget would be

$$
\begin{aligned}
10\log(P_F/P_R) &= L_{\text{pwr split}} + \alpha(2L) + 2L_C + L_{\text{insert}} + L_M \quad (8.1)\\
&= 10\log N + \alpha(2L) + 2L_C + L_{\text{insert}} + L_M,
\end{aligned}
$$

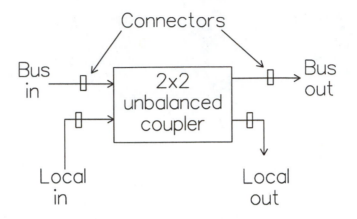

Figure 8.3 Tee coupler parameters.

where L_M is the link margin. From the right-hand side of Eq. 8.2 on the page before, we note that, as additional stations are added, the loss increases as $10 \log N$, where N is the total number of stations.

Linear Bus Power Budget

We now consider a linear data bus made up of tee couplers. Each station has two couplers. One coupler (the "tap out" coupler [shown in Fig. 8.3]) couples a fraction C_T of the incident light into the "local out" arm. The other "tap in" coupler adds the "local in" signal to the data bus. The losses include the following:

- The transmission for light in the fiber bus passing either tap will be $1 - C_T$; expressed as a loss, this is $L_{\text{thru}} = -10 \log(1 - C_T)$. Since each coupler has two taps, the total transmission loss for light passing through the coupler on the data bus will be $2L_{\text{thru}}$.

- Light that is coupled from the data fiber to the receiver arm will have a transmission factor of C_T; the loss is given by $L_{\text{tap}} = -10 \log(C_T)$.

- Light that comes from the "tap in" arm is assumed to add its power to the bus with C_T coupling efficiency. The loss it encounters is also $L_{\text{tap}} = -10 \log(C_T)$.

- There is a connector loss of L_C for each pair of connectors associated with coupler.

- Finally, there is an insertion loss L_{insert}, associated with the device to account for the excess losses.

We want to consider a total of N stations in the linear bus, each separated by a distance L by fibers with losses given by α (in dB/km). The smallest losses are encountered when transmitting from one station to its nearest neighbor station that is located a distance L away. The largest losses will occur when a station transmits to the N-th station located a distance of $(N-1)L$ away. We want to write a power budget for each of these cases.

We begin by calculating the power transmitted from adjacent stations, say from station #1 to station #2. We let P_F be the power available into the fiber from the source and neglect any coupling loss from the source fiber (typically a fiber pigtail) into the coupler fiber. The power path in this case is the following.

- We begin with the power P_{F1} being coupled into the transmitter arm of coupler #1.

- There is a connector pair when entering coupler #1 (loss of L_C).

- The power couples to the data bus with C_T efficiency (loss of L_{tap}).

- There is a connector pair when leaving coupler #1 (loss of L_C).

- There is an insertion loss associated with the "local in" to "bus out" path (loss of L_{insert}).

- There is a fiber loss between adjacent stations (loss of αL).

- There is a connector pair upon entering coupler #2 (loss of L_C).

- The power couples into the receiver arm of coupler #2 (loss of L_{tap}).

- There is a connector pair on leaving the receiver arm of the coupler on the way to the detector (loss of L_C).

- There is an insertion loss associated with the "bus in" to "local out" path (loss of L_{insert}).

Thus, the power budget for this path is written as,

$$10\log(P_{F1}/P_{R2,1}) = \alpha L + 2L_{\text{tap}} + 4L_C + 2L_{\text{insert}}\,, \tag{8.2}$$

where $P_{R2,1}$ is the power received at station #2 from station #1.

The power budget for the case of going from transmitter #1 to the receiver on station #N is determined from the following path analysis:

- We begin with power P_{F1} in the fiber at the transmitter.

- There is a connector pair upon entering coupler #1 (loss of L_C).

- The power couples to the data bus with C_T efficiency (loss of L_{tap}).

- There is a connector pair when leaving coupler #1 (loss of L_C).

- There is an insertion loss associated with the "local in" to "bus out" path (loss of L_{insert}).

- There are $(N-1)$ pieces of fiber, each of length L (loss of $[N-1]\alpha L$).

- There are $(N-2)$ couplers between the first and last station, and *each* coupler will introduce a transmission loss (losses of $L_{\text{insert}} + 2L_C + 2L_{\text{thru}}$).

- There is a connector pair upon entering coupler #N (loss of L_C).

- The power couples into the receiver arm of coupler #N (loss of L_{tap}).

- There is a connector pair on leaving the receiver arm of the coupler on the way to the detector (loss of $2L_C$).

- There is an insertion loss associated with the "bus in" to "local out" path (loss of L_{insert}).

The power budget for this link is

$$
\begin{aligned}
10\log(P_{F1}/P_{RN,1}) \;=\;\; & (N-1)\alpha L \quad \text{(the fiber losses)} \\
& +(N-2)(2L_C)+4L_C \quad \text{(the connector losses)} \\
& +(N-2)(2L_{\text{thru}}) \quad \text{(the coupler losses)} \\
& +(N-2)L_{\text{insert}}+2L_{\text{insert}} \quad \text{(the insertion losses)} \\
& +2L_{\text{tap}} \quad \text{(data bus and coupling loss)} \\
\;=\;\; & N\left(\alpha L+2L_c+L_{\text{insert}}+2L_{\text{thru}}\right) \\
& -\alpha L+2L_{\text{tap}}-4L_{\text{thru}}.
\end{aligned}
\tag{8.3}
$$

From this last equation (with the exception of the constant terms), we see that the losses of the linear bus increase *linearly* with an increasing number of stations, N.

The linear increase for the linear bus compares with a $\log N$ increase for the star bus, implying that systems with a large number of stations should use the star configuration. The actual number of stations for which the star configuration becomes advantageous depends on the relative value of the different losses of the buses.

———————————————— ∞ ————————————————

Example: (a) Consider a star network with a connector loss L_C of 1.5 dB per connector pair and an insertion loss (for each channel) of 0.75 dB. Calculate the system losses for $N = 3$ and $N = 50$ stations on the fiber. Ignore the fiber loss and the system margin.

Solution: The system loss for a star network is

$$
\begin{aligned}
L_{\text{total}} \;&=\; 10\log(N)+\alpha(2L)+L_{\text{insert}}+2L_C+L_M \\
&=\; 10\log(N)+0+0.75+2(1.5)+0 \\
&=\; 10\log(N)+3.75 \\
&=\; 8.52 \text{ dB} \quad \text{(for } N = 3\text{)}, \\
&=\; 20.7 \text{ dB} \quad \text{(for } N = 50\text{)}.
\end{aligned}
\tag{8.4}
$$

(b) Consider a linear data bus that taps 10% of the light into the arms of the tee couplers that are used. The insertion loss per tee coupler is 0.5 dB. Calculate the system losses for $N = 3$ and $N = 50$ stations on the fiber. Ignore the fiber loss and the system margin.

Solution: For the linear network, we first find

$$
L_{\text{tap}} = -10\log(C_T) = -10\log(0.10) = 10 \text{ dB}
\tag{8.5}
$$

and

$$
L_{\text{thru}} = -10\log(1-C_T) = -10\log(0.90) = 0.458 \text{ dB}.
\tag{8.6}
$$

The system loss for a linear network is

$$
\begin{aligned}
L_{\text{total}} \;&=\; N\left(\alpha L+2L_c+L_{\text{insert}}+2L_{\text{thru}}\right) \\
&\qquad -\alpha L+2L_{\text{tap}}-4L_{\text{thru}} \\
&=\; N\left[0+2(1.5)+(0.5)+2(0.458)\right] \\
&\qquad -0+2(10.0))-4(0.458) \\
&=\; 31.4 \text{ dB} \quad \text{(for } N = 3\text{)}, \\
&=\; 239 \text{ dB} \quad \text{(for } N = 50\text{)}.
\end{aligned}
\tag{8.7}
$$

(c) Plot the losses of both networks as a function of the number of stations on the network.

Solution: Figure 8.4 compares the losses of the star configuration vs. linear configuration for all values of N between 3 and 50. We note that the losses of the linear network continually increase as stations are added to the network, but the losses of a star configuration level off and increase at a much slower rate. This tendency is generally true of linear and star networks and provides an advantage to star configurations.

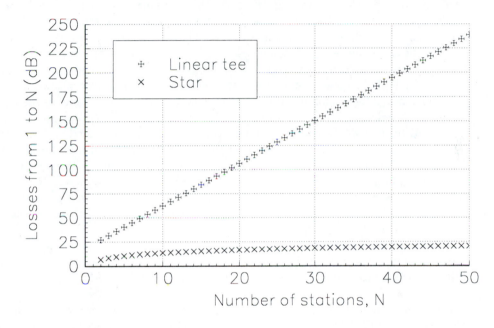

Figure 8.4 Comparison of losses of representative star and linear data bus configurations. (Parameter values are from the example problem.)

∞

Dynamic Range of Receivers in Networks

Local area networks or data buses are susceptible to dynamic-range limitations because of the wide range of possible distances between stations. Since $P_{R2,1}$ (the power received at station #2 from its next neighbor, station #1) is the maximum power received at a receiver, and $P_{RN,1}$ (the power received at station #N from station #1) is the minimum power between stations, we can compute the required *dynamic range DR* of the receiver from the ratio of these quantities,

$$DR = 10 \log \left(\frac{P_{R2,1}}{P_{RN,1}} \right) = 10 \log P_{R2,1} - 10 \log P_{RN,1}. \tag{8.8}$$

We now can find the dynamic range required by the receivers of a linear tee network. Subtracting Eq. 8.3 from Eq. 8.2, we find

$$DR = -2\alpha L + 4L_{\text{insert}} + 4L_{\text{thru}} + 4L_C - N(\alpha L + 2L_C + L_{\text{insert}} + 2L_{\text{thru}}). \tag{8.9}$$

Assuming that all receivers on the network are the same, each must have the dynamic range of Eq. 8.9 on the preceding page to avoid being saturated by a nearby transmitter or underdriven by a distant transmitter.

8.4   Standard Fiber Networks

Telecommunications standards-setting: Prior to the breakup of AT&T in 1984, the Bell system was such a dominant force in US telecommunications that it set *de facto* standards. Since then, setting of standards for telecommunications networks in the United States has followed a different path, involving many operating companies, long-distance service suppliers, and equipment manufacturers. (In most other countries the telecommunications industry is administered by the post, telegraph, and telephone government organizations [the "PTTs"].) International standards are set by the International Telecommunications Union–Telecommunication Standardization Sector (ITU–TSS). This organization was formerly called the International Telegraph and Telephone Consultative Committee (the CCITT).) The primary US standards are proposed and discussed in Committee T1 (and its technical subcommittees) of the American National Standards Institute (ANSI). Administrative support comes from a trade association, the Exchange Carrier Standards Association (ECSA). Committee T1 has been charged by ANSI with developing the North American Telecommunications standards and with developing reports and American recommendations for ITU. The committee has developed the Integrated Services Data Network (ISDN) standard, the SS7 switch services, and the Synchronous Optical NETwork (SONET) standard (discussed later in this chapter). A standard can take from one to three years to evolve due to multiple layers of discussion and compromise among the many parties to the committee. Once approved by the committee, the proposed standard is sent to the governing body of ANSI for approval and dissemination.

In the international arena, ITU deals only with governmental bodies, so the United States Department of State has a US National Committee for ITU (US ITU) to present the T1 Committee proposals for international standards. A T1 Committee proposal is sent to one of US ITU's four study groups. Once approved, the State Department submits them to ITU headquarters in Geneva as an official US proposal. ITU deliberates on the proposal, discusses pros and cons, and, if approved, issues its official Recommendations. These Recommendations can have a legal status in member countries that varies from mandatory observance to voluntary observance. (In the United States, observance of ANSI standards and ITU standards is voluntary, but failure to follow a standard is at the company's risk.)

Computer network standards-setting: Computer standards in most countries are set by the private industry without interaction with the government. Data networks, traditionally the realm of the computer industry, are starting to merge with the telecommunications network, leading to new problems. In the United States, computer-network standards are developed and set by Committee X3 of the Computer Business Manufacturers Association. Local area network standards, however, are set by the Institute of Electrical and Electronic Engineers (IEEE) Committee 802. It has set the standard for the Ethernet LAN and the Token Ring Network. Committee 802, the IEEE, and ANSI work with the International Standards Organization (ISO) to develop international computer standards.

Potential conflicts between telecommunications standards and computer-network standards call for close cooperation between the standards-setting groups, usually obtained by the overlapping membership of the groups.

Two fiber-optic data networks have been proposed as international standards. The Fiber Distributed Data Interface (FDDI) network is a wide-bandwidth local area network (LAN) operating at 100 Mb/s and the Synchronous Optical NETwork (SONET) is a proposed wide area network (WAN) operating at a base rate of 155 Mb/s (with expansion capability to achieve data rates of several Gb/s). We now want to consider these proposed standard networks in some detail.

8.5 FDDI Networks

In the 1980s it became apparent that the 10 Mb/s data rate associated with the Ethernet standard was limiting some wideband applications. A faster standard bus was required for two primary applications:

- high-performance interconnections between mainframes, peripherals, and other mainframes, and

- high-speed backbones for use in interconnecting lower speed LANs and devices.

Efforts to specify a standard were first initiated in 1982 to exceed links that were then being standardized (i.e., Ethernet and Token Ring Network) and to incorporate the data handling requirements of digital PBXs (private branch exchanges) that were widely introduced into the business environment to handle voice, fax, video, and sensor data streams. At the same time, the Open Systems Interconnection (OSI) standard was formalized. The primary advantage of the OSI standard was that it defined the layers of the interconnection so they could be *separately* designed and standardized.

The standard for FDDI is specified by the X3T.9 working group of ANSI (American National Standards Institute). In terms of the OSI layer nomenclature [18], it specifies the MAC (Media Access Control) layer, the PHY (Physical) layer protocol, the PMD (Physical Medium Dependent) layer, and the SMT (Station Management) Document (see Fig. 8.5 on the next page). The function of each of these is as follows:

- SMT (Station Management) provides the overall control of a ring by monitoring, managing, and configuring the ring. It determines the logical connection between the stations and automatically reconfigures the ring in case of a failure.

- MAC (Media Access Control) performs packet interpretation and controls token passing and packet framing. It interfaces with the local data through a standard Logical-Link Control (LLC).

- PHY (Physical) layer protocol performs clock recovery as well as encoding and decoding input signals from the PMD layer.

- PMD (Physical Medium Dependent) defines and characterizes the optical sources and detectors, cables, connectors, optical bypass provisions, and physical hardware characteristics. A detailed view of the separate functions implemented by the PHY and PMD layers is shown in Fig. 8.6 on the following page.

- The FDDI unit is placed on a data bus conforming to the IEEE P802.2 Logical-Link Control (LLC) data-bus standard.

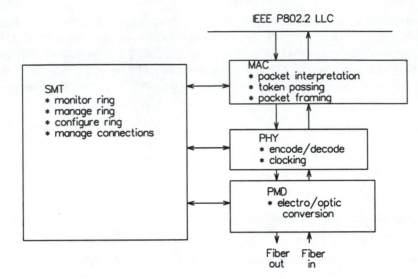

Figure 8.5 Block diagram of the FDDI layers.

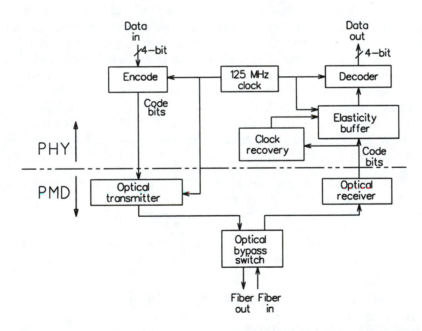

Figure 8.6 Detail of functions performed by the PHY and PMD layers of the FDDI network.

The outcome of the standards work was the Fiber Distributed Data Interface (FDDI). The topology of the network is primarily a dual counter-rotating ring. Concentrators will allow star arrangements of single-unit stations off of the ring. Properties (discussed in more detail in later subsections) of the proposed standard include the following:

- Data transmission rate: 100 Mb/s.

- Optical source: 1300-nm LEDs.

- Data encoding: The data are encoded using a 4-out-5 code into a *code group*. (This implies an actual transmission rate of 125 Mb/s to achieve the desired *data* rate of 100 Mb/s.) Each code group is called a *symbol*. There are thirty-two symbols possible:

  - Sixteen of the symbols represent data sets (i.e., 4 bits of ordered data representing values from 0,1,2,...,15).

  - Three symbols (called J, K, and L) are used for starting delimiter (JK).

  - One symbol (called T) is used for the ending delimiter (TT).

  - Two symbols (called R [reset] and S [set]) are used for control indicators.

  - Three symbols (called QUIET, IDLE, and HALT) are used for handshaking between stations.

  - The remaining seven symbol combinations are invalid.

- Signal Format: This data coding does not allow an unbalanced code (i.e., a long string of **1**s or **0**s with an average value that increases with the length of the string). (One of the advantages of Manchester encoding is that it is a balanced code with a constant DC value.) The code chosen for FDDI use has to lie within a $\pm 10\%$ range to minimize this DC effect on the receiver electronics. Empirical measurement indicates that the use of these codes adds only about a 1 dB power penalty to the link under worst-case conditions [19].

- Number of stations: Up to 1000 connections are allowed. (This implies that up to 500 dual-connectored stations are allowed.)

- Total fiber path: Up to 200 km of total fiber length are allowed. (Incorporating dual counter-rotating rings, up to 100 km of fiber would be allowed between stations.) The maximum distance between stations is 2 km in the early proposal; efforts have begun to offer a long-distance FDDI link.

- Frame size: The frame size is be ≤ 4500 bytes. Also, the frame time cannot exceed the minimum time that it takes a token to make a trip around the ring (i.e., with no data traffic on the ring).

It is expected that the standardized FDDI network will be popular for both commercial [20] and military [21, 22] applications.

8.5.1   Frame and Token Formats

The data and overhead bits are transmitted as a frame. A *frame* contains the data and ancillary information; a *token* is used to signify the right to transmit on the link. Only the possessor of the token is allowed access to the link.

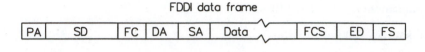

Figure 8.7 Structure of an FDDI data frame and token.

FDDI Data Frame

The contents of a *data frame* (Fig. 8.7) are

- *Preamble (PA) field*: consists of IDLE line-state symbols, which occur at the maximum frequency and are used to establish and maintain clock synchronization.

- *Preamble Starting Delimiter (SD) field*: a sequence of two delimiter sequences (JK). This field is recognized as a unique boundary of a frame.

- *Preamble Frame Control (FC) field*: defines the type of frame and its characteristics, such as

  - synchronous or asynchronous frames,
  - the length of the address field (i.e., 16 or 48 bits)
  - the kind of frame (i.e., a Logical Link Control [LLC] or a Station Management [SMT] frame), or
  - whether the frame is a token.

- *Preamble Destination Address (DA) field*: a 16-bit or 48-bit representation of the destination. (A group address that is recognized by more than one station is possible.)

- *Source Address (SA) field*: a 16-bit or 48-bit representation of the source, originating the message.

- Information field: contains the data symbols.

- *Frame Check Sequence (FCS) field*: a 32-bit field that contains the information from a cyclic redundancy check. This information is used at the destination to check the message for errors.

- *Ending Delimiter (ED) field*: one delimiter symbol (T).

- *Frame Status (FS) field*: a minimum of three control symbols. These symbols are modified by a receiving station and indicate

  - whether the addressed station has recognized its address,
  - whether the frame has been copied, and
  - whether any station has detected an error in the frame.

FDDI Token

The contents of a *token* (see Fig. 8.7 on the facing page) are

- *Preamble (PA) field* (as discussed in the frame description),

- *Preamble Starting Delimiter (SD) field* consisting of two delimiters,

- *Preamble Frame Control (FC) field* with a token description,

- *Ending Delimiter (ED) symbol* consisting of two end-of-token symbols (TT).

Two types of tokens are possible, depending on the contents of the FC field. They are the *restricted token* and the *unrestricted token*.

8.5.2   Network Operation

Each station repeats the message to its downstream neighbor. The receiving station has to decide which station has control of the medium and when to place information on the network. These ring control functions are controlled by the MAC.

Upon arrival of a message from its upstream neighbor, a station must first decide if the message is a token or a data frame, based on the frame contents.

- If the message is a token, the station then decides if it has data to transmit.

  - If the station has no data to transmit, it rebroadcasts the token to its downstream neighbor.

  - If the station does have data to transmit, it must decide if it is authorized to capture the token. This authorization scheme is a mechanism to allocate priority assignments to stations. Because their data are critical, some stations will have higher priority than others to capture the token; a means of setting and recognizing this priority must be incorporated in the network. The mechanism for setting the priority is in the rules for capturing the token. (More information on priority implementation is found in a later section.)

    * If the station is not authorized to capture a token when it receives the token, it must repeat the token to its downstream neighbor (while keeping track of the number of times repeated). If no other station wants to transmit after the number of repetitions set by the priority scheme, the station is free to capture the token and transmit its message.

    * The station captures the token by removing it from the network (i.e., failing to rebroadcast the token to the downstream neighbor). The MAC generates the station's data frame, and the station transmits the frame to its downstream neighbor. When receipt of the message has been successfully acknowledged by the receiving station (as described below), the transmitting station regenerates the token and transmits it to the downstream neighbor. Following transmission of a token, the transmitter sends a string of IDLE symbols until the starting delimiter (SD) of a new frame is received. At this point the MAC analyzes the frame and proceeds as described above.

In an alternative scheme, the MAC appends the token at the end of the frame for the use of the receiving station (only). For this *early-release token*, the transmitting station does not wait for the data frame to return with an acknowledgment and the network efficiency is increased (if there are not significant errors).

- If the arriving message is a data frame (i.e., it is not a token), the station checks the destination address (DA) portion of the frame to see if the message is addressed to it. (Note that a station can have more than one address.)

  - If the message is not addressed to the station, it transmits the unchanged frame to its downstream neighbor.

  - If the destination address (DA) of the message matches one of the receiving station's addresses, the frame is copied into the station's buffer, error checking is performed, and the MAC notifies the LLC (or SMT) that a message has arrived for processing. The MAC changes the frame status (FS) symbols of the data frame to show that the address has been detected and that the frame has been copied, or that the received frame was in error. The modified frame then proceeds around the ring until it arrives at the transmitting station.

  - After receiving a message whose source address (SA) fields match its own address (i.e., after receiving a message that it has originated), the transmitting station examines the contents of the FS symbols.

    * If an error message is indicated, then the original message is regenerated and retransmitted.
    * If the message is acknowledged (i.e., received without error), then the transmitting station removes all of its frames from the ring. This process is called *frame stripping*. During the stripping, the source transmits IDLE symbols to the network. It should be noted that, by the time that a transmitter decides that the frame has been received and acknowledged, the transmitter has already been repeating the beginning of the frame. This results in a remnant of a stripped frame appearing on the network. This remnant will not contain valid ending delimiters, so it will not be recognized as a valid message by any of the network stations and will be removed from the network. Following the frame stripping, the station regenerates the token and transmits it to its downstream neighbor.

8.5.3   Station Types

Three types of stations (see Fig. 8.8 on the next page) can be used on the FDDI network:

- A *dual-attachment station* (also called a *Class A node*) has two physical layer entities and two MAC entities to accommodate a dual counter-rotating ring arrangement. (These are called "P" and "M" in Fig. 8.8.) These stations can allow a wrap-around configuration in case of damage to the ring as discussed in a later section on FDDI reliability. Figure 8.9 on page 272 shows a representative ring configuration after the network has reconfigured itself to avoid a broken fiber cable.

  This wrap-around is configured by the SMT layer, which also determines the presence of the problem on the ring. Failure of the station allows an optical bypass mode to be initiated. In this mode an optical switch is activated, which routes the optical signal around the

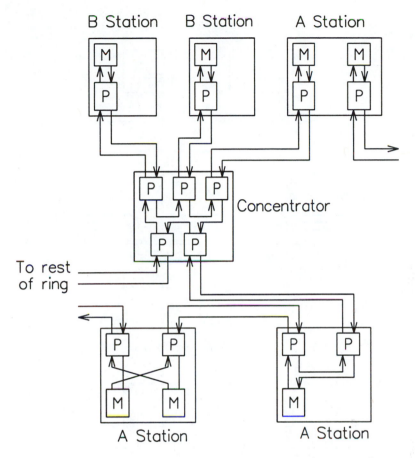

Figure 8.8 Representative interconnection of dual-attachment stations (A stations), a concentrator, and Class B stations (B stations) attached to the ring through the concentrator.

receiver and transmitter of the station. (The optical bypass also allows these stations to be powered down for maintenance without disrupting ring operation. This maintenance function is optional in the proposed standard.)

- A *single-attachment station* (also called a *Class B node*) has only one P entity and one M entity and is connected to the concentrator which is connected to the ring. In this fashion, failure of a Class B node will not affect the network operation; only that node is affected.

- A *concentrator* allows single-attachment stations to be joined together in a star configuration and then the combination is attached to the ring. It also serves as a bridge to slower networks, such as an Ethernet network, and allows single-attachment stations to be added to the ring without compromising the speed of the main ring.

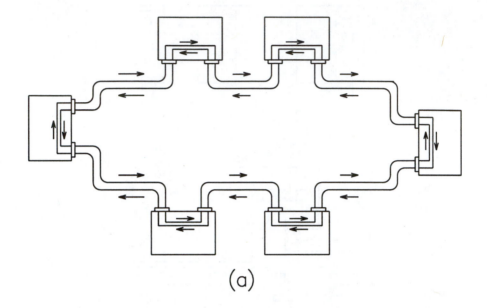

(a)

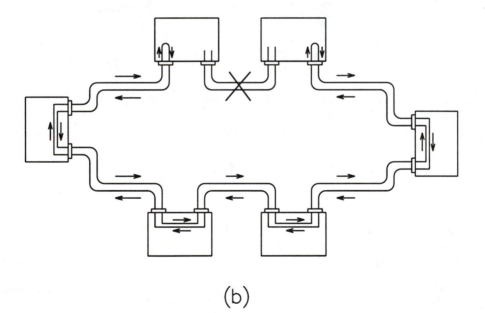

(b)

Figure 8.9 Example of link reconfiguration. (a) Original link. (b) Reconfigured link with failure point isolated.

8.5.4 Prioritization Schemes

One feature of token ring networks is the need for a *prioritization scheme* to ensure that stations with important traffic have a chance to obtain control of the token, rather than having the token controlled by a station or stations transmitting data of lesser priority. The control mechanism is implemented as part of the MAC. This is done by determining *a priori* the priority that a station will have. The implementation of the priority is controlled by the use of a *Timed Token Rotation (TTR) protocol.*

Generally two types of traffic are envisioned on the network:

- *Synchronous traffic* is controlled by a master synchronization scheme (e.g., the telephone voice system). The data must be transmitted during certain prescribed time periods because it is time-sensitive (e.g., a telephone conversation requires that the channel be revisited every 20 ms).

- *Asynchronous traffic* is "bursty" by nature; the communications occur sporadically. When an element has data to transmit, it can wait to assume network control to transmit its message. Small, reasonable delays (which depend on the message's importance and the station's priority) are allowable. Most computer data transmissions can be asynchronous.

In a network mixing synchronous and asynchronous traffic, the synchronous traffic is usually guaranteed a certain minimum amount of the network bandwidth (or, equivalently, a minimum data delay). Asynchronous traffic is allowed only in the bandwidth that exceeds this minimum that is allocated to the synchronous traffic. This idea is implemented by introducing the idea of *token timers* to keep the *token rotation time* within specific bounds. The token rotation time is simply the time that it takes the token to make one round trip on the network. The *minimum token rotation time* is the time that it takes the token to circulate one round trip when each node has no data to send. It is a function of the length of the ring, the token processing time at each node, and the number of nodes on the network. The *maximum token rotation time* is the time that we want to control; we want to avoid very long rotation times.

The timing of the token possession needs to be carefully regulated. Each station is guaranteed a minimum amount of time and uses this time to transmit synchronous traffic, if it has any. Each station has two timers associated with it. One is the *token rotation timer* (TRT) which is used to measure the time that it takes the token to circulate around the ring, and the other is the *token holding timer* (THT) which measures the time that the station has controlled the token. The THT is used with TRT to control the transmission of asynchronous traffic.

During network initialization, a *Target Token Rotation Time (TTRT)* is assigned to the ring, based on the expected traffic. This TTRT is set to the minimum time that will be allocated to each station. (The setting of the TTRT is a nontrivial exercise, since it is critical to network throughput and delay times.) The choice of TTRT depends on the length of the ring and the number of stations.

Using its TRT, each station measures the amount of time since the last token was received. The two classes of services are implemented as follows:

- The *synchronous class* allows the MAC to capture the token whenever the MAC has a string of synchronous frames ready to transmit. The station can transmit up to the maximum transmission time allocated to it.

- The *asynchronous class* allows the capture of a token *only if* the time since the last token was received (TRT) does not exceed the TTRT. If the TRT value exceeds the TTRT, transmission of asynchronous traffic is postponed. (Multiple priority levels can be implemented by adding more restrictive TRT requirements for token acceptance.) In the worst case, the time for the reception of two consecutive tokens will never exceed twice the TTRT.

Jain [23] describes some of the trade-offs that are necessary in setting the TTRT. Suppose that we have a ring with N stations attached to it and that the total length of fiber in the ring is L. The *latency D* of the ring is the amount of time that it will take the token to pass through the ring when there is no traffic. This is given by

$$D = \frac{L}{(c/n)} + NT_s = (5.085 \times 10^{-6})L + N(1 \times 10^{-6}) \text{ s}, \tag{8.10}$$

where L is in kilometers and T_s is the token processing time (in microseconds) at a station. Here, the first term on the right represents the fiber propagation time and the second term represents the processing time of N stations. A representative value that has been assumed is $T_s = 1$ μs. For longer links, we note that the propagation delay (the first term on the right side of Eq. 8.10) is dominant (unless there are a very large number of attached stations).

$$\infty$$

Example: We will consider three network configurations described by Jain [23]. The first is a "typical" network with 20 stations on the network and a total of 4 km of fiber in the ring. The second is a "big" network with 100 stations attached to a ring with 200 km of fiber (the most fiber allowed in an FDDI ring). The third configuration is the "largest" network which contains the maximum of 500 dual-attachment stations (i.e., the equivalent of 1000 stations) attached to 200 km of fiber.

We want to calculate the ring latency D for these cases.

Solution: For the "typical" network, we have

$$
\begin{aligned}
D &= \frac{L}{(c/n)} + NT_s = (5.085 \times 10^{-6})L + N(1 \times 10^{-6}) \text{ s} \tag{8.11}\\
&= (5.085 \times 10^{-6})(4) + (20)(1 \times 10^{-6}) = 40.3 \times 10^{-6} \text{ s} = 40.3 \ \mu\text{s}.
\end{aligned}
$$

The latency for the "big" network is

$$
\begin{aligned}
D &= \frac{L}{(c/n)} + NT_s = (5.085 \times 10^{-6})L + N(1 \times 10^{-6}) \text{ s}\\
&= (5.085 \times 10^{-6})(200) + (100)(1 \times 10^{-6}) = 1.117 \times 10^{-3} \text{ s} = 1.117 \text{ ms}.
\end{aligned}
$$

Finally, the latency of the "largest" network is

$$
\begin{aligned}
D &= \frac{L}{(c/n)} + NT_s = (5.085 \times 10^{-6})L + N(1 \times 10^{-6}) \text{ s}\\
&= (5.085 \times 10^{-6})(200) + (1000)(1 \times 10^{-6}) = 2.02 \times 10^{-3} \text{ s} = 2.02 \text{ ms}.
\end{aligned}
$$

$$\infty$$

Two parameters are used to characterize the ring performance. The first is the *efficiency* of the ring, given by Jain [23] as

$$\text{Efficiency} = \frac{N(T - D)}{NT + D},\tag{8.12}$$

where T is the TTRT assigned to the ring. The second is the *maximum access time delay*, $T_{\max}$, which is given by Jain [23] as

$$T_{\max} = (N - 1)T + 2D.\tag{8.13}$$

We want to maximize the ring efficiency and to minimize the access time delay (or at least not have too large a value). Both of the above expressions assume that the link is carrying asynchronous traffic only.

———————————————— ∞ ————————————————

Example: We continue the previous example which began with our latency calculations. Plot ...

(a) ... the efficiency as a function of assigned TTRT for the three network configurations.

Solution: For the "typical" network, we have

$$\text{Efficiency} = \frac{N(T - D)}{NT + D} = \frac{(20)(T - 40.3 \times 10^{-6})}{20T + 40.3 \times 10^{-6}}.\tag{8.14}$$

For the "big" network, we have

$$\text{Efficiency} = \frac{N(T - D)}{NT + D} = \frac{(100)(T - 1.117 \times 10^{-3})}{100T + 1.117 \times 10^{-3}}.\tag{8.15}$$

For the "largest" network, we have

$$\text{Efficiency} = \frac{n(T - D)}{nT + D} = \frac{(200)(T + 2.02 \times 10^{-3})}{200T - 2.02 \times 10^{-3}}.\tag{8.16}$$

Figure 8.10 on the following page shows the desired plot of network efficiency for these three cases.

(b) ... the maximum access delay time.

Solution: Performing the similar calculations for the maximum access delay, we have, for the "typical" network,

$$T_{\max} = (N - 1)T + 2D = (20 - 1)T + 2(40.3 \times 10^{-6}).\tag{8.17}$$

For the "big" network, we have

$$T_{\max} = (N - 1)T + 2D = (100 - 1)T + 2(1.117 \times 10^{-3}).\tag{8.18}$$

For the "largest" network, we find

$$T_{\max} = (N - 1)T + 2D = (1000 - 1)T + 2(2.02 \times 10^{-3}).\tag{8.19}$$

Figure 8.11 on page 277 shows the desired plot of the maximum access delay time for the three network configurations.

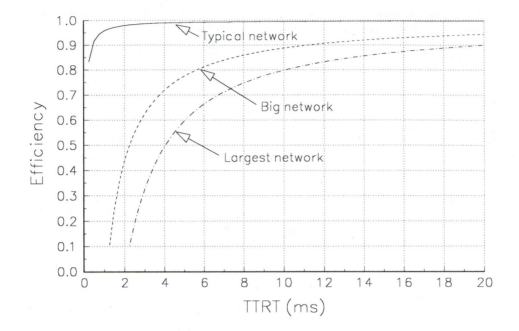

Figure 8.10 Plot of FDDI network efficiency vs. assigned TTRT for three network configurations.

Looking at these plots, we see the penalties of setting TTRT too small or too large. From the plot of the network efficiency, we see that setting TTRT too small lowers the efficiency, especially as the load on the network grows. In the second figure, we note that setting TTRT too big will cause the access delay to grow too long, especially if the network is heavily loaded.

We have seen from the example that setting the TTRT too big or too small will affect network performance. The FDDI specification establishes a minimum TTRT of 4 ms and a maximum of 165 ms to avoid extreme cases. The network manager, however, gets to select the actual value between these extremes. Jain [23] recommends a starting value of 8 ms as the default value of TTRT. If synchronous traffic is to be carried, then the TTRT needs to be at least one-half the revisit time requirement. For example, voice traffic (that requires a revisit every 20 ms) will require an allocation time T_{synch} of 10 ms. The ring latency D, the token time T_t (19 bytes for a total of 0.88 μs), and the maximum frame time $T_{\text{f max}}$ (4500 bytes for a total of 0.360 ms) are added to this to ensure

$$TTRT \geq T_{\text{synch}} + D + T_t + T_{\text{f max}}. \tag{8.20}$$

With a maximum latency of $D = 1.73$ ms, the TTRT should equal or exceed $2.3 + T_{\text{synch}}$ ms.

An additional scheme to establish priorities in FDDI is the use of *restricted and nonrestricted tokens*. Restricted tokens allow only certain stations, designed to recognize the special token, to receive and add data. Cooperating stations can, after prior agreement, enter into a restrictive

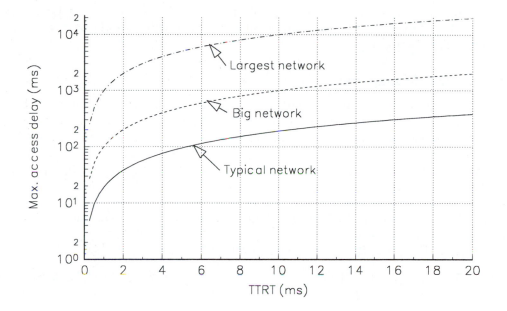

Figure 8.11 Plot of FDDI maximum access delay time vs. assigned TTRT for three network configurations.

token mode of operation. Use of this mode with multiple restricted tokens allows the stations to vie for available channel bandwidth adaptively, rather than on a pre-allocated basis.

Four kinds of traffic have been identified for use in this system of prioritization; all can be used on an FDDI network.

- The highest priority is given to circuit-switched data in an *isochronous channel*. Isochronous traffic is data that *must* be delivered in fixed units at fixed time intervals. The destination and source must be established by an *a priori* agreement. This traffic can be handled only by FDDI-II stations (i.e., stations built around a specialized version of FDDI that is optimized to handle only synchronous traffic) on the network and only after an isochronous channel has been assigned.

- The second highest priority is for *synchronous data traffic* (i.e., data that are delivered in fixed units at regular time intervals). In the FDDI network, data must be delivered at intervals that will not exceed twice the TTRT value to avoid exceeding the TTRT limit. The address and sender are encoded in the frame and either a restrictive or a nonrestrictive token can be used.

- The third highest priority is for asynchronous data after reception of a restrictive token. As described earlier, *asynchronous traffic* is data that is delivered with no time regularity. (The receiving station is expected to buffer the data until the complete message is received.) The bandwidth of the FDDI link is allocated first to isochronous and synchronous demands. The remaining bandwidth is allocated to asynchronous data. These data are transmitted in packets after capture of a token.

- The lowest priority is for asynchronous traffic transmitted after reception of a nonrestrictive token. The station vies for bandwidth on a packet basis with all of the other asynchronous traffic on the network.

8.5.5 Station Management

The SMT controls station functions including initialization, performance monitoring, activation and deactivation, error control, and maintenance. It also is in communication with the other SMTs on the network, controlling network administration in such areas as address assignment, bandwidth allocation, and network configuration and reconfiguration in case of a failure. A special function of the SMT is management of the Connection Management (CMT) function. The CMT controls the logical interconnection of the PHY and MAC functions within a station and establishes the logical connection with the adjacent stations upstream and downstream.

To establish a connection with a neighbor after a request from the SMT, the CMT checks for the existence of an operating link to its neighbor. To do this, it has the PHY transmit a sequence of continuous QUIET, HALT, and IDLE signals [24].

- A return stream of QUIET symbols (or the lack of any response at all) indicates that no link is available, due to either a network partition that has occurred or a failure in the link or station.

- A return string of HALT symbols or a string of alternating HALT and QUIET symbols indicates a willingness of a neighboring concentrator to make a connection.

- A stream of IDLE symbols indicates the acceptance of the requested connection.

Once the physical connection is made and acknowledged through these handshake signals, the CMT needs to logically configure the paths between the MAC and the PHY. There are several possibilities (depending on the complexity of the station); this flexibility allows the station manufacturer to adapt the station to various applications. The A stations in Fig. 8.8 on page 271 show three alternative internal connections between the MAC and PHY available to the CMT.

8.5.6 FDDI Optical Components

The FDDI specification gives the dimensions and tolerances of the components to ensure interoperability.

Fiber Media

In the original FDDI specification, only optical fiber was allowed, although the size of the fiber can vary. (Work has begun to specify a shielded twisted-wire version of FDDI for short-distance transmission.) The recommended fiber size is 62.5/125 μm, although it is expected that 50/125, 85/125, and 100/140 μm fibers will also be used. The best compromise between fiber dispersion and power coupling is in 62.5/125 [25] and 85/125 fibers. Typically, the fibers will be required to have bandwidth-distance products of 400 MHz·km or more and attenuations of less than 2.5 dB/km [19]. Fibers of other sizes will change the maximum distance between stations.

Sources

The operating wavelength is specified to be 1300 nm. The data rate is specified at 100 Mb/s with an expected maximum transmission distance of 2000 m. This combination can be met most economically with 1300 nm LEDs for the distance specified. (Long-range FDDI systems under development could use lasers.) While short-wavelength sources and receivers would be more economical for short-distance use, the standards designers decided to avoid the confusion that a second wavelength would bring, so the standard allows 1300 nm sources only [19]. The minimum allowable power for the transmitter (into the fiber) is −16 dBm.

Receivers

The standard calls for pin diodes to be used in the link. While avalanche photodiodes would be more sensitive and would result in more link margin, the designers felt that pin diodes would be a more mature technology and would result in a lower cost receiver. (This decision to use LEDs combined with pin-diode receivers reduces the acceptable losses of the system to a fairly low value of 11 dB, as described in the next section [25].) To achieve the required bit-error rate with a pin diode requires a receiver power level of at least −27 dBm.

Losses

The maximum optical loss, including cable loss, connector loss, and optical bypass devices, over the 2-km maximum distance is 11 dB. This loss-budget goal can be distributed among the components in any fashion desired [25]. Short links within buildings will typically have mostly connector losses; long-distance links will have appreciable cable loss. (The cable must also be designed to meet the 100 Mb/s data rate for the 2 km distance specified.) A worst-case cable loss of 2.5 dB/km will give 5 dB of cable loss for the 2 km maximum transmission distance, allowing a 6 dB loss allocation for the rest of the link, including splices, connectors, and bypass switches.

FDDI Connectors

The cable receptacle defines the boundary between the cable and the station. The cable connector is a duplex plug (see Fig. 8.12 on the next page). The design of the single-station connector is different than that of the dual-attachment station since they should never be connected together; the single-attachment station should be connected only to the concentrator. The connector is a dry, lensless connector with the fiber typically epoxied into a ferrule. The grind-and-polish method is used to smooth the ends. Connectors will exhibit losses in the 0.2 to 1 dB range [19]. While connectors at the station are specified in the standard, other connectors in the cable system are unspecified and left for the user to choose, as long as the 11 dB total loss budget is met.

An additional specification is made on the optical bypass switch included in the dual-port stations. The losses of this switch in the closed (or bypass-activated) position cannot exceed 3 dB. (Recall that this switch was required to allow us to bypass a faulty station.)

Jitter

The phenomenon of *jitter* is movement of the data transitions slightly from their theoretically predicted positions in time. Jitter can be caused by random components, such as thermal noise

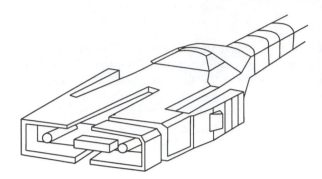

Figure 8.12 FDDI connector with two fibers.

in the receiver. Sources of random jitter can be combined to form an rms jitter J_{rms} as

$$J_{\text{rms}} = \sqrt{\sum_i \Delta\tau_i^2}\,, \tag{8.21}$$

where $\Delta\tau_i$ is the individual jitter component from any given source. There is also a component of jitter that is data-dependent. An example of this is the recognition of the fact that a logical **1** voltage only asymptotically approaches its value, it never quite gets there. The voltage level where a downward transition begins, therefore, depends on how many **1**s precede it. The more **1**s there have been, the higher is the initial starting voltage. This small difference in starting voltage will cause a small difference in the time when the transiting voltage achieves the voltage threshold, causing some jitter. This data-dependent jitter cannot be added together in an rms fashion, since it is dependent on the previous data.

The FDDI standard specifies the maximum allowed jitter at the optical output. (The designer decides the trade-offs that must be made within the station to achieve less than the required amount of jitter.) The amount of jitter is difficult to measure and to design. Systems designers will need more work and experience to gain familiarity with how to meet a jitter specification. New instrumentation is also becoming available to measure and characterize the various jitter components.

Other Specifications

The bit-error rate (BER) of the network is 4×10^{-11} [26], allowing for high data-transmission integrity. The maximum number of nodes is 500 [26].

8.5.7 FDDI Reliability Provisions

Reliability of a data network is of paramount importance. Token-ring networks are, by their nature, susceptible to reliability problems, since any defect that prevents transmission of the token can cause the net to fail. (Accidental loss of the token due to, for example, data transmission errors will cause the network to eventually reinitialize itself through the SMT functions.) The architecture of the FDDI net and its built-in reconfigurability strengthen the network's reliability. Features that increase the reliability include the following:

- Station-bypass switch: This switch allows an optical bypass of a station. Broken and powered-down stations can be taken off the network for maintenance or replacement without affecting the network operation.

- Counter-rotating ring connection: All stations directly attached to the link must be dual-connectored. The second counter-rotating link can be used as a standby link or for concurrent transmission. If a station fails or the link fails, the two rings will automatically be reconfigured into one ring (with twice the original length) to maintain link continuity (see Fig. 8.9 on page 272). In this reconfiguration, the stations on either side have noted the break in the ring and have used their bypass switches to reroute traffic over the secondary ring, maintaining continuity on the network. It should be emphasized that this reconfiguration is done automatically by the network (using the control path) and that the link will automatically return to the original configuration once the repair has been made. Several breaks in the link at different locations will cause the ring to segment into several independent sub-rings.

- Concentrators: These are used to bring single-connected stations into the link. Failure of a single-connected station plugged into a concentrator will not affect the network. Failure of a concentrator will drop all attached stations, but the network will either bypass the failed unit (with the optical bypass switch) or reconfigure around it.

8.6   SONET/SDH

The proliferation of optical-trunk networks has given rise to the requirement for a standard format for these signals so that they may be shared between networks (e.g., so that a signal can be passed from an MCI trunk to an AT&T trunk) without any need for an interface. (This is the so-called "mid-fiber meet.") The Synchronous Optical NETwork (SONET) standard has been proposed by the T1 committee of the American National Standards Institute (ANSI) and is also being considered by the ITU as an international standard [27]. The international version is called the *Synchronous Digital Hierarchy (SDH)*. (A *synchronous network* has a master clock that controls the timing of events all through the network.) The extension of SONET to become an international standard (SDH) holds the promise of breaking down an incompatibility between North American data rates, European data rates, and Japanese data rates (Fig. 8.13 on the next page).

The features of the standard include [27]

- families of standardized digital interfaces to accommodate future data rate increases (see Fig. 8.14 on page 283);

- a base rate of 51.84 Mb/s to carry all signals in the North American hierarchy up to DS3 (44.7363 Mb/s);

- expandability to higher data rates;

- synchronous multiplexing and demultiplexing for simple combining of channels, for simplified access to data payloads on the network, and for increasing synchronization of the network; and

- inclusion of sufficient overhead channels to accommodate the maintenance of the network.

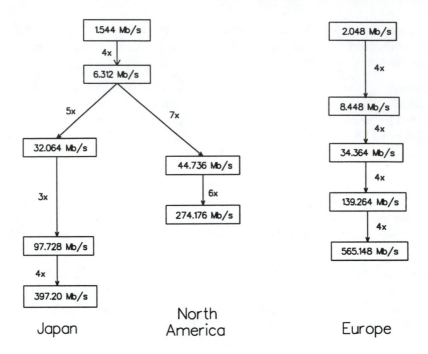

Figure 8.13 Current telecommunications multiplexing schemes. (Products are not exact because of the addition of overhead bits to the data stream.)

The standard has several parts [27]: the optical interface that specifies the operating wavelengths, power levels, etc.; operations specifications; and the rate and format specifications that give the data format, the frame size, etc.

8.6.1   SONET Optical Specifications

Some of the optical parameter specifications [28] are

- nominal wavelength: 1310 nm (within a range of 1280 to 1340 nm),

- maximum spectral width: 9 nm,

- signal optical extinction ratio: 10:1,

- coding: NRZ,

- pulse shape: to be determined.

8.6.2   SONET Rate and Format Specification

The key challenge that arose in setting the proposed standard was to allow SONET to work in a pleisiochronous environment and *still* to maintain the synchronous nature of the network. (Signals are considered to be *pleisiochronous* if their "significant instants" occur at nominally the same rate; any allowed variations in the rate are constrained to be within predetermined

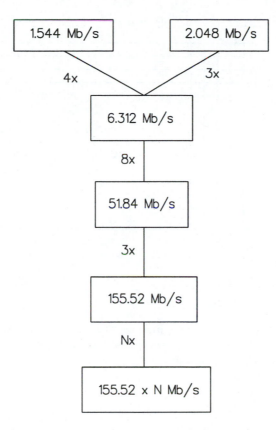

Figure 8.14 Proposed international standardized telecommunications multiplexing scheme. (Products are not exact because of the addition of overhead bits to the data stream.)

limits.) The SONET solution was to incorporate a feature that gives the network a high degree of flexibility, the use of payload "pointers" to indicate the phase of the data payload within the SONET frame.

The basic building block for a US telecommunications system is the Synchronous Transport Signal–Level 1 (STS-1) transmitted at 51.84 Mb/s. The frame structure can be modeled as a 90-column by 9-row structure of 8-bit bytes (Fig. 8.15 on the next page). One frame is transmitted every 125 μs (set by the 64 kb/s sampling rate for voice signals, which produces one byte every 125 μs). The first three columns are section overhead and line overhead bytes. The remaining 87 columns $\times$ 9 rows contain the *Synchronous Payload Envelope* (SPE) which carries the data plus nine bytes of path overhead. The format of the frame is set up so that the frame can carry a DS3 signal (at 44.736 Mb/s) or a mixture of lower-rate signals such as DS1 and DS2.

8.6.3 SONET Overhead Channels

The *overhead channels* are divided into section overhead, line overhead, and path overhead and are used to manage the communications network. Each overhead channel is designed to be used by different pieces of equipment in the network.

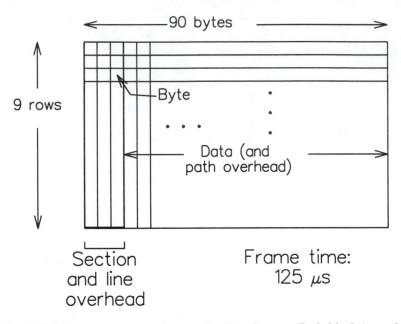

Figure 8.15 The STS-1 frame structure of 9 rows by 90 columns. Each block is an 8-bit byte. The first three columns contain overhead data for section and line management. The remaining 87 columns contain data and path layer management data.

Section Overhead

The *section overhead* contains information used by all of the SONET equipment and breaks down as follows (see Fig. 8.16 on the facing page):

- two framing bytes (A1 and A2) that locate the start of a frame,

- an STS-1 identification byte (C1),

- an 8-bit Bit-Interleaved-Parity (BIP-8) byte (B1) that is used to check for errors in the section-overhead information,

- an orderwire byte (E1) used by the network system to control the channel,

- a byte reserved for user applications (F1), and

- three bytes for section-overhead data (D1, D2, and D3), such as maintenance data or supply support data.

Line Overhead

The *line overhead* bytes are processed by all SONET equipment except the signal regenerators (i.e., the repeaters). The bytes are allocated as follows (see Fig. 8.16 on the next page):

- an STS-1 pointer (H1, H2, and H3) that provides information about the location of the data relative to the STS-1 frame (explained in more detail later in this section),

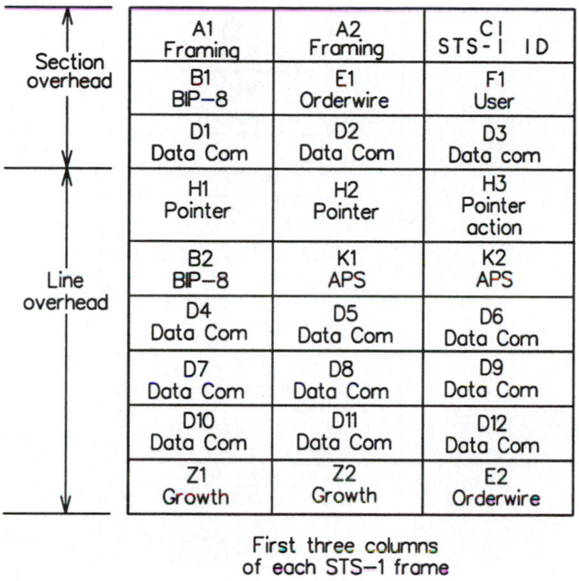

A1 Framing	A2 Framing	C I STS-1 I D
B1 BIP-8	E1 Orderwire	F1 User
D1 Data Com	D2 Data Com	D3 Data com
H1 Pointer	H2 Pointer	H3 Pointer action
B2 BIP-8	K1 APS	K2 APS
D4 Data Com	D5 Data Com	D6 Data Com
D7 Data Com	D8 Data Com	D9 Data Com
D10 Data Com	D11 Data Com	D12 Data Com
Z1 Growth	Z2 Growth	E2 Orderwire

Section overhead (spans first three rows) — Line overhead (spans remaining rows)

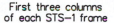

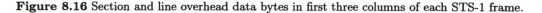

First three columns
of each STS-1 frame

Figure 8.16 Section and line overhead data bytes in first three columns of each STS-1 frame.

- an 8-bit Bit-Interleaved-Parity (BIP-8) byte (B2) that is used to check for errors in the line-overhead information,

- a two-byte autoprotection switch (APS) message protection channel (K1 and K2) that will switch to backup components or otherwise implement the user's protection scheme,

- nine bytes of line overhead data (D4-D12),

- two bytes reserved for future uses (Z1 and Z2), and

- an orderwire channel (E2) for control of the channel by the SONET equipment.

Path Overhead

The *path overhead* is processed by the SONET payload terminating equipment at the destination and is associated with the data package. Its byte assignments are as follows (see Fig. 8.17 on the following page):

- a trace byte (J1) used to trace the location of the signal through the various elements of the network,

- an 8-bit Bit-Interleaved-Parity (BIP-8) byte (B3) used to check for errors in the path-overhead information,

- a signal-label byte (C2) used to identify the type of payload being carried,

- a path-status byte (G1) used to carry path-maintenance signals,

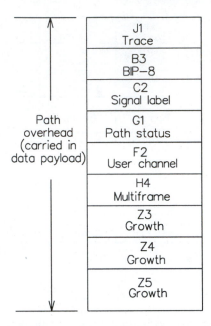

Figure 8.17 Path-overhead data bytes in first column of a Synchronous Payload Envelope (SPE) frame.

- a byte reserved for user applications (F2),

- a multiframe byte (H4) used to indicate the DS0 signaling-bit phase when carrying DS0 signals, and

- three bytes (Z3, Z4, and Z5) reserved for future growth of the standard.

8.6.4   SONET Payload Pointer

The *payload pointer* carried in the line overhead is a new feature introduced in SONET. Its purpose is to provide multiplexing synchronization of the pleisiochronous frames and to align the frames in an STS-N signal.

Low-data-rate signals are combined (multiplexed) into higher data-rate signals for trunk transmission. The usual multiplexing strategy of the telecommunications industry is shown in Fig. 8.18 on the next page [29]. The incoming data streams have already been multiplexed and might represent a DS3 signal (at 44.736 Mb/s) in the United States, a CEPT-4 signal (at 139.264 Mb/s) in Europe, or a fifth-level signal (at 397.20 Mb/s) in Japan. The *multiplexer* provides synchronization and the addition of bits (*bit stuffing*) for performance monitoring and maintenance functions. The *multiplexing function* is a simple interleave of the tributary channels. The *high-speed signal conditioner* balances the statistics into the data stream (e.g., to avoid long strings of **0** values or **1** values or provide for the insertion of block codes). The conditioned signal is then passed on to the optical source for high-speed transmission. Two types of conventional multiplexing have evolved, bit stuffing and fixed-location mapping.

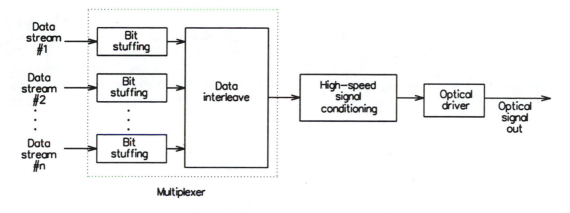

Figure 8.18 Typical multiplexing technique.

Bit Stuffing

In the *bit-stuffing method*, extra bits are stuffed into the higher-speed channel to fill up the data stream to capacity, as the signals are multiplexed together (e.g., four 1.544 Mb/s DS1 *tributary channels* being multiplexed into a 6.312 Mb/s DS2 stream). Bit-stuffing indicators are included in the frame at fixed positions and tell whether the stuffing bits contain some useful system information or are just dummy bits. The method has the advantage of being able to accommodate signals with widely varying clock speeds, but has the disadvantage of requiring de-stuffing of the bits at the demultiplexer end.

Fixed-Location Mapping

In the second multiplexing method, the tributaries are mapped into fixed, preassigned locations in the main data stream. Fixed-location mapping requires that synchronized data arrive at the multiplexer at just the right instant for inclusion in the channel. In networks that are highly synchronized (e.g., long-distance carriers), this required synchronization of the tributaries is no problem, as it is already included in the network. Even in these networks, however, there can be phase misalignments due to propagation delays and other effects, so 125 μs buffers (called "slip buffers") are included in the multiplexers to align the signals with their respective slots. In systems that are to multiplex signals from sources with separate synchronizing clocks, this is a major problem. The floating payload in the SONET frame is a potential solution.

There are two ways to align the signal within the STS-1 frame—to "float" the payload at any arbitrary location within the frame (and to use the payload pointer to record the payload starting position) or to rigidly fix the location of the payload and require that the system timing be stringent enough to meet the subsequent requirements.

Floating Payload

The floating payload is an innovation of SONET and has an advantage when dealing with signals that are not set up by a master clock. In SONET the DS3 signal is mapped into the STS-1 frame directly. No data buffers or master synchronization of the signals are required. The payload pointer is contained in the H1 and H2 bytes of the STS-1 line overhead and indicates the starting-byte location of the payload. A late-starting payload is allowed to overlap into the following STS-1 frame [27] (see Fig. 8.19). The payload is allowed to float both vertically and horizontally within the STS-1 frame [27] (also portrayed in Fig. 8.19). While offering high flexibility to data stream handling, the technique does have the disadvantage of requiring pointer data generation, reading, and interpretation. This new manipulation hardware and software must be designed and integrated into SONET equipment.

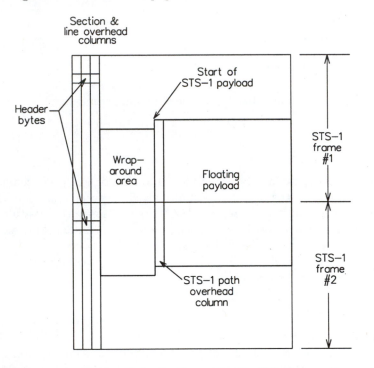

Figure 8.19 Representative alignment of a floating payload within two contiguous STS-1 frames.

The payload pointer also has the function of accommodating slippage in the data due to slight differences in the data clock and the STS-1 clock. Figure 8.20 on the facing page shows the case where the data clock is slightly faster than the frame clock. The payloads in frames #1 and #2 are at a given location. In frame #3, the faster clock of the data places the payload at a location that is one byte earlier than the previous frames. A stuff byte is moved into the H3 location of the payload pointer in this frame and the payload pointer location (in bytes H1 and H2) is decreased by one to point to the new location of the payload. In frame #4 of Fig. 8.20 on the next page, the payload has adjusted to its new position. This method of handling slippage in the data without lengthy data delays is one of the primary advantages of the floating payload structure.

Figure 8.21 on page 290 [27] shows a similar adjustment when the data clock is slightly slower than the frame clock. In frame #3 the payload has moved to one byte later in the frame. A stuff bit has been added immediately after the H3 location and the payload pointer is increased by 1.

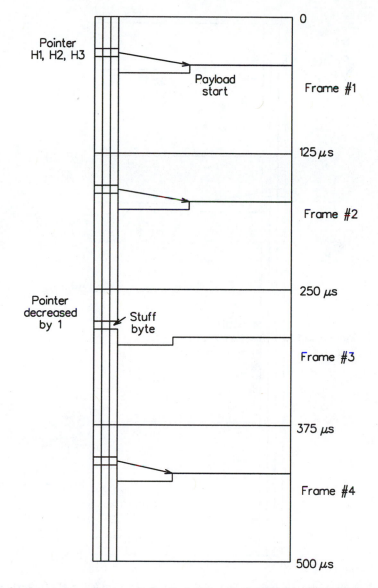

Figure 8.20 Payload pointer adjustment for fast data clock. (After R. Ballart and Y.-C. Ching, "SONET: Now it's the standard optical network," *IEEE Communications Magazine,* pp. 8–15, March 1989, ©1989 IEEE.)

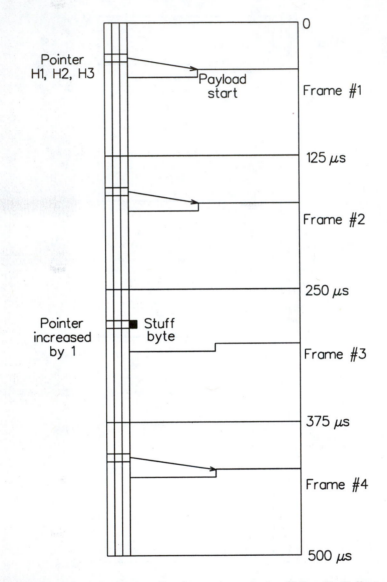

Figure 8.21 Payload pointer adjustment for slower data clock. (After R. Ballart and Y.-C. Ching, "SONET: Now it's the standard optical network," *IEEE Communications Magazine*, pp. 8–15, March 1989, ©1989 IEEE.)

Fixed Payload

An alternative use of the frame is to assign the payload to a rigidly defined, fixed location in the STS-1 frame. The payload is not allowed to float, but must fit into exactly the same location in each and every frame. This is accomplished by rigid synchronization of the data clock with the frame clock, as sometimes found in single-provider telecommunications networks. The data are synchronized to be at the input at the proper time for inclusion at the proper location. As long as the synchronization is correct, the data will be assigned their proper locations.

8.6.5 Broadband Signal Handling

One of the advantages of SONET is its capability in handling higher data-rate signals than today's standard DS3 signals. (In fact, the European standard transmission of 139.264 Mb/s means that the European network cannot use the STS-1 frame structure alone for its signals; the STS-1 frame does not have a high enough data rate. This upward compatibility allowed SONET to be proposed as the SDH international standard.) Higher data-rate signals are obtained by byte interleaving N STS-1 frames into a so-called STS-N frame [27] (see Fig. 8.22).The frame is now $N \times 90$ columns (bytes) wide by 9 rows high. The time duration of the frames is $155N$ μs. For example, three STS-1 frames can be multiplexed to reach the 139 Mb/s rate preferred by the European networks.

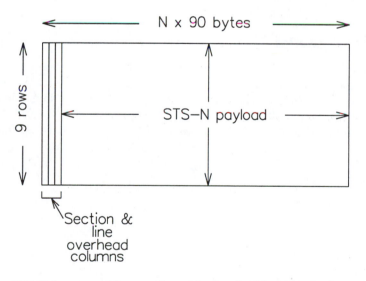

Figure 8.22 An STS-N frame used for carrying wider bandwidth payloads than the STS-1 frame.

The payload pointers are used to multiplex the N STS-1 frames to STS-N levels. When multiplexed, the section and line overhead bytes of the first STS-1 frame are used for the overhead information; the overhead bytes in the rest of the multiplexed frames are not used. The STS-N frame is then scrambled (to avoid the possibility of long strings of **0**s or **1**s) and converted to an Optical Carrier-Level N (OC-N) signal. The line rate of the OC-N signal will be exactly N times the rate of the OC-1 signal rate. The OC-N levels allowed by the standard are shown in Table 8.3. These higher data-rate signals provide upward compatibility of future systems.

Level	Line rate (Mb/s)
OC-1	51.84
OC-3	155.52
OC-9	466.56
OC-12	622.08
OC-18	933.12
OC-24	1244.16
OC-36	1866.24
OC-48	2488.32

Table 8.3 SONET signal rates.

Name	No. of columns	No. of bytes	Application
VT1.5	3	27	DS1
VT2	4	36	CEPT-1
VT3	6	54	DS1C
VT6	12	108	DS2

Table 8.4 Virtual tributary descriptions and carrying-capacity.

8.6.6 Virtual Tributaries

Just as the STS-1 frames can be combined for higher data-rate signals, so they can also be subdivided for signals that are low in data rate. SONET allows for a variety of fixed-sized smaller payload units (called *virtual tributaries* or *VTs*). There are four allowed sizes of VTs as shown in Table 8.4; this table also shows the carrying-capacity of each VT in terms of the present communications standards. A *VT group* is a 12-column × 9-row payload structure that can carry

- four VT1.5s *or*

- three VT2s *or*

- two VT3s *or*

- one VT6.

Seven VT groups (i.e., 84 columns), one path overhead column, and two unused columns are byte interleaved to form one STS-1 payload. This payload is then placed into the STS-1 frame. Figure 8.23 on the next page shows 28 VT1.5s put together to form a payload. When added to the three columns of section and line overhead, they form the complete STS-1 frame. VT groups can also be mixed when they are combined to form a frame; they do not all need to be of the same type.

The VT payload can either float within the VT or be assigned a fixed position (called a "locked" position).

- The *floating VT payload* is allowed to float in a fashion similar to the STS-1 payload; the VT pointer is used to show the starting byte location within the VT payload. Again, the

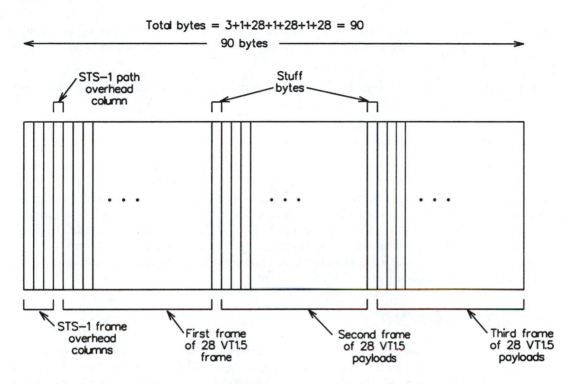

Figure 8.23 Combination of 28 VT1.5 packages to form an STS-1 payload. (After Now it's the standard optical network," *IEEE Communications Magazine*, pp. 8–15, March 1989, ©1989 IEEE.)

primary advantage is that no data buffers (with subsequent data delay) are required to handle asynchronous data streams.

- The *locked VT payload* is assigned a fixed location within the STS-1 payload (which can, itself, be floating or fixed within the STS-1 frame). Again, the locked mode provides a simple scheme for handling the multiplexing of channels *if* a master synchronization is maintained. This scheme is primarily useful in the telecommunications industry where several DS0 channels (in the United States) can be handled with this approach.

It should be noted that floating VTs may not be mixed with locked VTs within the STS-1 payload. All have to be either floating or fixed.

The bundling of information channels within the VTs is called *payload mapping*. Five types are possible

- Asynchronous mapping into a floating VT group uses conventional bit stuffing to multiplex asynchronous channels into the VT payload.

- Byte-synchronous mapping into either a floating VT group or a locked VT group (counts as two separate mapping methods) is proposed for synchronous mapping of the DS0 channels (and their overhead).

- Bit-synchronous mapping into either a floating VT group or a locked group (also counted as two separate mapping methods) has been proposed for carrying unframed synchronous signals.

The most useful mappings will probably be the asynchronous mapping and the byte-synchronous mode, both used with the floating VT group mode.

8.6.7 SONET Compatibility

Although proposed as a method of ensuring intercompatibility of equipment, the SONET standard does not completely specify *all* of the possible permutations and combinations. For example, some of the overhead is left for future user applications. Users must ensure that the equipment will follow their own application but disregard the application data from "foreign" systems. How can this be ensured? Protective codes are included in the SONET overhead, but each manufacturer will handle a protection-provoking event differently (e.g., one might power-down its source while another may activate an optical switch). Much work still needs to be done to ensure and demonstrate intercompatibility of SONET equipment.

8.7 ATM and SONET

One of the goals of the convergence of the data communications and telecommunications fields is to have a common communications format that would allow "seamless integration" of the local area network (LANs) with the wide area networks (WANs) of the telecommunications industry. The *Asynchronous Transfer Mode (ATM)* [30–33] technique offers the promise of achieving a common format from the PC or workstation on the desktop through the international switches of the global telecommunications giants. ATM is a WAN technology that can be used for LAN links as well. Besides its seamless integration of WANs and LANs, it offers the promise of the following features:

- It has a scalable network performance that allows it to run at design rates from 25 Mb/s up to multi-gigabits per second.

- It offers low delay times through a network due to minimum packet overhead and the elimination of the need of the intermediate node to read and process the entire packet.

- It is a point-to-point *connection service*. This means that *every* ATM cell in an ATM transmission travels over the same path. This path is specified either in the call setup or, in a private network, is wired in. The header of the ATM cell contains all of the needed information to relay the cell from station to station. (A *connectionless service*, on the other hand, allows the packets to follow many paths to the receiver. The receiver reassembles the packets in the correct order. These services are applicable for data; they are not used for real-time voice and video since delays or packet sequence scrambling cannot be tolerated.)

- It has an architecture that allows users to share the available bandwidth (rather than allocating the entire bandwidth to a single user for a finite amount of time).

- It allows transmission of a mixture of synchronous signals (voice, video) and asynchronous signals (data) to share the bandwidth. This feature is becoming especially important as transmission of multimedia data files and video team teleconferencing reach fruition.

Before beginning our discussion of ATM, we want to consider the differences between WANs and LANs. WANs typically have the following characteristics. They cover larger geographic region and are usually configured as a series of point-to-point nodes that are connection oriented and support some type of switching to add and drop signals. WANs typically omit source address from their packet because of their point-to-point nature. WANs typically also have had a higher error rate than LANs (although, this is changing). This necessitated error-checking at each node, leading to increased overhead, increased processing time (leading to longer end-to-end latencies), and increased node buffer requirements. The error-checking at every node is required in order to have reasonable throughputs in high-error systems. Without it, the relatively high error rate ensures that many packets will require retransmission. (Low-error fiber-optic WANs are reducing this requirement. For example, the new frame relay systems handshake only between the ends of the link; the older X.25 packet network nodes check every frame at every node.) The first widely support WAN were packet networks based on X.25 link access protocol (LAP-B) standard. The X.25 standard defines a series of interchange protocols and specifies the framing at the physical, data link, and network levels. The standard requires handshaking and error checking between each pair of nodes as data frame traverses network.

LANs, on the other hand, encompass a smaller geographic region and usually feature connectionless service. They typically operate at a lower error rate; hence, error checking is done only between the source and user rather than at every node pair in between. This allows for simpler design of the intermediate nodes, resulting in lower link overhead, lower intermediate storage, and less delay. The cost is a longer time to detect and correct errors, but, since errors are less likely to occur, the trade-off advantage is in favor of end-to-end error correction.

One of the major advances in telecommunications and data communications is the ability to assign bandwidth on demand. Unlike many present systems where the user "rents" all of the bandwidth for the duration of the connection, the desire is to share the available bandwidth adaptively among different users. The first attempt at this is SMDS (Switched Multimegabit Data Service) in the United States and CBDS (Connectionless Broadband Data Service) in Europe. This service allows users to request only the required bandwidth for their application rather than paying for fixed bandwidth service. Several users share the payload space on a given hop; the user payloads are switched off and on at the intermediate nodes depending on their final destination. This services is threatened to be overtaken by ATM due to its increased advantages.

ATM is a packet-based system with fixed-length cells that are 53 bytes in length. These bytes are distributed as shown in Fig. 8.24 on the following page [34]. Five bytes are assigned to the cell header with the following bit distribution [31]:

- 4-bit generic flow control (GFC),

- 8-bit virtual path identifier (VPI),

- 16-bit virtual channel identifier (VCI),

- 3-bit payload-type indicator (PTI),

- 1-bit cell loss priority (CLI) indicator, and

- 8-bit header-error control (HEC) field.

The VCI and VPI contain routing information for the cell. The VPI identifies the major path (*virtual path*) and the VCI identifies the individual paths (the *virtual channels*) within the trunk.

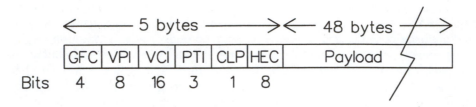

Figure 8.24 Structure of an ATM cell.

(The "virtual" descriptor is used to distinguish the connections from the dedicated circuits of the synchronous transfer mode [STM] systems.) The rest of the header bits maintain the message flow and ensure the integrity of the addresses as the cell traverses the network. The remaining 48 bytes are assigned to the payload. The short cell length was chosen to reduce network latency and also allows better statistical multiplexing. A fixed cell length was chosen to simplify switching the cells on and off the link since buffer and other device capacities are standardized (rather than allowed to have variable lengths) It also ensures that the cell lengths are reasonable (e.g., 3 μs at a 155 Mb/s data rate) so that audio and video signals will have almost-immediate access to the link. The fixed cell length also allows a scalable network performance.

A *virtual path* may consist of several virtual trunks. The VPI can represent the trunk between two locations and the VCIs can be used to identify the individual calls. The nodes in the trunk need only look at the VPI portion of the cell to transfer the cell; they do not need any further information from the contents of the cell. (The connections are described as "virtual" to distinguish them from fixed connections such as are used with the synchronous transfer mode (STM) in current use in the telecommunications networks.)

The network performance of an ATM system is determined (almost) by switch speeds. For the duration of the connection, user stations can have access to the full data rate provided since they are not sharing the bandwidth with other users.

In order to implement ATM transmission, the application must divide up its data into ATM cells at the transmitter and reassemble the data from the cells at the receiver. This task is handled in accordance with the ISO layer structure. This layer structure is a standardized way to break down the responsibilities for intercommunication in software and hardware designs. The assignment allows software engineers, for example, to design their application to send and receive data to/from other locations without having to worry about the details of the link between them (e.g., whether the link is fiber or twisted pair). Table 8.5 on the next page shows the definitions of the layer assignments and their functions.

In ATM, the network-layer frame must be divided up into ATM cells at the transmitter and, then, at the receiver, the data within the ATM cells has to be recombined into a network-layer frame. The layers above the network layer know how to handle the network-layer frame and do not know or care that ATM or FDDI is being used to transmit the data. In ATM, the assembly and disassembly of the ATM cells is done by a part of the ATM standard called "ATM adaptation layer 5 (AAL5)" [31, 34].

Figure 8.25 on the facing page shows a typical ATM LAN setup [32]. A group of workstations, servers, and any other network elements are direct-connected to an ATM switch. Using plug-in cards, each connection can be tailored to the user's bandwidth requirements. One user might have a high-rate optical fiber connection while another user might have 1.5 Mb/s coax link. The

Layer	Function
Application	Provides service elements to process the data; e.g., resource sharing, file transfer, database management
Presentation	Services necessary to format exchange data and manage session
Session	Establishes and terminates connections; arbitrates rights to channel; synchronizes data exchanges
Transport	Provides functions for error-free delivery of messages, e.g., flow control, error recovery and acknowledgment
Network	Provides transparent routing of messages between two transport entities
Data link	Provides rules of transmission on physical link, e.g., packet formats, access rights, error detection and correction
Physical	Mechanical and electrical interconnection

Table 8.5 ISO layers and their functions.

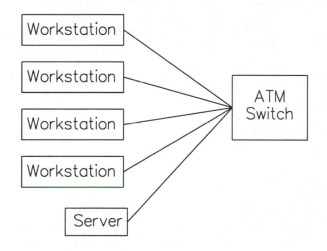

Figure 8.25 Representative ATM LAN setup.

ATM switch is responsible for receiving and delivering ATM cells from and to the users. Since each user has a dedicated line there is no contention problem, as exists in techniques using a shared medium. The network is easily added to a wide-area network through the addition of a WAN port to the switch.

On the other hand, Fig. 8.26 on the next page illustrates a typical private WAN that might be operated by a company with nationwide (or worldwide) offices. The local area networks of the office are tied into routers. The company leases data lines from a public service provider (e.g., a dedicated 1.544-Mb/s T1 data line leases for about $6,500 per month [32] in 1994). The user pays for the entire bandwidth whether it is used or not. For example, most data lines are used most intensely during business hours; after-hour usage is usually lower. This structure can be replaced by that shown in Fig. 8.27 on page 299. Here the routers are connected to ATM switches owned by a public service provider. The switches are interconnected by a mesh of broadband public lines; the service provider has the task of filling up these lines with traffic, not the end-user. An interconnection is set up by the ATM switches for the duration of the complete ATM transmission; this ensures the timely, sequential arrival of ATM cells as required for audio and video signals.

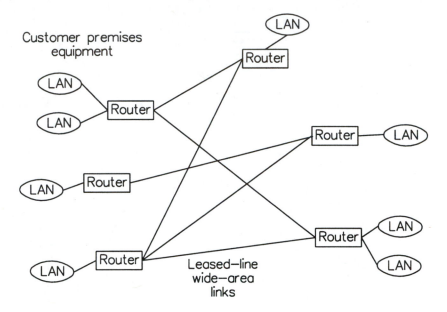

Figure 8.26 Structure of a typical leased-line WAN.

The ATM cell was designed to run on SONET/SDH links at 155.42 Mb/s, 622.08 Mb/s, and 2.48832 Gb/s. ATM cells fit seamlessly within the SONET payload STS frame as described earlier. As mentioned before, ATM LANs can run from 25 Mb/s up to 2 Gb/s or more. This easy incorporation of the ATM cell into the SONET frame is one of the most powerful features of the ATM approach, as it is integral to the upward compatibility of the ATM architecture into the worldwide telecommunications network. It is an example of the synergistic effects that can occur with open standards that allow compatible combinations to be made.

Hence, we have seen that ATM technology is a connection-oriented scheme. The packets in a message all traverse the same path. Unlike synchronous transfer mode technology, however, the path is not reserved exclusively for one user. The path is available to other users if the original user is not using it. This sharing of resources is one of the key elements of ATM's efficiency. Since most of the communications capacities today are unused. For example, a typical voice channel uses only 30 to 40% of the bandwidth due to silences in the communication and times of listening [32]. Currently, it is expected that LAN application of ATM technology will be implemented in the near future, followed by more intensive use of ATM WAN technology in later years.

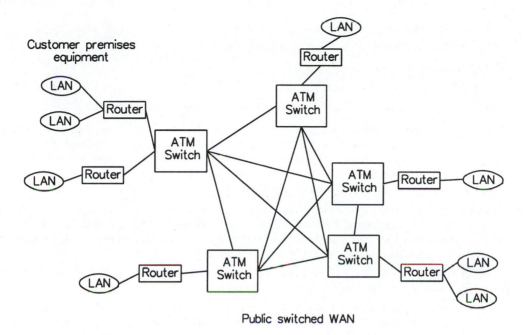

Figure 8.27 Structure of an ATM WAN using public broadband ATM switches and lines.

8.8 Summary

In this chapter we have considered fiber optics used in data networks. As the long-haul applications for fiber optics have become saturated, data networks offer opportunities for fiber-optic systems. We have seen that star and linear data buses have proved to be popular configurations. For fiber-optic networks, an analysis of the power budget showed that the star configuration has advantages over the linear bus for applications with many stations on the network. The FDDI and SONET standards represent efforts by the local area network community and the telecommunications community to establish standards that incorporate fiber optics as the media of choice. It is important to realize, however, that optical transmission is only part of the breakthroughs represented by these standards. The novel architectures proposed and being adopted offer equally revolutionary concepts. ATM standards have been proposed that offer the prospect of allowing the seamlessly merging of the telecommunications applications with computer network applications. The marriage of the ATM architecture with the modular capabilities of SONET offer promise of providing an upwardly scalable path for data transmission that allows video and audio applications to coexist with data traffic on LANs, MANs, and WANs. Testbed demonstrations of networks are currently being implemented and studied [35–38].

8.9 Problems

1. Consider a 20 station linear data bus with connector losses of 1 dB per connector, coupling fractions of 5% at each arm of the coupler, insertion losses of 2 dB per coupler, fiber losses of 3 dB/km, and station spacings of 1 km.

(a) Calculate the transmissivity of each connector.

(b) Calculate the transmission loss (in dB) of each arm of the coupler (L_{arm}).

(c) Calculate the ratio of $P_{1,2}/P_f$ in dB.

(d) Calculate the ratio of $P_{1,20}/P_f$ in dB.

(e) Calculate the dynamic range required of the receiver used in this network.

2. Consider a star network that operates with sources that produce +3 dBm of output power in a fiber. Assume that the fiber loss is 0.6 dB/km and that the station-to-star distance is 2 km. The required receiver sensitivity is −30 dBm. Calculate the number of stations that can be on this network if the connector loss is 1 dB per connector, the total insertion loss of the star is 3 dB (from any input to any output) and the link margin is 0 dB.

3. Consider the same transmitter and receiver as the previous problem used in a linear network. Assume that 5% of the light is coupled into the arm of the tee coupler and that insertion loss is 1 dB per tee coupler. Calculate the number of stations that can be on this network if the connector loss is 1 dB per connector and the link margin is 0 dB.

4. Consider the network of Fig. 8.9a on page 272. Suppose that the node on the left side of the ring fails. Draw a sketch of the ring after reconfiguration has occurred.

5. Consider an FDDI link with a transmitter power of −16 dBm and a receiver sensitivity of −27 dBm. The loss in the fiber has a worst-case value of 2.5 dB/km and the loss of the connectors is 1 dB per connector pair. The losses of the bypass switch in the station is 2 dB when bypassed. The separation between the stations is 2 km.

(a) Ignoring all other losses, how many consecutive stations can be shut down (i.e., bypassed) in the network between two active stations?

(b) It is desired to allow one station to be bypassed between two active stations. Find the sum of the fiber loss, connector loss, and bypass switch loss that would allow this to happen.

(c) Assuming that fiber losses, the connector loss, and the bypass switch loss stay as originally stated, find the station separation that can be supported if one station is allowed to be bypassed between two active stations.

(d) Repeat the calculation of part (c) if the connector losses are reduced to 0.2 dB per connector pair (i.e., were typical of the losses associated with splices).

(e) Repeat the calculation of part (c) if the fiber loss is reduced to 0.6 dB/km.

(f) Repeat the calculation of part (c) if the bypass switch loss is negligible (i.e., 0 dB).

References

1. P. Cochrane and M. Brain, "Future optical fiber transmission technology and networks," *IEEE Communications Magazine*, vol. 25, no. 11, pp. 45–60, November 1988.

2. Paul E. Green, Jr., *Fiber Optic Networks*. Englewood Cliffs, NJ: Prentice Hall, 1993.

3. R. L. Cruz, G. R. Hill, A. L. Kellner, R. Ramaswami, G. H. Sasaki, and Y. Yamabayashi, "Optical networks," in *Selected Areas in Communications*, IEEE Communications Society, 1996.

4. S. Su, L. Jau, and J. Lenart, "A review on classification of optical switching systems," *IEEE Communications Magazine*, vol. 24, no. 5, pp. 50–55, 1986.

5. J. R. Lee and C. K. Un, "A code-division multiple-access local area network," *IEEE Trans. on Communications*, vol. COM–35, no. 6, pp. 667–671, 1987.

6. G. J. Foschini and G. Vannucci, "Using spread-spectrum in a high-capacity fiber-optic local network," *J. Lightwave Technology*, vol. 6, no. 3, pp. 370–379, 1988.

7. J. A. Salehi, "Code division multiple-access techniques in optical fiber networks—Part I: Fundamental principles," *IEEE Trans. on Communications*, vol. 37, no. 8, pp. 824–833, 1989.

8. J. A. Salehi and C. A. Brackett, "Code division multiple-access techniques in optical fiber networks—Part II: Systems performance analysis," *IEEE Trans. on Communications*, vol. 37, no. 8, pp. 834–842, 1989.

9. M. Fujiwara, M. Goldman, M. O'Mahony, O. Tonguz, and A. Willner, "Special issue on multiwavelength optical technology and networks," in *Journal of Lightwave Technology*, 1996.

10. S. Personick, "Protocols for fiber–optic local area networks," *J. Lightwave Technology*, vol. LT–3, no. 3, pp. 426–431, 1985.

11. J. E. Midwinter, "Current status of optical communications technology," *J. Lightwave Technology*, vol. LT–3, no. 5, pp. 927–930, 1985.

12. M. Gerla, P. Rodrigues, and C. W. Yeh, "Token-based protocols for high-speed optical-fiber networks," *J. Lightwave Technology*, vol. LT-3, no. 3, pp. 449–466, 1985.

13. E. G. Rawson, "The Fibernet II Ethernet-compatible fiber-optic LAN," *J. Lightwave Technology*, vol. LT-3, no. 3, pp. 496–501, 1985.

14. J. W. Reedy and J. R. Jones, "Methods of collision detection in fiber optic CSMA/CD networks," *IEEE J. on Selected Areas in Communications*, vol. SAC–3, no. 6, pp. 890–896, 1985.

15. M. Barnoski, "Design considerations for multiterminal networks," in *Fundamentals of Optical Fiber Communications* (M. F. Barnoski, ed.), pp. 329–351, New York: Academic Press, 1981.

16. G. Keiser, *Optical Fiber Communications, Second Edition*. New York: McGraw-Hill, 1991.

17. R. Ramaswami and K. Liu, "Analysis of effective power budget in optical bus and star networks using erbium-doped fiber amplifiers," *J. Lightwave Technology*, vol. 11, no. 11, pp. 1863–1871, 1993.

18. J. Walrand, *Communications Networks: A First Course*. Homewood, IL: Aksen Associates and Richard D. Irwin, Inc., 1991.

19. W. Burr, "The FDDI optical data link," *IEEE Communications Magazine*, vol. 24, no. 5, pp. 8–23, 1986.

20. M. J. Strohl, "High performance distributed computing in FDDI networks," *IEEE Lightwave Telecommunications Systems (LTS)*, vol. 2, no. 2, pp. 11–15, 1991.

21. R. J. Kochanski and J. L. Paige, "SAFENET: The standard and its application," *IEEE Lightwave Communications Systems (LCS)*, vol. 2, no. 1, pp. 46–51, February, 1991.

22. F. Halloran, L. A. Bergman, E. G. Edgar, R. Hartmayer, and J. Jeng, "An FDDI network for tactical applications," *IEEE Lightwave Communications Systems (LCS)*, vol. 2, no. 1, pp. 29–35, February, 1991.

23. R. Jain, "Performance analysis of FDDI token ring networks: Effect of parameters and guidelines for setting TTRT," *IEEE Lightwave Telecommunications Systems (LTS)*, vol. 2, no. 2, pp. 16–22, 1991.

24. F. E. Ross, "FDDI – A tutorial," *IEEE Communications Magazine*, vol. 24, no. 5, pp. 10–17, 1986.

25. M. J. Hackert and G. D. Brown, "System design methodology for power budgeting of 62.5–μm multimode fiber systems," *J. Lightwave Technology*, vol. 11, no. 2, pp. 205–211, 1993.

26. I. Chlamtac and W. R. Franta, "Rationale, directions, and issues surrounding high speed networks," *Proc. IEEE*, vol. 78, no. 1, pp. 94–120, 1990.

27. R. Ballart and Y.-C. Ching, "SONET: Now it's the standard optical network," *IEEE Communications Magazine*, vol. 26, no. 3, pp. 8–15, March 1989.

28. R. J. Boehm, Y.-C. Ching, C. G. Griffith, and F. A. Saal, "Standardized fiber optic transmission system — a synchronous optical network view," *IEEE J. on Selected Areas in Communications*, vol. SAC–4, no. 9, pp. 1424–1431, 1986.

29. I. Jacobs, "Design considerations for long-haul lightwave systems," *IEEE J. on Selected Areas in Communications*, vol. SAC-4, no. 9, pp. 1389–1395, 1986.

30. J. Bryan Lyles and D. C. Swinehart, "The emerging gigabit environment and the role of local ATM," *IEEE Communications Magazine*, pp. 52–58, April, 1992.

31. S. M. Walters, D. S. Burpee, and G. H. Dobrowski, "Evolution of fiber access systems to ATM broadband networking," *Proc. IEEE*, vol. 81, no. 11, pp. 1588–1593, 1993.

32. J. Lane, "ATM knits voice, data on any net," *IEEE Spectrum*, pp. 42–45, February, 1994.

33. A. Miller, "From here to ATM," *IEEE Spectrum*, pp. 20–24, June, 1994.

34. J. J. Bae and T. Suda, "Survey of traffic control schemes and protocols in ATM networks," *Proc. IEEE*, vol. 79, no. 2, pp. 170–189, 1991.

35. G. R. Hill et al., "A transport network layer based on optical network elements," *J. Lightwave Technology*, vol. 11, no. 5/6, pp. 667–679, 1993.

36. Stephen B. Alexander et al., "A precompetitive consortium on wide-band all-optical networks," *J. Lightwave Technology*, vol. 11, no. 5/6, pp. 714–735, 1993.

37. Paul E. Green, Jr. et al., "All-optical packet-switched metropolitan-area network proposal," *J. Lightwave Technology*, vol. 11, no. 5/6, pp. 754–763, 1993.

38. L. Kazovsky and P. Poggiolini, "STARNET: a multi-gigabit-per-second optical LAN utilizing a passive WDM star," *J. Lightwave Technology*, vol. 11, no. 5/6, pp. 1009–1027, 1993.

Chapter 9

Wavelength-Division Multiplexing

9.1   Introduction

So far, we have been concerned only with links that carry a single wavelength. Despite the high capacity of an optical channel, most current systems use only one wavelength for carrying information. In order to increase the capacity of a link or to use innovative wavelength addressing ideas in networks, links utilizing multiple wavelengths are becoming increasingly popular [1, 2].

We have seen, as illustrated in Fig. 9.1 on the next page, that optical fibers have relatively broad regions of low loss that could support the operation of the link with more than one source. Since we have the ability to control the operating wavelength of the semiconductor emitters through the alloy composition or through the grating spacing in the distributed-feedback (DFB) lasers or distributed Bragg reflector (DBR) lasers , it is possible to hypothesize a link that would carry several wavelengths, as illustrated in Fig. 9.2 on the following page. At the receiving end, filters, or other wavelength-sensitive elements, such as gratings, would separate the wavelengths and each carrier would fall on a separate receiver for detection. Since the wavelengths are used to share (or multiplex) the channel, the technique is called *wavelength-division multiplexing (WDM)* [3, 4]. Early efforts separated the channels by about 10 nm; current efforts have separations of 1 nm or less. These current efforts are called *dense WDM* to distinguish them from the earlier, wider separations.

The key elements in the link are the devices that combine the separate source emissions (the *couplers* or *multiplexers*) at the transmitter end and the elements that separate the channels (the *demultiplexers*) at the receivers. We also need techniques to set and maintain the wavelength of the sources, so that they are fairly close together.

Bidirectional links are also possible in WDM systems by intermixing sources and receivers at each end, as shown in Fig. 9.3 on page 305. The optical fiber passband extends roughly from 800 nm to 1600 nm. (There are significant losses associated with the OH$^-$ ions at 1400 nm and 1250 nm. These losses are not important for short-distance links [up to 10 km or so] but become increasingly important as one tries to go maximum distances.) Three fiber-optic transmission windows are identified near 850, 1300, and 1550 nm.

- Multimode systems operating at 850 nm have bandwidth-distance products of 1 GHz·km (attained with graded-index fibers).

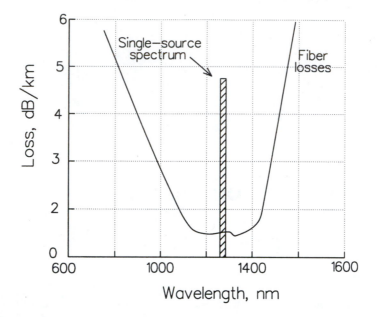

Figure 9.1 Underutilized optical bandwidth in a single-source multimode system.

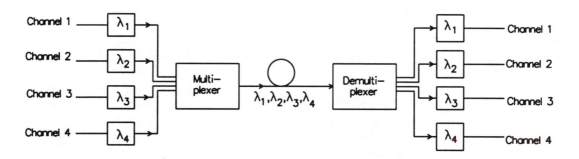

Figure 9.2 System using wavelength-division multiplexing.

- As seen in Fig. 9.4 on the facing page, there are about 14,000 GHz of low-loss bandwidth in the 1300 nm region of a typical single-mode fiber. Single-mode fibers operating at 1300 nm have losses of about 0.4 dB·km^{-1}. The dispersion of a single-mode fiber with a 10 μm core and a Δ of about 0.5% is zero at about 1300 nm, with a representative dispersion in the 1260 nm to 1360 nm window of about 4 ps·km^{-1}·nm^{-1}. (Recall that the effects of dispersion become apparent when the dispersion exceeds about 20% of the bit period.)

- As also seen in Fig. 9.4 on the next page, there are about 15,000 GHz of low-loss bandwidth in the 1550 nm region of a typical single-mode fiber. Single-mode fibers operating at 1550 nm have losses of about 0.2 dB·km^{-1} with a typical dispersion of 18 ps·km^{-1}·nm^{-1}. This dispersion can be reduced if dispersion-shifting techniques are applied to move the

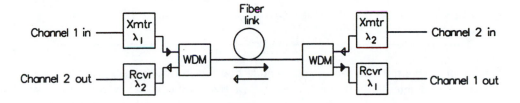

Figure 9.3 Bidirectional WDM system.

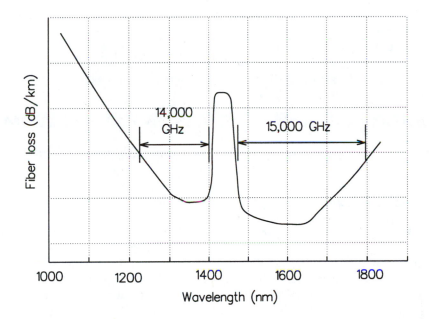

Figure 9.4 Spectral windows in single-mode fibers.

minimum-dispersion wavelength up to the 1500 nm window or by decreasing the source linewidth.

For long-distance high-data-rate links, then, the usable windows are from 1250 to 1350 nm and from 1450 to 1600 nm, allowing a combined optical window of 250 nm.

─────────────────── ∞ ───────────────────

Example: Calculate the width of the 1300 nm window in nm.

<u>Solution:</u> The frequency and wavelength are interrelated by $\nu = c/\lambda$; so,

$$\frac{d\nu}{d\lambda} = -\frac{c}{\lambda^2} \tag{9.1}$$

$$|\Delta\nu| = \frac{c}{\lambda^2}\Delta\lambda$$

$$|\Delta\lambda| = \frac{\lambda^2}{c}|\Delta\nu| = \frac{(1300 \times 10^{-9})^2(14,000 \times 10^9)}{3.0 \times 10^8}$$

$$= 7.89 \times 10^{-8} \text{ m} = 78.9 \text{ nm}.$$

Similarly, you can show that the 15,000 GHz window at 1550 nm is 120.1 nm wide.

––––––––––––––––––––––––––––––– ∞ –––––––––––––––––––––––––––––––

Other factors can limit the use of single-mode fibers in multi-wavelength applications. The fiber ceases to be single-mode if the wavelength of the source is below the cutoff wavelength of the fiber. At longer wavelengths, a fiber becomes more susceptible to microbending losses. Additionally, commercial semiconductor laser sources do not exist over much of the band but probably could be engineered.

The transmission window can be exploited by wavelength-division multiplexing. In these systems, sources operating at various wavelengths within the window are combined for transmission over the channel and are separated for reception, as in Fig. 9.2 on page 304. The prime advantage is that synchronous, asynchronous, and analog signals can all be carried simultaneously over the fiber channel with full utilization of the bandwidth offered by the fiber.

9.2   Wavelength-Selective WDM vs. Broadband WDM

Wavelength-division multiplexing can be divided into wavelength-selective techniques and broadband techniques.

9.2.1   Wavelength-Selective WDM

The *wavelength-routed WDM technique* is illustrated in Fig. 9.5a. Each source operates at a separate, assigned wavelength. The power from all sources is combined (ideally) without loss. The demultiplexer at the receiving end is wavelength-sensitive; it separates each wavelength into a different route to a receiver; the route being unique to that wavelength. (The separation is, again, ideally lossless.) The multiplexing and demultiplexing can be achieved with low insertion loss, as all of the power at a given wavelength is (in theory) directed along only one path; the other wavelengths go along separate paths. Neglecting the insertion loss of the couplers, then,

$$P_{\text{out}}(\lambda_j) = P_{\text{in}}(\lambda_j)\,. \tag{9.2}$$

The amount of light that leaks from one channel into another is called the *crosstalk*. In incoherent systems, such as we are describing, the crosstalk CT from channel j into channel i is expressed (in dB) as

$$\text{CT}_{\text{ij}} = 10\log\left(\frac{P_{ij}}{P_{ii}}\right)\,, \tag{9.3}$$

where P_{ij} is the power measured in channel i when only channel j is active and P_{ii} is the power in channel i when it is the only channel that is active. The crosstalk can be expressed either pair-wise, as just described, or in total as

$$\text{CT}_{\text{total}} = 10\log\left(\frac{\sum_j P_{ij}}{P_{ii}}\right)\,, \tag{9.4}$$

where the numerator is the sum of the power from all of the other channels as measured in channel i.

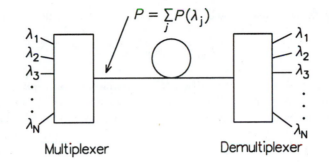

(a)

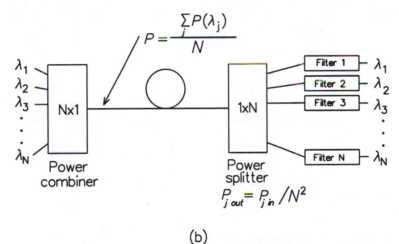

(b)

Figure 9.5 (a) Wavelength-routed multiplexing. (b) Broadcast-and-select multiplexing.

9.2.2   Broadcast-and-Select Techniques

The *broadcast-and-select technique* (Fig. 9.5b) combines the powers of the sources at the transmitting end and then divides this total, combined signal power at the receiver end.

In the arrangement shown each transmitter is preassigned a fixed wavelength. All wavelengths are broadcast to all receivers; each receiver requires its own filter to separate the channels. The divided power reaching each receiver is filtered for the desired wavelength, which is then passed to the detector (similar to commercial TV broadcasts). It should also be noted that the source end and receiver end may, in fact, be co-located in a star configuration. (Alternatives allow a tunable transmitter that can operate at any of the allowed wavelengths and a fixed receiver, or both tunable transmitters and tunable receivers.)

In these systems, simple splitters and combiners are used to combine and split the signals, resulting in a loss of $1/N$ for each device. The overall loss, then, is $1/N^2$, as illustrated. Hence, we have a significant power loss in each channel, expressed by (neglecting insertion losses)

$$P_{\text{out}}(\lambda_j) = \frac{P_{\text{in}}(\lambda_j)}{N^2}.$$

(9.5)

The losses of the splitter are $1/N$ or $10 \log N$ dB (excluding any excess losses). While this loss penalty can be severe without the use of compensating optical amplifiers, the broadband technique has some advantages. The sources need only to be tuned within the passband of the spectral filter, a fairly generous spectral width. The minimum channel spacing depends on the steepness of the passband filter characteristics and the amount of crosstalk that can be tolerated. Suitable receiver filtering elements include Fabry-Perot etalons, tuned semiconductor laser amplifiers, integrated filters, and heterodyne receivers, as discussed later. In addition, the broadband system is an agile system for military applications, since the source wavelengths and receiver filters can be easily switched without disturbing the distribution system.

We now want to consider some of the optical components used in WDM systems.

9.3   Multiplexers

Broadband WDM systems use broadband versions of the power splitters and combiners that were described Chapter 4. The bandwidth of a wavelength-flattened fused-fiber combiner or splitter can be as much as 400 nm [4], which is more than adequate for the postulated window extending from 1250 nm to 1600 nm.

Wavelength-selective WDM systems require multiplexing and demultiplexing that can be carried out by a variety of techniques including diffraction gratings, spectral filters, or directional couplers. The elements that separate the wavelengths are of two types, angularly dispersive devices and filter devices.

9.3.1   Angularly Dispersive Devices

Angularly dispersive devices (e.g., prisms and gratings) are optical devices that transmit or reflect light at an angle that depends on the wavelength of the incident light (assuming a collimated input light beam) [5]. The main parameters that describe such a device are the excess loss and the angular dispersion of the device. The *excess loss* is the loss (in dB) incurred by passing through the device in excess of the expected $1/N$ splitting loss. Since these devices are, by their nature, wideband devices, the excess loss should be specified as a function of the wavelength [6].

The *angular dispersion* of the device is the measure of the angular spread $d\theta$ between two beams that are spatially coincident at the input with wavelengths separated by $d\lambda$. The angular dispersion is given by $d\theta/d\lambda$. Angular dispersion can be converted into lateral displacement by the addition of a lens, as shown in Fig. 9.6 on the next page. Here, the dispersive device is placed in the front focal plane of the rear lens and the detector inputs are located in the back focal plane.

The dispersive element requires a collimated beam of light incident on it. Figure 9.6 on the facing page shows a collimating lens intercepting the beam from the input fiber. The end of the input fiber is in the front focal plane of the lens; the dispersive device is in the back focal plane of the lens.

In most angularly dispersive systems, a grating is the device that is combined with the lenses (Fig. 9.7 on the next page). A grating is a reflecting or transmitting element with a series of

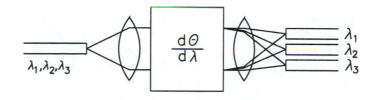

Figure 9.6 Wavelength separation system using an angular disperser and lenses.

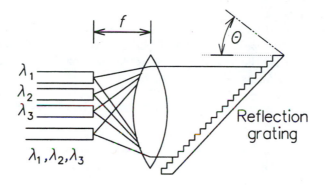

Figure 9.7 Geometry for reflecting grating WDM receiver.

close parallel lines engraved or etched into the surface. Excess losses of 1 to 3 dB are typical for these demultiplexing devices; the typical channel separation is 1 to 10 nm.

9.3.2   Filtering Devices

Filtering devices reflect or transmit light depending on the wavelength. Several types of filtering devices can be used to separate the wavelengths [6]. Table 9.1 shows representative bandwidths that can be achieved with each type of filter.

1. Interference filters—Most optical filters used in this application are multilayer dielectric stacks (rather than the absorbing filters used in photographic applications) to provide a sharp change in reflectivity at the desired wavelength and to minimize the insertion loss. The filters typically pass one wavelength (with a typical resonance response) and reflect all

Method	Typical channel bandwidth
Interference filters	5 nm
Fabry-Perot filter	0.1–10 nm
Tuned semiconductor amplifier	1–10 GHz
Heterodyne receiver	1–10 GHz

Table 9.1 Typical channel bandwidths for different filtering techniques.

other wavelengths more than 10% (typically) away from the center wavelength of the filter. (With more complex designs, the spacing can be made to be 1% or even 0.1% of the center wavelength.) For operation with more than two channels, a series arrangement of several filters becomes necessary [7], with increasingly large insertion losses for the combination of filters. Only cascades of two or three filters are practical for use, due to their relatively high insertion loss, limiting broadband systems operation to five or six separate wavelength channels [6].

2. *Fabry-Perot resonator filters*—The Fabry-Perot resonator is a bandpass optical filter that transmits a narrow band of wavelengths and reflects others [8]. The transmission frequencies of the filter occur every $c/2L$, where L is the spacing between the mirrors of the resonator. This period is called the *free spectral range*, FSR, of the interferometer. The 3 dB frequency widths of the filter passband (i.e., the full-width at the half-maximum points) is Δf and is related to the FSR by the *finesse* F of the filter by

$$\Delta f = \frac{\text{FSR}}{F} \tag{9.6}$$

where $F = \pi\sqrt{R}/(1-R)$ and R is the reflectivity of the resonator mirrors (assumed the same for both mirrors of the resonator filter). The minimum transmission between the peaks is $(\pi/2F)^2$ [9].

3. Semiconductor amplifier filters—A diode laser that is pumped below threshold (i.e., it is not pumped sufficiently for lasing) is a tuned bandpass amplifier with a narrow passband [10]. The center frequency is determined by the composition of the material. Using a distributed-feedback (DFB) laser or a distributed-feedback reflector (DFR) laser with multiple electrodes (as shown in Fig. 9.8 on the next page) allows one to adjust the amplified wavelength. In the distributed-feedback amplifier (Fig. 9.8a), one current input can be used for the pump while the other is used to control the nominal center wavelength of the amplifier. In the distributed-feedback-reflector amplifier, the current I_1 varies the optical periodicity of the reflector (and, hence, the center wavelength of the amplifier), the current I_2 controls the phase of the amplifier, and the current I_3 controls the pump power or gain of the amplifier.

4. Coherent detection with electronic filters—Once a source with a linewidth smaller than the signal bandwidth is possible, *coherent detection* techniques become possible. In these techniques, the signal is mixed with a frequency-shifted reference signal from a frequency-locked local laser. The signals mix (i.e., interfere) coherently on the detector, generating a difference frequency (called the *intermediate frequency*). The channel spacing of the detected signals is determined by the bandwidth of the modulated signal (dependent on the type of modulation used and the data rate) and the receiver arrangement used. This detection method allows the use of narrow channels that are separated in the electronic processing subsystem of the receiver where precision electronic filters can be used, rather than filtering in the optical regime where separation filters are still fairly crude and primitive in terms of the minimum channel width.

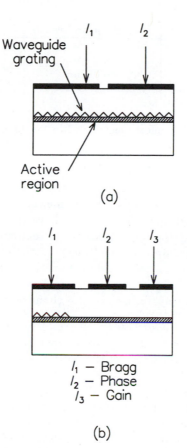

Figure 9.8 (a) A distributed-feedback (DFB) amplifier (shares its feedback region with the current pump). (b) A distributed-reflector (DFR) amplifier with multiple control electrodes (separates the reflector region from the current-pumped region).

9.4 Sources

To make WDM systems possible, we would ideally have a powerful source with sufficient stability and tunability to allow rapid selection of the desired wavelength. Table 9.2 on the following page illustrates the spectral width of some candidate sources that we want to discuss.

- **LEDs:** The wide bandwidth of an LED is usually considered a disadvantage, due to the resulting fiber dispersion. It is possible, however, to pass the output through a spectral filter to narrow the spectral width of this source; this technique is called *spectral slicing* [4]. This method has the advantage of allowing use of several of the same relatively inexpensive LEDs as the sources and, using different filters, selecting the operative wavelengths. This selective filtering further reduces the already meager power level available from the LED source, however. The reduction in power is proportional to the bandwidth reduction, making this technique practical only for short-distance, low-data-rate links.

Type	Typical spectral width	Typical spectral width
LED	50–100 nm (or more)	5000–10000 GHz
Fabry-Perot laser	3–6 nm	300–600 GHz
DFB laser	<0.01 nm	10–100 MHz
External cavity laser		<1 MHz

Table 9.2 Typical spectral widths for different sources.

- **Fabry-Perot Diode Lasers:** The *Fabry-Perot (FP) laser* will typically have 6 to 8 modes oscillating with a spacing of $c/2L$ Hz between them. For these lasers the center wavelength has a typical tolerance of ± 3 nm and a typical linewidth of 6 nm. A representative tolerance on the multiplexer/demultiplexer is ± 1 nm [4], so the best available channel spacing is about 14 nm. (Note that the laser needs to be temperature-controlled, since the center wavelength has a temperature coefficient of about 0.4 nm·°C^{-1}.) Allowing a guard band between channels that is equal to the channel width itself, we find that we need to allocate about 28 nm per channel and that our 100 nm and 150 nm wide windows can accommodate 3 and 5 channels, respectively, for a total of 8 channels.

- **DFB Diode Lasers:** The *distributed-feedback (DFB) laser* [8, 11] allows single-mode operation of the diode laser and offers a lower temperature sensitivity of about 0.08 nm·°C^{-1}, as well. The typical spectral width of the output is 10 to 100 MHz. This is smaller than the signal spectral width (for high data rates) or the wavelength "chirp" that accompanies the output of these lasers when pulsed. (The chirp can be as much as several 10s GHz [or several tenths of a nm]. This chirp can be removed by using a cw laser as a source and using an external modulator to modulate the data onto the light.) The channel width can be estimated from a source center wavelength tolerance of ± 0.5 nm (assuming a temperature-controlled laser), a chirp tolerance of ± 0.2 nm, and a filter tolerance of ± 1 nm [4]. The actual channel width is about 3.5 nm and, adding a guard band equal to the channel width, the allocated channel width would be about 7 nm. Therefore, we could have 14 and 21 channels in our respective fiber transmission windows.

- **Frequency-Locked Lasers:** A laser can be phase-locked to an atomic resonance and, with temperature and current feedback controls, can achieve a linewidth as small as 5 MHz. This small linewidth allows the channel bandwidth to be determined only by the signal bandwidth, independent of the source linewidth. In this case, the demultiplexing cannot be done by fixed filters but must be done by a tunable, narrow-bandwidth element to ensure matching the passband of the filter element precisely to the wavelength of the source. Once frequency-locked, the laser cannot be easily tuned. Other techniques can be used to offset a second laser from the reference-locked laser by an adjustable (or preset) amount.

9.5   Nonlinear Effects on WDM Links

Nonlinear effects, as discussed in Chapter 3, can be deleterious on multichannel data links [12, 13]. Chraplyvy [12] analyzed the effects of nonlinearities on both single-signal and multisignal fibers.

- For multisignal fiber links, he found that *stimulated Raman scattering* would have a negligible effect.

- The effects of *carrier-induced phase noise* (i.e., changes in the optical phase of one signal due to changes in the power in itself or in other light waves in the link) will present problems for phase-modulated signals in a WDM link. Broadband multiplexed links will need to keep their power levels below 20 mW to avoid this effect; selective-wavelength links will need to keep their power below a few milliwatts to successfully operate over 10 or so channels.

- The effects of *Brillouin scattering* depend only on the power level and are independent of the number of channels in the link. Power limits of 10 mW are suggested [12] for maximum-performance links. (Generally, broadband systems require that each source power level be degraded by a factor of $1/N$ from the maximum dictated by the nonlinearity [where N is the number of channels in the link]; wavelength-selective links allow *each* source to achieve the power limit dictated by the nonlinearity.)

- *Four-photon mixing* occurs when two propagating light waves mix nonlinearly as they travel through the fiber and produce frequency components at $2\nu_1 - \nu_2$ and $2\nu_2 - \nu_1$. These sidebands grow and, through further nonlinear interaction, produce nine new optical frequencies at $\nu_{ijk} = \nu_i + \nu_j - \nu_k$ with $i, j, k = 1, 2, 3$. The appearance of these new signals and the power reduction of the original signals degrade the performance of the link. The strength of the effect depends on the channel separation and the fiber dispersion. To minimize the effect, Chraplyvy [12] calculates that the channel spacing will have to be larger than 50 GHz and that the region of zero-dispersion should be avoided (a surprising result that reduces the data rate-distance product of the link). Source power levels of less than a few milliwatts are recommended. Waarts et al. [13] perform a similar analysis of the effects of nonlinearities on *coherent* frequency-division multiplexed links where the channels are desired to be very close to each other.

9.6 WDM and Optical Amplifiers

Optical amplifiers have the ability to amplify several wavelengths simultaneously. Hence, their incorporation into WDM links is a natural fit to achieve long-distance multichannel transmission or to implement a multichannel large-ring network. The advantage of offering gain is not without potential penalties, however. Two problems to be discussed here are:

- The nonuniform gain spectrum of amplifiers can result in a variable SNR at the end of a link with cascaded amplifiers [14]. The effect of this nonuniform gain is further accentuated because the strongly amplified channels rob gain from weaker channels through process of gain saturation. The degree of mismatch has been modeled [14] with the critical parameter being identified as the losses between the amplifiers (assumed wavelength independent). An optimum loss of 25 dB was determined using representative link parameter values. The possible solutions include

  - preskewing the optical signals at the transmitter,

  - inserting a notch filter cantered at the 1531-nm peak of the gain,

  - using only the relatively flat gain region near 1557 nm, and

  – cooling the amplifiers so that the gain is based on wavelength-independent inhomogeneous broadening rather than wavelength-dependent homogeneous broadening.

  – We can also add spectrally shaped filters after several amplifiers to remove the effect. Reference [14] used a simple notch filter with center wavelength of 1560 nm (i.e., the location of the spectral gain peak in a cascade of amplifiers) and 3-dB width of 2 nm. The filter was placed after every 20 amplifiers in a simulated experiment and successfully removed the nonuniform gain effects.

- We cannot add a narrow spectral filter after each amplifier to reject some of the ASE generated. The resulting ASE is larger than occurs in a single-channel system; it can saturate the gain from the amplifiers and reduce the signal-to-noise ratio.

- Under some circumstances, a low-dispersion fiber of long length and modest power signals (i.e , 1 mW per channel) can allow significant four-wave mixing to occur [15], generating new signals close to or atop the channels. (This mixing was described earlier in our discussion on nonlinear effects in fibers.) In low-dispersion fibers, such as dispersion-shifted fibers, that operate in links with optical amplifiers, this four-wave mixing is *the* dominant mechanism that limits the link performance. These mixing signals may interfere with the desired signal and make its detection difficult. The mixing signals are especially emphasized if operated at zero-dispersion wavelength. Some possible solutions are to use unequal channel separation and to use *dispersion management* [15].

  – Unequal spacing of channels: In general, N channels launched will cause M mixing products, where

  $$M = \frac{N^2(N-1)}{2}. \tag{9.7}$$

  Due to the coherent nature of the mixing signals and signals, even small mixing signals can cause major impairment of received signal if the mixing signal frequency falls atop the signal frequency (e.g., mixing signal with 1% of signal power can impose a ± 1 dB fluctuation in the received power). We can avoid this effect with unequal channel spacing, with the separation in frequency spacing being larger than the bit rate to avoid interference [16, 17]. (We note that equal spacing in wavelength is unequal spacing in frequency, but it is *not* enough separation to remove the effect.) A frequency allocation proposed in Ref. [17] allowed the input power levels to be increased by as much as 7 dB. This increase in power can be used for higher bit rates, increased distance, or some combination of desirable link properties. The allocation scheme has several solutions and some trade-offs in link performance can be made [17]. (It should be noted that the generation of the mixing products will still deplete the signal power, reducing the signal-to-noise ratio at the receiver.) This technique should prove useful for upgrading links utilizing dispersion-shifted fiber and optical amplifiers from single-channel use to multichannel applications.

  – Dispersion management. In links with conventional fiber, the dispersion at 1550 nm may be significant. It is possible to remove the effects of four-wave mixing in a multichannel link in these fibers by managing the dispersion in the link. We want the *total* dispersion to be low in order to have long dispersion-limited distance. To eliminate the mixing effects, though, we want the *local* dispersion to be higher valued. We can obtain the combined effects if we concatenate fiber lengths with alternating signs of their dispersion. illHigh-value dispersion (i.e., 1 to 2 ps·nm^{-1}·km^{-1}) is desirable to avoid four-wave mixing. The techniques are the same as in our discussion of dispersion compensation in Chapter 7.

9.7 Multipoint, Multiwavelength Networks

We now want to discuss some samples of network implementations that use broadband transmission and wavelength-selective techniques.

9.7.1 Broadband Transmission Systems

In the *broadcast-and-select network* (also called a broadband network), as shown in Fig. 9.5b, any signal transmitted by a fixed-frequency transmitter is broadcast to *all* receivers (allowing the maximum in network flexibility). The receivers tune to the desired transmitter. If N is the number of stations in the array, then the power is split into $1/2^N$ and the overall splitting loss is $10 \log 2^N$ dB. The network can be configured as either a linear tree or as a star configuration (as seen in Fig. 9.9 on the next page).

9.7.2 Wavelength-Routed Transmission Networks

These networks combine wavelength-division multiplexed systems in various ways. The power transmitted at a particular wavelength is routed to particular receiver through the network through the use of wavelength-selective elements. Once a wavelength is dropped to the desired receiver, it can be reutilized to add another signal to the mix, thereby conserving the number of wavelengths required. Some networks even envision the use of wavelength conversion devices to change wavelengths of a signal within the network routing device. This also has the effect of increasing the network efficiency. Both the wavelength spectrum and the optical power can be utilized more efficiently, but at a cost of decreased network flexibility.

Examples of these networks include the star network, the chain network, and the ring network.

WDM Star Network

In one version of the wavelength-selective WDM *star network* (Fig. 9.10 on page 317), the receivers have a set wavelength and the sources are tunable. (A configuration with set source-wavelengths and tunable receivers is also possible.) Any source can talk to any receiver by tuning to the receiver's preassigned wavelength. Two possible receiver configurations are shown, an optical WDM direct-detection receiver that uses optical filters to pass the proper wavelength, and a coherent FDM receiver that uses heterodyne detection to tune to the proper frequency of the receiver. The N×N coupler in this arrangement is a wideband device that combines the source power and splits it equally among the receivers.

WDM Chain Network

In a wavelength-selective WDM *chain network* (Fig. 9.11 on page 317), each node has a wavelength-selective splitter that separates the wavelengths. In this *add/drop configuration*, certain wavelengths are routed to the receiver(s) (i.e., dropped from the network to their receivers). New signals are added to the network at these wavelengths. Additional wavelengths can also be added

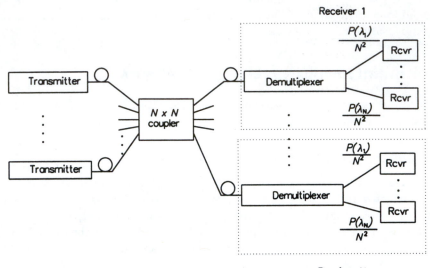

(a)

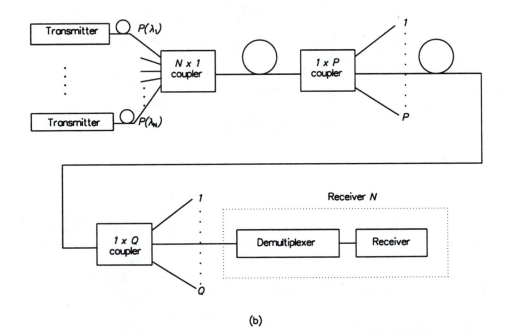

(b)

Figure 9.9 (a) A broadcast-and-select, broadband-transmission, star network. (b) A broadband-transmission tree network. (After G. R. Hill, "Wavelength domain optical network techniques," *Proc. IEEE*, vol. 77, no. 1, pp. 121–132, ©IEEE, 1990.)

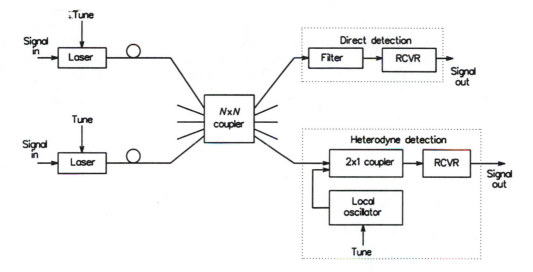

Figure 9.10 Example of a WDM network arranged as a star network. Receivers can be either direct-detection receivers with spectral filters or heterodyne detection FDM receivers, as shown. (After G. R. Hill, "Wavelength domain optical network techniques," *Proc. IEEE*, vol. 78, no. 1, pp. 121–132, ©IEEE, 1990.)

to the network (provided that the output coupler of the node has inputs for them). Data can be added and dropped at each wavelength channel as the information traverses the network. In the figure, we note that nodes 1 and 5 should be co-located to have full transmit/receive capabilities.

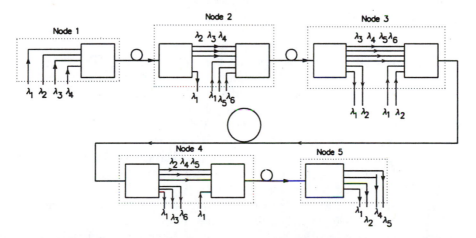

Figure 9.11 A wavelength-division multiplexed chain network illustrating wavelength-dependent routing to accomplish signal add/drop functions. Nodes 1 and 5 should be co-located to have full transmission and reception capabilities in this unidirectional link. (After G. R. Hill, "Wavelength domain optical network techniques," *Proc. IEEE*, vol. 78, no. 1, pp. 121–132, ©IEEE, 1990.)

WDM Ring Network

In the wavelength-selective WDM *ring network* (Fig. 9.12), the configuration is similar to the chain network with its ends tied together. In simplest form, the number of wavelengths added at a node are equal the number received, so that all of the couplers and splitters are the same.

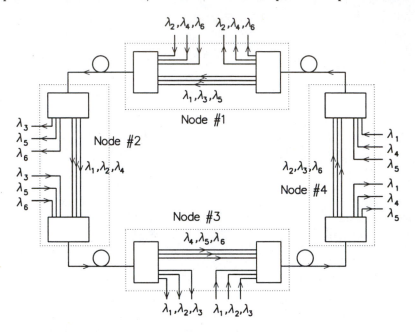

Figure 9.12 A wavelength-division multiplexed ring network with wavelength-routed add/drop functionality. (After G. R. Hill, "Wavelength domain optical network techniques," *Proc. IEEE*, vol. 78, no. 1, pp. 121–132, ©IEEE, 1990.)

In these configurations the basic idea is that a set of identical WDM elements provides a fixed optical path between transmitter and receiver. The wavelengths can be reused in another section of the network, thereby decreasing the number of wavelengths required to implement the network. It can be shown that a set of N nodes will require only $N(N-1)$ optical channels, rather than the N^2 wavelengths that might be expected to fully interconnect the nodes (i.e., so that any node can talk to any other node). Figure 9.13 on the next page illustrates a WDM star network that can interconnect four nodes with eight identical WDM units but requires only three wavelengths. The wavelength assignments are also shown. (Note that it is assumed that transmitter #1 does not want to connect to receiver #1, etc.) Table 9.3 on the facing page shows the minimum number of wavelengths required [4] to implement a WDM network between N nodes; Fig. 9.14 on page 320 plots the relations shown in the table.

When multiplexing devices are cascaded, it should be noted that the bandwidth of the combination is reduced from the bandwidth of a single element. For example, if the individual bandwidth of an element were Gaussian-shaped with a bandwidth value of BW, then a cascade of N units would be Gaussian-shaped with a bandwidth value of $BW/\sqrt{N}$, thereby reducing the bandwidth of the combination and placing more emphasis on maintaining the source wavelength within a tighter tolerance.

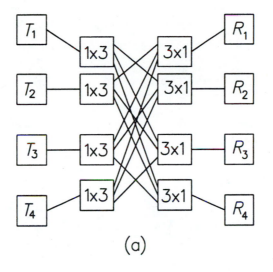

(a)

$$
\begin{array}{c|cccc}
 & R_1 & R_2 & R_3 & R_4 \\
\hline
T_1 & - & \lambda_1 & \lambda_2 & \lambda_3 \\
T_2 & \lambda_3 & - & \lambda_1 & \lambda_2 \\
T_3 & \lambda_2 & \lambda_3 & - & \lambda_1 \\
T_4 & \lambda_1 & \lambda_2 & \lambda_3 & - \\
\end{array}
$$

(b)

Figure 9.13 (a) A wavelength-division multiplexed star network with four nodes and three wavelengths. (b) Wavelength assignment table. (After G. R. Hill, "Wavelength domain optical network techniques," *Proc. IEEE*, vol. 78, no. 1, pp. 121–132, ©IEEE, 1990.)

Method	Minimum number of wavelengths
Star	$N-1$
Chain	$\left(\frac{N}{2}\right)^2$ (if N is even) $\frac{(N-1)(N+1)}{4}$ (if N is odd)
Ring	$\frac{N(N-1)}{2}$

Table 9.3 Minimum number of wavelengths required to connect N channels with a WDM network.

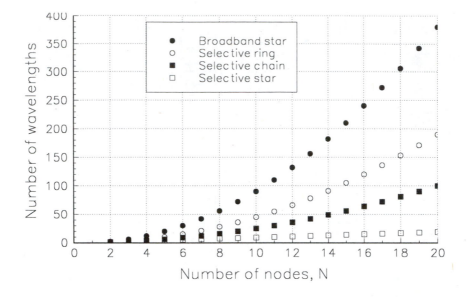

Figure 9.14 Minimum number of wavelengths required to connect N nodes of a network.

9.7.3   Switched Networks

Switching in a WDM network can be accomplished by changing the source or the receiver to use a different transmitter-receiver wavelength and, hence, a different path through the network. (We note that a broadband optical path is needed, unless we want to subdivide the network into smaller channels.)

Figure 9.15 on the facing page shows an all-optical switched optical network. The hub is totally passive, consisting only of $1 \times N$ and $N \times 1$ couplers. The receiver of each node is assumed to be co-located with the transmitter of that node, although shown separately on the figure for clarity. The switching is done by choosing a wavelength of the transmitter to match the appropriate receiver and then routing the traffic to that transmitter of the node. For example, the traffic at node #1 that is to be sent to node #2 is routed into the laser with wavelength λ_{12}. The WDM coupler at node #1 combines this wavelength with all of the other wavelengths (i.e., all of the traffic for the other nodes) and sends it to the hub. At the hub, the $1 \times N$ wavelength-sensitive coupler divides the wavelengths and sends the traffic at λ_{12} to the $N \times 1$ which combines the power at all wavelengths that are to be sent to node #2. At node #2 the channels are split by the wavelength-sensitive $N \times 1$ coupler and received simultaneously. The individual messages are time-demultiplexed and routed to the appropriate user via the switch at node #2. It can be shown [3] that only N wavelengths are required to completely crossconnect N nodes with this topology (rather than the N^2 wavelengths that might seem to be required).

We note from this arrangement that, while the hub has been simplified, the nodes require several sources and receivers. The number of these sources and receivers can be reduced to one if we can tune the source and the receiver filter and limit ourselves to transmitting and receiving only one message at a time. A trade-off is made between the number of components and the number of simultaneous channels desired.

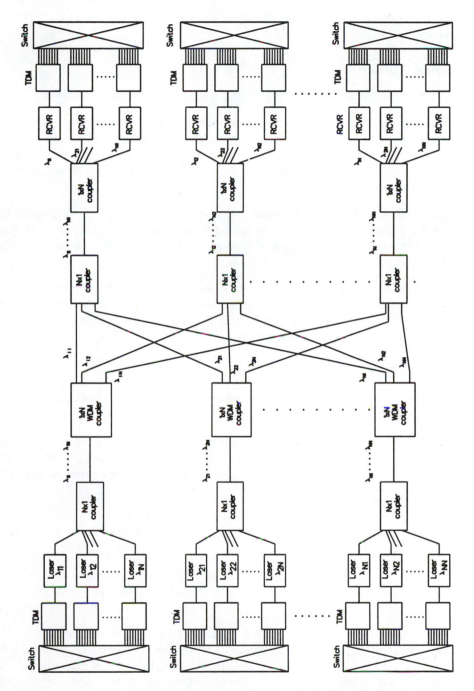

Figure 9.15 Example of a switched WDM network. (After S. S. Wagner and H. Kobrinski, "WDM applications in broadband telecommunications networks," *IEEE Communications Magazine,* pp. 22–30, March 1989, ©IEEE 1989.)

Various methods have been proposed to tune the sources or the receivers (of both) in a switched network.

9.7.4   Source Tuning

Some *source tuning* mechanisms for rapidly tuning a source have been identified.

- Temperature/current tuning—The source wavelength can be changed by a modest amount by changing the temperature of the source (a resulting change of about 0.08 nm·°C^{-1} for a distributed-feedback laser, making a 1 nm change feasible) or by changing the laser current (a resulting change of about 200 MHz·mA^{-1} for switching rates higher than 10 MHz, making major tuning infeasible) [4].

- *DFB and DFR tuning*—Broader tuning ranges can be achieved [4] by combining the DFB laser with a phase-control section and an active section or by fabricating an integrated structure that controllably combines the power of different laser structures oscillating at various wavelengths.

- External element tuning—Additionally, it is possible to use a Fabry-Perot laser with an external cavity that has a movable grating controlled by a piezoelectric pusher. Tuning with this technique has been demonstrated over 55 nm [18], although rapid, accurate tuning of this device still needs to be demonstrated.

9.7.5   Receiver Tuning

To tune the receivers, tunable filters can be based on technology using multilayer filters, Fabry-Perot interferometers, or integrated optic devices.

- A tunable *multilayer filter* device can be made whose transmission properties vary spatially across the face of the filter (e.g., by varying layer thickness). Tuning is accomplished by translating the filter. A filter that uses this technique has reported coverage of 270 nm with a passband of 7 nm [7].

- *Fabry-Perot interferometer filters* are tuned by changing the length of the cavity using a piezoelectric pusher. Devices with 30-channel operation have been demonstrated [4]. Tuning time can be a few microseconds.

- Integrated optic tunable filters that use TE-to-TM mode convertors (i.e., that convert transverse electric modes to transverse magnetic modes), have been demonstrated in the laboratory [4].

- Distributed feedback lasers operated below threshold (i.e., operated as amplifiers rather than oscillators) present narrow passbands. The devices can be tuned by incorporating a phase-control section. The response times are fast (on the order of nanosecond switching times) and tuning ranges of 71 GHz (for 10-channel operation) have been demonstrated [4].

9.8   Summary

In this chapter, we have seen that current fiber systems utilize only a fraction of the spectral width available in both multimode and single-mode fibers. Wavelength-division multiplexing (WDM) [1] offers a method to fill up the fiber's passband with several sources of different wavelengths. WDM techniques hold much promise for allowing vast amounts of data to be transferred over fiber links.

Wavelength-division multiplexers and demultiplexers can be either wavelength-selective or broadband. The wavelength-selective demultiplexers typically use an angularly dispersive element, such as a diffraction grating, to separate the wavelengths. Filters of various types have better control of the wavelength passband but have an insertion loss that limits the number of filters that can be cascaded to just a few filters.

WDM techniques require sources with a controllable operating wavelength and, having been set at a specific wavelength, require long-term stability to maintain that wavelength. Various techniques have been described including filtered LEDs, Fabry-Perot resonator lasers, DFB diode lasers, and frequency-locked diode lasers.

When operated with modest power levels in fibers with near-zero dispersion, four-wave mixing can become the mechanism that dominates the link's performance. Such links can best be operated with non-periodic frequency spacings between the channel wavelengths (although the power depletion from the signal channels that goes into the mixing channels can still limit the link performance). Another technique is to use concatenated lengths of fiber having dispersions of alternating signs, so that the total combined dispersion is near-zero, but the local dispersion is large enough to keep the four-wave mixing from being significant.

Various sample architectures have been described which allow tunable point-to-point communications or broadcast communications. The architecture for an optical-switched network has also been described [1, 2].

9.9   Problems

1. Consider the windows in a single-mode fiber.

   (a) Calculate the width in Hz of the optical window extending from 1250 to 1350 nm.

   (b) Calculate the width in Hz of the optical window extending from 1450 to 1600 nm.

   (c) Find the total bandwidth in Hz in both windows combined.

2. Consider the geometric arrangement of Fig. 9.6 on page 309.

   (a) If two wavelength are separated in wavelength by $\Delta\lambda$, show that the focussed spots at the receiving fiber locations are separated by $w = f\,\Delta\lambda\,d\theta/d\lambda$.

   (b) From the result of the previous part, calculate an expression for the maximum outside diameter of the receiving fibers.

3. (a) The typical channel separation of an interference filter is 5 nm. Calculate the equivalent frequency linewidth $\Delta\nu$ if the operating wavelength is 1300 nm.

   (b) Calculate the ratio of $\Delta\nu$ for the filter of part (a) to the 10 GHz frequency linewidth achievable with a heterodyne receiver.

4. Calculate the value of reflectivity required for a Fabry-Perot filter to have a ΔF of 0.1 nm if the mirror separation is 1 cm and the operating wavelength is 1.5 μm.

5. Make a wavelength assignment table similar to the one shown in Fig. 9.13 on page 319 for the five-node chain network shown in Fig. 9.11 on page 317. Note that each transmitter can talk only to its downstream neighbors (e.g., transmitter 2 can talk to receivers 3, 4, and 5 but it cannot talk to 1).

6. Make a wavelength assignment table similar to the one shown in Fig. 9.13 on page 319 for the four-node ring network shown in Fig. 9.12 on page 318.

7. Table 9.3 on page 319 gives the minimum number of wavelengths required to connect N channels in a WDM star network. Sketch the star network configuration for $N = 3$.

References

1. M. Fujiwara, M. Goldman, M. O'Mahony, O. Tonguz, and A. Willner, "Special issue on multiwavelength optical technology and networks," in *Journal of Lightwave Technology*, 1996.

2. R. L. Cruz, G. R. Hill, A. L. Kellner, R. Ramaswami, G. H. Sasaki, and Y. Yamabayashi, "Optical networks," in *Selected Areas in Communications*, IEEE Communications Society, 1996.

3. S. S. Wagner and H. Kobrinski, "WDM applications in broadband telecommunication networks," *IEEE Communications Magazine*, pp. 22–30, March 1989.

4. G. R. Hill, "Wavelength domain optical network techniques," *Proc. IEEE*, vol. 77, no. 1, pp. 121–132, 1990.

5. G. Keiser, *Optical Fiber Communications, Second Edition*. New York: McGraw-Hill, 1991.

6. H. Ishio, J. Minowa, and K. Nosu, "Review and status of wavelength-division-multiplexing technology and its application," *J. Lightwave Technology*, vol. LT-2, no. 4, pp. 448–463, 1984.

7. J. McCartney, P. Lissberger, and A. Roy, "Recent developments in the production of narrowband position tuned interference filters for wavelength multiplexed optical fiber systems," *Optics Communications*, vol. 64, pp. 338–342, 1987.

8. A. Yariv, *Optical Electronics, Fourth Edition*. New York: Holt, Rinehart and Winston, 1991.

9. I. P. Kaminow, "Non-coherent photonic frequency-multiplexed access networks," *IEEE Network*, pp. 4–12, 1989.

10. S. Suzuki, M. Nishio, T. Numai, M. Fujiwara, M. Itoh, S. Murata, and N. Shimosaka, "A photonic wavelength-division switching system using tunable laser diode filters," *J. Lightwave Technology*, vol. 8, no. 5, pp. 660–666, 1990.

11. T.-P. Lee, "Recent advances in long-wavelength semiconductor lasers for optical communication," *Proc. IEEE*, vol. 79, no. 3, pp. 253–276, 1991.

12. A. R. Chraplyvy, "Limitations on lightwave communications imposed by optical-fiber nonlinearities," *J. Lightwave Technology*, vol. 8, no. 10, pp. 1548–1557, 1990.

13. R. G. Waarts, A. Friesem, E. Lichtman, H. H. Yaffe, and R.-P. Braun, "Nonlinear effects in coherent multichannel transmission through optical fibers," *Proc. IEEE*, vol. 78, no. 8, pp. 1344–1368, 1990.

14. A. E. Wilner and S. Hwang, "Transmission of many WDM channels through a cascade of EDFA's in long-distance links and ring networks," *J. Lightwave Technology*, vol. 13, no. 5, pp. 802–816, 1995.

15. R. W. Tkach, A. R. Chraplyvy, F. Forhieri, A. H. Gnauk, and R. M. Derosier, "Four-photon mixing and high-speed WDM systems," *J. Lightwave Technology*, vol. 13, no. 5, pp. 841–849, 1995.

16. F. Forghieri, R. W. Tkach, and A. R. Chraplyvy, "Reduction of four-wave mixing crosstalk in WDM systems using unequally spaced channels," *IEEE Photonics Technology Letters*, vol. 6, pp. 754–756, 1994.

17. F. Forghieri, R. Tkach, and A. Chraplyvy, "WDM systems with unequally spaced channels," *J. Lightwave Technology*, vol. 13, no. 5, pp. 889–897, 1995.

18. R. Wyatt and J. Devlin, "10 kHz linewidth 1500 nm InGaAsP external cavity laser with 55 nm tuning range," *Electronics Letters*, vol. 19, pp. 110–112, 1983.

Index